Zoophysiology and Ecology
Volume 1

Endocrines and Osmoregulation

A Comparative Account
of the Regulation of Water and Salt
in Vertebrates

by
P. J. Bentley

With 29 Figures

Springer-Verlag
New York   Heidelberg   Berlin   1971

P. J. Bᴇɴᴛʟᴇʏ
Professor of Pharmacology
Mount Sinai School of Medicine
of the City University of New York

ISBN   0-387-05273-9   Springer Verlag New York · Heidelberg · Berlin
ISBN   3-540-05273-9   Springer Verlag Berlin · Heidelberg · New York

To S. E. D.
For Friendship

Preface

It is an old and trite saying that a preface is written last, placed first and read least. It may also be an explanation or justification for what follows, especially if the author feels that the reader may not agree with him. This is thus, something of an apologia.

When I first agreed to write this book, I felt that it was to be directed principally towards the endocrine system of vertebrates especially as there were available some excellent accounts of their osmoregulation. However, it was soon apparent to me that in order to appreciate the role of the endocrines one must strongly emphasize non-endocrine functions involved in the regulation of the animals water and salt balance. In addition several years have elapsed since a full account of vertebrate osmoregulation has been given. I have thus felt free to take an overall look at the animals water and salt metabolism, but have especially emphasized more recent contributions. Possibly the book should now be called 'Osmoregulation and the Endocrines'.

A rather conventional format has been retained for, despite many claims to the contrary, I do not feel that all things new, experimental and innovative are necessarily desirable. In the first two chapters I have attempted to introduce, and summarize, at a general level, the two major topics, osmoregulation and endocrines, while emphasizing the diversity of such processes in the vertebrates. In the second chapter I have also included a section on the 'Criteria, methods and difficulties' of experimental comparative endocrinology. I felt that this information may be useful for the reader who desires to make a critical appreciation and evaluation of many of the experiments that are described. It may, hopefully, also be useful to the younger workers interested in comparative endocrinology and help them to glean something about the sort of work that they may expect to do. The vertebrate groups have been described within the framework of their customary classification, a decision arrived at primarily because this also makes a little evolutionary sense. Some have questioned the wisdom of placing the mammals first, and traveling down, rather than up, the phyletic scale. However, each chapter is reasonably complete in itself so that those who disagree can resort to the subterfuge of reading the book backwards.

I have attempted to relate the physiology of the various beasts to that of the environments that they normally occupy. At the same time I have not hesitated to descend to the more currently fashionable 'molecular level'. I hope that this may help to remind both some of the more 'mature' readers and emphasize to the younger ones the very broad scope of interest that animals have to offer. It is principally for the younger group that this book has been written.

New York   P. J. Bentley
1970

Acknowledgements

The observations of many people, only some of whom are quoted, made this book possible. I would like to extend my thanks to all who have contributed to these subjects and only wish that it would have been possible to acknowledge the work of all.

The latter part of this book was written while I was a guest of the Commissariat à L'Energie Atomique. I would like to thank them for their hospitality, particularly Professeur J. Coursaget, Chef du Departement de Biologie, and Dr. Jean Maetz and his Groupe de Biologie. The help and guidance of Jean Maetz made the latter part of this book possible. Dr. Erik Skadhauge kindly gave me some advice on avian matters.

The Mount Sinai School of Medicine of the City University of New York kindly granted me leave so that I could join the Commissariat à L'Energie Atomique. A number of nice ladies have helped me during the various stages in preparation of this manuscript; special thanks are due to Mrs. Jean Homola, Mrs. Norma Beasley, Miss Marguerite Murphy and Mrs. Wolina Shapiro.

My wife showed considerable forbearance and gave much needed help in the final preparation of the proofs. The editorial help and advice of Dr. D. S. Farner is gratefully acknowledged as well as the cooperation of the staff of Springer-Verlag in Heidelberg and Mr. B. Grossmann in New York.

Contents

XI

Chapter 1

Osmotic Problems of Vertebrates

General Introduction

Water provides the principal physical framework in which the energy transformations characteristic of life take place; it is the solvent for life as we know it. Chemical reactions take place in a variety of media, but the physico-chemical properties of water make it uniquely suitable for such a role. Its mobility and plasticity provide it with structural advantages suitable to living systems, its high heat capacity and latent heat of evaporation buffer the organism against thermal changes in the environment, while its remarkable capacity as a solvent, for both organic and inorganic materials, provides a vehicle for the interactions of many different substances.

Living organisms contain 40 to 99% water, the content in vertebrates ranging from 65 to 85% of their body weight. This water is nearly all in a free uncombined state and contains solutes that participate in transformations of energy, together with those which are structural components contributing to the stability of the system. Thus, the solubility of proteins can be facilitated by the presence of certain concentrations of ions, while metals may act as co-factors in enzyme reactions. Solutes may also be utilized to provide chemical gradients which act as stores of potential energy, as in nerve and muscle cells, where electrical depolarization represents a release of stored energy due to earlier ion accumulation. In addition solutes influence the osmotic mobility of water, restricting its movements across structural barriers within the organism or between it and the external environment.

1. Geographical and Ecological Distribution of the Vertebrates (Osmotic Distribution)

It is uncertain whether the vertebrates originated in the sea or fresh water, but subsequently both environments, as well as the dry land, have been occupied by representatives of most of the major groups. The geographical location(s) for the transition to a vertebrate is also unknown, but subsequent dispersal has been widespread. The major groups of fishes, the Agnatha, Chondrichthyes and Osteichthyes, have representatives in the sea and fresh water. The tetrapods are primarily terrestrial, but the Amphibia, reptiles, birds and mammals all have species that normally live in marine or freshwater environments. Osmoregulatory processes play a major role in such dispersal, and it is not surprising to observe that the physiological mechanisms by which such adaptations take place, often differ greatly. Even among the fishes, an apparently novel occupation of a certain medium may represent a secondary, tertiary or even further removed alteration of the situation that was primary in the group. DARLINGTON (1957) has traced the probable historic origins of the endemic catfishes of Australia: teleost fish originated in the sea from whence a fish related to the order Isospondyli moved into fresh water to give rise to the order Ostariophysi. Some of this group probably re-entered the sea and returned subsequently to fresh water in Australia. Among the

1

tetrapods at least one amphibian, the crab-eating frog, *Rana cancrivora*, has returned to a marine environment, and even more numerous examples of such a transition can be seen among the reptiles that have marine and freshwater as well as terrestrial representatives. The biological adaptations accompanying such transitions are sometimes very similar in different groups, providing examples of parallel evolution. In other instances, more novel and unique mechanisms have evolved. Thus both the marine frog, *Rana cancrivora*, and the sharks and rays, maintain their body fluids hyperosmotic to sea-water by accumulating urea. On the other hand marine reptiles, like the loggerhead turtle, *Caretta caretta*, have body fluids similar in concentration to their terrestrial relatives which is hypotonic to sea-water, but they can excrete salt as a highly concentrated solution from a modified orbital gland. *Rana cancrivora* has evolved a parallel mechanism to that of the sharks and rays, while the reptiles have utilized a novel mechanism not seen in their phyletic forbears. Both similarities and diversities in osmoregulatory mechanisms are found within, and between, the major phyletic groups of vertebrates.

The vertebrates have occupied most of the earth's geographic areas, being sparse in the cold terrestrial polar regions, and not as numerous in hot dry deserts as in the wetter tropical zones. The desert regions, where the supply of water may be limiting to life, make up about one third (50 million square kilometers) of the land surface of the earth (Schmidt-Nielsen, 1964a). Despite the potential osmoregulatory problems, vertebrates do live in even the extremely dry parts of such desert areas, and exhibit interesting physiological and behavioural patterns consistent with their life there. The seas, even in polar regions, have a diverse vertebrate fauna of fish, birds and mammals. Major geographic limitations exist for the various vertebrate groups and will be discussed later, but in the instance of the Amphibia and reptiles, they are mainly dictated by temperature rather than by water.

Geographic dispersal, followed by genetic isolation, has played an important role in the process of evolution. Both the ability and inability to osmoregulate in different situations have played a part in breaking and maintaining the physical barriers involved. Deserts can retard the dispersal of animals (Darlington, 1957), as shown by the distribution of the Amphibia in Australia. Darwin (1859) considered the sea to be the major physical barrier to dispersal, a view that is still current (Darlington, 1957). The sea constitutes the most effective barrier to the movement of the freshwater fishes, and Darwin noted the absence of Amphibia and terrestrial mammals from most oceanic islands, where reptiles are frequent inhabitants. This probably reflects the osmoregulatory powers of the different groups. Reptiles with their less permeable integument, considerable tolerance to internal osmotic change, extrarenal salt excretion and marked independence of environmental temperature would appear to be better suited to prolonged oceanic voyages than freshwater and terrestrial Amphibia and mammals. At Duke University I kept a freshwater turtle, *Pseudemys scripta*, for 30 days in sea-water. At the end of this time it was sluggish and had a plasma sodium concentration of nearly 300 m-equiv/l, but on being returned to fresh water it resumed its normal life. Prolonged oceanic voyages are thus conceivable, even in contempory species of freshwater-terrestrial reptiles, a tribute to their osmoregulatory capabilities.

2. The Osmotic Anatomy of the Vertebrates

The normal osmotic concentrations of the body fluids in vertebrate species lie between 200 and 1100 m-Osmole/l of water (Table 1.1). There is sometimes a considerable tolerance to changes in concentration without vitally serious effects. The Australian lizard, *Trachysaurus rugosus*, can tolerate an increase in its plasma sodium level from 160 to 240 m-equiv/l (BENTLEY, 1959a). Some Australian frogs (*Cyclorana*) may lose water by evaporation equivalent to 50% of their body weight and survive if they are then allowed to rehydrate, but other species (*Hyla*) die after losing half as much water (MAIN and BENTLEY, 1964). Man is intolerant to such change, and may die after losing water equivalent to 12% of his body weight, or if his plasma sodium rises from 150 to 170 m-equiv/l (ADOLPH, 1947).

The predominant osmotically-active solutes in vertebrates are sodium, potassium and chloride. Proteins make up only a small part of the total osmotic concentration, but, due to their restricted movements across membranes, play an important role in the regulation of fluid movements across capillaries and cell membranes. Other solutes, such as urea and trimethylamine oxide, may, in certain osmotic circumstances, make a substantial contribution to the osmotic composition of the animals. Significant concentrations of such organic solutes occur in marine chondrichthyeans and the frog, *Rana cancrivora*, in which they contribute to maintain hypertonicity to the sea-water. Other solutes, such as calcium, magnesium, bicarbonate and phosphate are not an important part of the osmotic anatomy. They are, however, indirectly essential as they influence the functioning of osmoregulatory organs like the kidney, and are necessary for the physiological function of all living processes.

The total sodium and potassium content of the body has been measured in man, and is 68 m-equiv/kg body weight for sodium and 55 m-equiv/kg for potassium (THORN, 1960). The amount that is readily exchangeable, and thus substantially active osmotically, is 42 m-equiv/kg for sodium and 46 m-equiv/kg for potassium, so that these two cations make a similar overall osmotic contribution.

Solutes (with the exception of urea) are not distributed uniformly in the body. The fluid inside the cells (*intracellular fluid*) contains mainly potassium and organic solutes, while that outside the cells (*extracellular fluid*) consists principally of sodium and chloride. This distribution is not mutually exclusive, as small amounts of sodium and chloride may be present inside cells, while low, but essential, concentrations of potassium and protein occur in the extracellular fluid (Table 1.1).

The total body water can be estimated by measuring the dilution of certain injected solutes, or by dessication. The extracellular fluid can similarly be estimated by measurements of the dilution of known amounts of injected inulin or sucrose. Evans blue is confined to the blood plasma, and so can also be used to measure this space. The volume of the intracellular fluid is derived arithmetically, as the difference between the total volumes of body water and extracellular fluid. Using such techniques, the intracellular fluid is found to be 45 to 55% of a vertebrate's body weight, while the extracellular fluid is 15 to 25% of this. The plasma varies from about 2% of the body weight in some bony fishes, to a more usual value of about 5% in other vertebrates (Table 1.2). The proportions of the various fluid

3

Table 1.1 *Principal osmotic constituents of plasma and intracellular fluid of some vertebrates*

| | Plasma mM/Kg water (or serum) (or mM/l plasma) | | | | | | Intracellular mM Kg/H_2O | | | | | |
	Na	K	Cl	Mg	Other	Osmolarity	Na	K	Cl	Mg	Other	Osmolarity
Sea-water	470	12	550	50		1070						
Agnatha												
Myxine glutinosa[1] (Hagfish)	549	11	563	19	No TMAO	1152	132	144	107	68	Amino acid 71 TMAO 211	990
Lampetra fluviatilis[2] (River lamprey)	120	3	96	4		270						
Chondrichthyes												
Raja clavata[3] (Ray)	289	4	311		Urea 444	1050						
Pristis microdon[4] (Freshwater sawfish)			170		Urea 130	540						
Chimaera montrosa[5] (rabbitfish)	360	10	380		Urea 265							
Osteichthyes												
Platichthys (Pleuronectes) flesus[6] (Flounder)												
Sea-water	194	5	166			364	10	157	30		Amino acid 44 and TMAO 14	
Fresh water	157	5	114			304	15	158	42		Amino acid 71 and TMAO 30	
Latimeria chalumnae[7] (Coelacanth)	181		199	29	Urea 355	1181						
Protopterus aethiopicus[13] (African lungfish)	99	8	44			238						

Table 1.1 *Principal osmotic constituents of plasma and intracellular fluid of some vertebrates*

	Plasma mM/Kg water (or serum) (or mM/l plasma)						Intracellular mM Kg/H$_2$O					
	Na	K	Cl	Mg	Other	Osmolarity	Na	K	Cl	Mg	Other	Osmolarity
Amphibia												
Bufo viridis[8] (European green toad)	113	5	99		Urea 16	279	30	95	17		Amino acid 100 Urea 17	
Rana cancrivora[9] (Crab-eating frog) (in 80% sea-water)	252	14	227		Urea 350	830						
Reptilia												
Trachysaurus rugosus[10] (Bobtail lizard)	168	5					50	125				
Aves												
Larus glaucescens[11] (Gull)	154	4										
Mammalia												
Rat[12]	150	6	119	16	Urea 7 Amino acid 30	324	16	152	5	16	Amino acid 3 Urea 7	350

References: [1]BELLAMY and CHESTER JONES (1961); [2]ROBERTSON (1954); [3]MURRAY and POTTS (1961); [4]H. SMITH (1936); [5]FÄNGE and FUGELLI (1962); [6]LANGE and FUGELLI (1965); [7]PICKFORD and GRANT (1967); [8]GORDON (1965); [9]GORDON *et al.* (1961); [10]BENTLEY (1959a); [11]HOLMES, BUTLER and PHILLIPS (1961); [12]HARRIS (1960); [13]SMITH (1930a).

Table 1.2   *Distribution of water in the body of various vertebrates*

| | | g/100 g body weight | | | | |
		Total body water	Plasma	Extracellular space interstitial	Total	Intra-cellular
Agnatha						
Petromyzon marinus[1] (Lake lamprey)		75.6	5.5	18.4	23.9	51.7
Chondrichthyes						
Carcharhinus nicaraguensis[2] (F. W. shark)	FW	72.1	5.1	14.6	19.7	52.4
Squalus acanthias[2] (Spiny dogfish)	SW	71.7	5.5	15.7	21.2	50.5
Osteichthyes						
Chondrostei *Polyodon spathula*[3] (Paddlefish)	FW	74.0	2.2	13.4	15.6	58.4
Holostei *Amia calva*[3] (Bowfin)	FW	74.5	2.2	16.7	18.9	55.6
Teleostei *Cyprinus carpio*[3] (Carp)	FW	74.1	1.8	13.7	15.5	55.9
Parrot fish	SW	73.1	2.4	14.2	16.6	56.5
Reptilia						
Amphibolurus ornatus[4] (Australian lizard)	T	73.6	5.4	20.3	25.7	47.9
Mammalia						
Man[5]	T	66.0	4.3	15.7	20.0	46
Camel[6]	T	69.8	5.3	13.9	19.2	50.6

References: [1]THORSON (1959); [2]THORSON (1962); [3]THORSON (1961); [4]BRADSHAW and SHOEMAKER (1967); [5]ELKINGTON and DANOWSKI (1955); [6] MACFARLANE (1964).
SW = sea-water   FW = fresh water   T = terrestrial

compartments thus vary in the different vertebrate groups, and THORSON (1961) has suggested this may follow a phylogenetic pattern in the fishes.

Physical separation between the fluid compartments of the body takes place at the cell membrane for intra- and extracellular compartments, while the walls of the capillaries separate the intercellular and plasma phases of the latter. The skin, and linings of the respiratory organs and gut separate the body fluids from the external environment. The volume and composition of the intracellular fluid is controlled by the cell, and more especially its outer 'plasma' membrane. The extracellular fluid is regulated principally by the kidney, but may be assisted in different species by the skin, gills, 'salt' glands and gut. The maintenance of the correct relationship of volume and composition between the intercellular fluid and the plasma is due to the blood vessels, particularly the capillaries.

3. Osmotic Exchanges with the Environment

Water, solutes and nutrients continually move in and out of animals in a pattern that results ideally in their 'steady state' with the environment. Many of the osmotic exchanges are obligatory for the animal and not under its direct control, though their magnitude determines the extent of the regulatory processes that are ultimately necessary in order to achieve the 'steady state'.

This obligatory osmotic 'exchange' or 'turnover' depends on several aspects of the animal's physiology.

a) Physico-chemical Gradients with the Environment

Differences between the chemical composition of an animal and its environment will influence water and solute movements in, or out, of the beast. Thus aquatic species in fresh water will tend to accumulate water and lose salts, while the opposite occurs in bony fishes living in the sea. Temperature gradients will influence evaporative water losses especially for cooling in homoiotherms.

b) Size and Surface Area

Exchange of water and solutes is related to surface area. Large animals have a relatively smaller surface area exposed to the environment than small animals. The gill surface in fishes varies considerably and, apart from a relation to the animal's size, may reflect the normal oxygen needs of the species.

c) Nature of the Integument

The barriers exposed to the environment vary in their permeability to water and solutes. Thus, the skin of the Amphibia is usually more permeable to water than that of reptiles, while the gills of the Chondrichthyes are less permeable than those of the teleosts.

d) Metabolism and Oxygen Consumption

The exchanges of oxygen and carbon dioxide in terrestrial animals are usually accompanied by water loss, due to saturation of the expired air with water vapour. Metabolism is accompanied by the production of heat and waste solutes; the first facilitates evaporative water loss, while the latter often requires body water for their excretion.

e) Feeding

Excess water and solutes may be taken in with nutrients. Animals on a succulent herbivorous diet containing 98% water will usually have a water intake in excess of their needs, while a diet of marine invertebrates will provide superfluous salt.

The magnitude of these effects varies considerably in different species. Water exchanges in a day may be equivalent to as much as 50% of the body water in the leopard frog, *Rana pipiens,* or less than 1% in a lizard like the chuckawalla, *Sauromalus obesus.* The daily sodium exchange of a marine fish, like the flounder, is more than 20 times as great as the total sodium content of the body, but in man this exchange only represents about 3% of the total present. In many instances differences in obligatory exchanges could be adaptive with respect to osmoregulation. Some of these variations will be discussed subsequently.

4. Forces Effecting Exchanges of Water and Solutes

The osmotic composition of the vertebrates differs greatly from that of their external environments and considerable differences in solute composition exist between the different solutions present within the animal. Various physico-chemical forces act towards the attainment of osmotic equilibrium of the animal with its environment. In vertebrates these are mainly the processes of diffusion, osmosis and evaporation.

a) Diffusion

Molecules in solutions, both solute and solvent, are in a continual state of motion dictated by their kinetic energy. Collisions will take place constantly at the borders of the system. The number of collisions with the border will depend on the concentration of the molecules and their average velocity. Such collisions may result in the transfer of molecules to an adjoining phase. This will be affected by the nature of the barrier separating the two regions and the velocity or energy of the particular molecule concerned. When two such systems are separated by a barrier through which the molecules can pass, there will be a transfer or *diffusion* in both directions. If the thermodynamic conditions on both sides of the barrier are identical, then the rates of collision with each side will be the same and, although an exchange will occur, there will be no net change. There will be a *flux* in each direction (*influx* and *outflux*) but no *net flux*. If, however, due to differences in concentration, temperature or electric charge, collisions occur more often on one side of the barrier than the other, there will be a net movement towards the solution with the lower thermodynamic energy. This will cease when physico-chemical equilibrium has been attained. The nature of the separating barrier will play a critical role in the rate of such transfer, and this will be influenced by such factors as its thickness, porosity and chemical composition. Apart from its concentration and kinetic energy, the nature of the diffusing molecule itself will affect its rate of diffusion, due to its size (in inverse proportion) and, in biological systems, interactions (such as solubility) with the interspersed barrier.

Measurements of the influx and outflux of molecules across membranes, with the aid of isotopes, have revealed that such transfers are not always strictly related to differences in electro-chemical activity between the two sides. Active transport is not necessarily involved in such circumstances. The 'uphill' and 'downhill' exchange of solutes, such as sodium, may be tightly coupled to each other so that a molecule moving towards one side of a membrane is directly exchanged for another at the opposite side so that no net flux results. Thus, reducing the concentration of the solute (by replacing it with an impermeant one) on the outside will similarly reduce the rate of its flux from the inside. Ideally, on a simple electro-chemical basis, the outflux should not change in these circumstances. This effect has been called *'exchange diffusion'* (USSING, 1947; KEYNES and SWAN, 1959). Solutes involved in this process would appear to interact with the membrane in some manner. This is usually conceived as involving the participation of a 'carrier' that moves passively in either direction across the membrane but only when combined with the appropriate solute. Influx and outflux are thus tightly coupled to each other and are the same in each direction. Ideally they do not follow classical electro-chemical gradients. This process plays an important role in controlling the permeability of the gills of some fish to sodium.

b) Osmosis

When two solutes of different concentrations are separated by a barrier which prevents (semipermeable), or strongly restricts, the movement of the solute, the solvent will flow from the side with the lower to that of the higher solute concentration. This flow of solvent is called *osmosis*. The application of an opposing pressure will impede the movement of the solvent. If the fluid on one side is pure solvent then the pressure that must be applied to the opposing solution in order to stop net water transfer is called its osmotic pressure. A pressure of 22.4 atmospheres is required to prevent movement of water across a semipermeable membrane into a 1 molal (or osmolal) solution (1 M or Osm.).

The osmotic movement of water can be viewed either as its molecular diffusion or as a 'bulk' flow in a continuous phase.

(i) Molecular diffusion. The solute is seen as reducing the number of favourable collisions of solvent molecules with the dividing membrane, by interacting with them and/or lowering their molecular concentration. Hydrostatic pressure may then influence the movement of the solvent molecules by altering their kinetic energy (HARRIS, 1960).

(ii) 'Bulk flow' of fluid may result from hydrostatic forces that arise in narrow channels, or 'pores', through which the solvent flows as a continuous phase. These forces arise in such channels at the junction of the membrane and solution (see MAURO, 1965).

9

The osmotic behaviour of a solution is a colligative property, related to the number of molecules of solute(s) present. It will be influenced by the degree of impermeability of the membrane to the solutes, its permeability to water and differences in hydrostatic pressure between the two sides.

Electro-osmosis is water movement that results from the application of an electric field to a solution containing electrically charged substances. If the charged materials are a fixed membrane, then water will move relatively to this structure and across it. Such water flow has been often considered in biological systems but does not appear to play a major role in substantial movements of water.

c) Evaporation

The change from a liquid (or solid) into a vapour or gas is called vaporization or evaporation. Kinetically this can be looked at as the escape of molecules, with a greater than average kinetic energy, across a boundary from a liquid to a gas phase. The excess kinetic energy of the molecules will be lost to the solution and this will be translated as heat loss. When water evaporates at 100 ° the loss of heat is 539 cal/g; at more physiological temperatures, it is considered for practical purposes to be 580 cal/g. Heat lost as a result of evaporation of water plays an important role in thermoregulation of birds and mammals.

Evaporation of water takes place more rapidly when it is at a higher temperature (higher kinetic energy), and when its concentration (vapour pressure) in the gas phase is low. Evaporation will cease when this phase is saturated with the vapour (saturated vapour pressure), a concentration that becomes greater with increasing temperature. If the vaporized water mixes slowly with the surrounding atmosphere, then the rate of evaporation will be slowed, while if mixing is hastened, the process will be facilitated. The disposition and arrangement of cutaneous structures such as scales, feathers and hairs may thus be expected to alter the rates of evaporative water loss. For more practical purposes the rate of evaporation of water from a surface, like the skin, will depend on the difference between the predicted saturated vapour pressure of the air at skin temperature and the actual vapour pressure of water that is present. This is called the 'saturation deficit' of the air.

d) Facilitated Diffusion

When molecules move down their electro-chemical gradients more rapidly than the properties and structure of the dividing membrane would indicate is likely to occur by simple diffusion, we may suspect that some interaction is taking place with the cell. This may be confirmed by such phenomena as a reduced rate of movement in the presence of other similar molecules (due to competitive antagonism) and, instead of strict adherence to the laws of diffusion, maximal rates of transfer in the presence of relatively small differences in concentration. In addition low concentrations of diverse substances, such as heavy metal ions, which are known to react with cell constituents, may reduce the rate of change. A definitive criterion is the absence of an energy requirement, except for the small amount that may be

10

needed to maintain the structure of the membrane. Facilitated diffusion has best been characterized in the processes of sugar transfer into cells, but may play a role in ion movement, and possibly in the movement of sodium across the outer border of the epithelial cells of amphibian skin and urinary bladder.

e) Active Transport

Active transport is described by USSING (1960) 'as a transfer which cannot be accounted for by physical forces'. Solute transfers down the sum of gradients of electrical potential, chemical concentration, temperature, pressure and as a result of 'drag' forces accompanying solvent movement *do not* constitute active transport. Transport *against* the sum of such gradients requires the expenditure of energy by the cell into or out of which solute moves.

To satisfy these criteria is not always an easy matter, for while differences in chemical concentration are relatively simple to determine *in vivo,* other contributing factors such as electrical differences may be more difficult to measure. In addition the transfer of several substances across a membrane may be linked in such a manner as to make it difficult to identify the primary process. Thus, in membranes that are permeable to both anions and cations, each ion species could conceivably move down an electrical gradient created by an active transfer of the other. While it is currently considered unlikely that water is transferred actively in vertebrates, there are instances, as in the intestine, in which it moves against an osmotic gradient. Such transfer of water has been linked to active sodium transport, and may be due to local osmotic and hydrostatic forces that arise in the membrane as a result of sodium movement; it does not constitute active transport (CURRAN, 1965). A linkage in transport of sodium and other solutes such as sugars and amino acids has also been described (CRANE, 1965).

Satisfaction of the criteria of active transport, and identification of the primary process, can often be obtained from information about the electrical conditions across membranes and the measurement of the transfer of each constituent solute in the absence of its normal associates. Electrical differences can usually be measured more easily *in vitro,* and if a difference in potential (p. d.) exists, its sign and magnitude in relation to the transfer of a cation or anion will often suggest the prime mover. Thus in the instance of the frog's skin, the inside (corium) may have a p. d. of 100 mV positive, relative to the outside solution. When the chemical concentrations are identical on both sides of the skin, sodium and chloride move towards the corial side. The charge present there would suggest that the cation is actively transported, though the possibility of a slower active transport of the anion in the same direction cannot be excluded without more information. An electrical p. d. cannot always be detected across membranes (such as the rabbit gall bladder) that transport sodium and chloride, so that such an analysis is not possible; measurements of the movements of the separate ion species with the aid of their radioactive isotopes is then necessary. Separate measurements of the transfers of each linked solute may aid the identification of the primary process. Such experiments can be difficult to perform and interpret, especially in cases where one of the solutes

11

is also an energy substrate, or if there are interactions with parts of the cell indirectly concerned with the transport process.

A very elegant, yet simple, demonstration of active sodium transport was made by USSING and ZERAHN in 1951 using the frog skin. These experiments have provided both a standard method and experimental design for the analysis of such processes. It has been known for 100 years that frog skin *in vitro* displays an electrical p.d. with the corium (inside) positive, relative to the solution on the outside (DU BOIS REYMOND, 1848). In 1937 KROGH showed that frogs can take up sodium and chloride through their skin against a chemical concentration gradient, from solutions as dilute as 10^{-5}M. The experiment of USSING and ZERAHN allowed a precise analysis of these processes to be made in this tissue. Pieces of the ventral skin of frogs were clamped between two lucite chambers each containing Ringer solution of identical composition. The electrical p.d. between the two sides was measured and found to be 40 to 50 mV, inside (corium) positive. An external current was applied from a battery in the opposite direction to the biological current, so as to reduce the p.d. to zero, thus effectively short-circuiting the skin. Under these conditions no electrical or chemical concentration difference existed across the membrane. Sodium transfer in either direction was measured with the aid of ^{22}Na(influx) and ^{24}Na(outflux) and it was found that the influx exceeded the outflux, clearly demonstrating the active transfer of sodium. In addition it was found that when the current used to short-circuit the skin (the short-circuit current, SCC), and the net sodium transfer, were both expressed as millicoulombs they were nearly equal. This rules out the active transfer of chloride, which in such a system carries the current through the external circuit. These experiments not only afford a clear demonstration of active sodium transport, but the methods have in addition been used for the analysis of the ion permeability of other membranes such as the intestine and anuran urinary bladder. Vertebrate hormones can alter sodium transport in such *in vitro* systems and have been widely used to examine the actions of such hormones more closely.

When more than one substance is moved simultaneously, arguments as to the prime mover are often of semantic as well as biological interest. Such co-transport, linked to active sodium transport, influences the movement of many sugars and amino acids across the plasma membrane as well as across epithelial membranes (transepithelial). Whether or not sodium transfer itself is an example of such co-transport linked to the primary movement of a metabolite in the cell is not clear, but affords an example of the difficulties in such a classification. Active sodium transport would appear to take place transepithelially in the gut, amphibian skin and urinary bladder, fish gills and in the tubular systems of the kidney and glands like the 'salt glands' of birds, reptiles and chondrichthyeans. Active chloride transport may also occur across some of these membranes but while the evidence for this is accumulating, its unequivocal elevation to such a status seems to be lagging among the more conservative 'transport workers'. For the purposes of physiology it is nevertheless important.

Transport of sodium across the plasma membrane itself is ultimately responsible for both transepithelial transfer and the normal functioning of the cell. Active extrusion of sodium, usually coupled to active accumulation of potassium, has been demonstrated in many tissues including the erythrocyte, muscle and nerve cells.

Biologically important substances such as calcium, magnesium and urea may be actively transported in certain circumstances but it is difficult to obtain satisfactory evidence that these are primary instances of such a process.

Water usually moves readily by osmosis across plasma membranes but is *not*, according to contemporary evidence, subject to active transport in vertebrates, though it may indirectly be influenced by such processes. The permeability of membranes to water may nevertheless be subject to alteration; in the tetrapod kidney and the skin and urinary bladders of some Amphibia, neurohypophysial hormones increase the osmotic permeability of the membranes.

The process of active transport requires energy which, in the cases of sodium and potassium transport, is now generally considered to be derived from a labile-high energy phosphate group present in adenosine triphosphate (ATP) and which becomes available when it is broken down to the diphosphate (ADP). A very elegant demonstration of the relationship of ATP to active sodium transfer was made by Caldwell, Hodgkin, Keynes, and Shaw (1960) using the squid axon. The supply of high energy phosphate in the axon was allowed to disappear under the influence of cyanide, sodium transport declining simultaneously. When ATP, or compounds that could be converted to ATP, were injected into the axon, active sodium extrusion was resumed.

ATP is formed in cells by the metabolism of carbohydrates, fats and proteins. Some of the interrelationships of the metabolism of such substrates and the formation of ATP is given in Fig. 1.1. One g-mole of glucose, during aerobic metabolism, gives with the aid of 6 g-moles of oxygen, 36 g-moles of ATP. Thirty of these result from the aerobic metabolism of the intermediary pyruvic acid in the

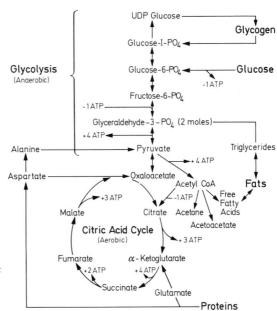

Fig. 1.1. Summary of the metabolic transformations involved in the production of ATP from carbohydrates, fats and proteins.

13

KREBS citric acid cycle. Fatty acids and amino acids may also enter the citric acid cycle. As the supply of ATP is limiting to sodium transport, the reactions and substrates from which it is formed may influence the rate of sodium movement, and it is at such sites that hormones which alter sodium transfer could be expected to act.

Estimates of the number of ions transferred in relation to each molecule of ATP hydrolysed depend principally on measurements of oxygen consumption in relation to the rate of sodium transport. Such estimates range from the transport of 4 to 28 sodium ions for each molecule of oxygen used, though 14 to 18 ions per molecule of O_2 is more usual. It will be recalled that six molecules of ATP are produced for each molecule of oxygen utilized so that we can expect 2 or 3 sodium ions to be transported for each molecule of ATP broken down.

The nature of the linkage of the phosphate bond energy provided by ATP, with the 'ion jump' from one electro-chemical phase to another is not clear. SKOU in 1957 noted the association and similar properties of the enzyme Na-K ATPase with the sodium 'pump' in peripheral nerves. This enzyme requires Mg^{2+} as well Na^+ and K^+ for its activation and accelerates the breakdown of ATP to ADP with the release of the energy and inorganic phosphate. The enzyme has been found in a wide variety of tissues noted for their ability to transport sodium actively and including, apart from nerve, muscle, red blood cells and tissues especially noted for transepithelial ion transport, such as the kidney, 'salt' glands, gills and amphibian skin and urinary bladder (see CSAKY, 1965). Histochemical and cell fractionation procedures indicate that the Na-K ATPase is localized in or near the cell membrane across which 'ion jumps' may be expected to occur. In addition, the sodium 'pump' and ATPase have a number of similar, rather specialized characteristics; both may be inhibited in a similar way by the drug ouabain (strophanthin G) and can be activated by sodium and potassium. It thus seems that ATPase and the Na 'pump' are closely associated and it is possible that the enzyme itself acts as the final 'carrier' for the ions, but the relationship and method of bringing about the final 'ion jump' is unknown.

Hormones have not been shown to influence Na-K ATPase directly but may do so indirectly by changing the local concentrations of the ionic substrates and possibly ATP. Levels of Na-K activated ATPase are somewhat labile depending on environmental conditions. Increasing temperature can facilitate the amount of this enzyme in the goldfish gut (SMITH, COLOMBO, and MUNN, 1968). Adaptation to salt water increases Na-K ATPase in the gills of the eel (UTIDA et al., 1966) and the killifish, *Fundulus heteroclitus*, (EPSTEIN, KATZ, and PICKFORD, 1967) and the intestine of the eel, *Anguilla japonica* (OIDE, 1967). Ducks adapted to drinking hypertonic saline solutions show an increased level of Na-K ATPase in their nasal 'salt' glands (FLETCHER, STAINER, and HOLMES, 1967). It is possible that hormones may be involved in such changes.

Conclusion. For the purpose of our theme it is primarily important to know *if* solutes and water can move across a membrane in the animal, in which *direction* this may take place, and whether such movement(s) require *some* metabolic intervention by the cell. The latter makes a 'control system' feasible, while precise information about such a metabolic link may suggest the nature of the 'signal' that activates such a system.

14

5. Biological Structures Participating in Fluid Exchange

The fluids of the body are separated from each other by the cell membrane and capillary, and from the external environment by the skin, the epithelia of the respiratory organs (gills and lungs) and the gut. The integrity of the intracellular fluid is dependent on the cell itself and its surrounding membrane, as well as the composition of its bathing (extracellular) fluids. The regulation of the components of the extracellular fluid is due principally to the kidney, but also to the gills, various 'salt' glands, the gut and even the skin in some species.

a) The Cell Membrane

The contents of cells resist mixing with their external bathing solutions. This property has resulted in the concept that the fluid contents of cells are contained behind a surface skin or membrane, called the *cell membrane* or *plasma membrane,* which represents the immediate barrier separating the intracellular from the extracellular fluid. Numerous properties, related to both the general homeostasis of the cell and to the special functions localized in certain types of cells, have been attributed to the cell membrane. This membrane concept has been developed from studies of the movements of water and numerous solutes between different cells and their immediate environment under divers conditions, including the presence and absence of metabolites, electrical stimulation and the presence of hormones. While it is not always clear which of the properties of the cells reside precisely with the external membrane, the presence of such a structural barrier can be demonstrated. This is shown very elegantly in the isolated axolemmal 'sheath' of the giant axon of the squid *Loligo.* BAKER, HODGKIN, and SHAW (1961) isolated this external membrane by pressing the internal axoplasm from the nerve fibre with a squeegee, leaving the axolemma intact. This 'sheath' could then be refilled with various artificial salt solutions and was shown under such conditions to conduct nerve impulses in a manner similar to the intact axon. Such conduction is associated with a regular pattern of sodium and potassium movements across the membrane. In addition, BAKER and SHAW (1965) have shown that the isolated perfused axolemma can breakdown ATP and contains ouabain-sensitive Na-K ATPase, the enzyme that is closely associated with the function of the sodium 'pump'.

The behaviour of the cell with respect to movements of various solutes and water has resulted in the following general views about its structure and properties.

(i) Sieve or pore-like nature. The movement of solutes between the cell and its environment are often related to their molecular size; large molecules move less rapidly than smaller ones. This contributed to the concept of a porous or sieve-like structure, which discriminates by reason of having holes of different sizes that will admit certain molecules while excluding others. The overall permeability of the cell will depend on the total area and length of the pores as well as their diameter.

(ii) Solvent-like nature. Lipid soluble substances such as ethers, aldehydes and ketones may enter cells very rapidly compared with more water soluble solutes. This suggests that the outer cell membrane has a lipid component, which can distinguish between materials by being able to act as a solvent for some of them.

15

(iii) Metabolically-oriented nature. The relative movements of many substances, in or out of the cell, cannot be explained by reason of their size or lipid solubility and indeed their direction of movement may, on simple physico-chemical grounds, be somewhat unexpected. This suggests that substances may interact chemically with the membrane in such a way as to facilitate or even promote their transfer.

(iv) Excitable nature. Many substances interact with cells to change their function and permeability. This may be manifested as an internal metabolic transmutation, the conduction of an electrical impulse, contraction of a muscle or the secretion of a gland. Such responses are usually highly specific, both with respect to the chemical structure of the excitant, and the cellular component in the reaction (the receptor).

The considerable diversity that exists in the transit of water and divers solutes between different cells and their bathing solutions suggests that the cell membrane is a mosaic of all these properties including pores, lipid solvents and metabolically mediated processes. This is combined with an ability to interact and respond to excitants, including metabolites and hormones.

The 'cell membrane' may be viewed as an array of physico-chemical and biochemical properties not necessarily located at the cell surface or even present in a discrete structure. However, biologists and especially anatomists, usually shrink from such ethereal concepts so that a considerable effort has been made to find if the molecular structure and physico-chemical properties at the cell surface correspond with those attributed to the 'cell membrane'. In the laboratory, the external cell membranes may be physically separated from other cell constituents by procedures that involve exposure to hypoosmotic solutions, as with red blood cells, or by carefully breaking up the tissue and separating the components by differential centrifugation. Such isolated pieces of membranes may be analyzed (see STEIN 1967) to determine their chemical composition and enzymatic activities. Mammalian red cell membranes are generally found to contain about equal proportions of protein and lipid, the latter consisting of similar amounts of cholesterol and phospholipids. The proportional composition is, however, found to vary in cells from different tissues and species. Such differences may contribute to the distinctive permeabilities of diverse cell types. The external cell membrane of the red cell is also found to possess Na-K activated ATPase and acetylcholinesterase activity.

Molecular arrangement of the cell membrane. The membrane at the border of the cell can be seen under the electron microscope but is too small to observe in detail by light microscopy. Even before the advent of the electron microscope DAVSON and DANIELLI, (see DAVSON and DANIELLI, 1952) on the basis of the then current chemical and physical information, suggested that the lipids in the membrane were arranged as a bimolecular leaflet oriented with the ends of their hydrophobic hydrocarbon chains facing each other inwardly, while the opposing hydrophilic heads of the molecules faced outwards. These hydrophilic 'heads' were both thought to be associated electrostatically with a globular protein layer. The cell membrane was thus conceived as being a sandwich: lipid betweeen protein.

The electron microscope reveals structures which are fairly consistent with such a model; two electron dense lines separated by an electron transparent region having a total width of about 100 Å. This structure has been called the unit membrane

16

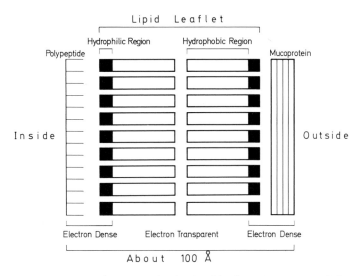

Fig. 1.2. Diagrammatic representation of the bimolecular lipid leaflet arrangement of the plasma membrane according to the DAVSON-DANIELLI-ROBERTSON model.

and was first described by J. DAVID ROBERTSON (1958 and 1964). ROBERTSON suggested that the electron transparent area represented the hydrophobic chains of the lipids, while the dark border on either side consists of their hydrophilic 'heads', together with a protein or polypeptide at the inner surface and a mucoprotein or mucopolysaccharide on the outside. This later hypothesis deviates from the DAVSON-DANIELLI model and was suggested by differences in the appearance of each border when 'fixed' in either osmic acid or potassium permanganate. This DAVSON-DANIELLI-ROBERTSON model has gained widespread acceptance (but see the repeating unit of GREEN and PERDUE, 1966) and is supported by evidence obtained by X-ray diffraction and optical birefringence techniques, as well as the behaviour of lipid-protein-water mixtures *in vitro*.

The detailed relationship of such a general molecular arrangement to the diverse special characteristics of various types of cells is not yet clear. Thus, how do such molecular specific functions as facilitated diffusion, active transport and excitant-receptor sites fit into such a structure? Unit membranes may contain numerous molecular species of proteins, phospholipids and cholesterol and it is conceivable that their characteristic properties reflect differences in their precise molecular composition and arrangement. Such specific information on the structure of cell membranes is not yet available.

b) Capillaries

The capillaries constitute the barrier separating the interstitial (or intercellular) fluid from the blood plasma. They are tubular vessels roughly 10^{-2} mm in diameter with walls about 10^{-3} mm wide and are connected at either end with arterioles and venules. Histologically they consist of three layers; an inner lining of flattened endothelial cells lying on a basement membrane surrounded by an adventitial layer of cells. The individual endothelial cells are separated by intercellular regions consisting of a 'ground substance' to which some of the special properties of capillaries have been attributed.

Fluids mainly move across capillaries by hydrodynamic or 'bulk' flow that is superimposed on diffusional movements of water and solutes. Water is transferred 1000 times more rapidly than across the red cell wall. Modern concepts about the process of fluid transfer across capillary walls were initiated by STARLING and have been called the STARLING-filtration-absorption principle (LANDIS and PAPPENHEIMER, 1963). Our current knowledge about such processes is substantially due to LANDIS and PAPPENHEIMER and the review cited should be consulted for a detailed account of the subject. The transfer of fluid between the plasma and the interstitial solutions is the result of the balances of hydrostatic and colloid osmotic pressures on either side of the capillary wall. Filtration is assisted by the hydrostatic pressure of the plasma and the osmotic pressure of the interstitial fluid, and is opposed by the corresponding pressures in the interstitial and plasma fluids respectively. The situation can be summarized as follows:

Fluid movement = filtration coefficient $(P_c - \pi_{pl} - P_{if} + \pi_{if})$
P = hydrostatic pressure π = osmotic pressure; c = capillary
if = interstitial fluid
Average corresponding pressures mm Hg in man; 32 − 25 − (1 to 9) + (0.1 to 5)

In the proximal (arteriolar) end of the capillaries the forces result in filtration. Water and small solutes pass out of the plasma while the larger protein molecules are retained, a process of *ultrafiltration*. After passage to the distal end of the capillary, the hydrostatic pressure drops sufficiently to permit a reabsorption of fluid to take place from the interstitial solution. The hydrostatic and osmotic pressures cited vary considerably in the different regions of the body. In reptiles, Amphibia and fishes the osmotic pressure of the plasma proteins is usually equivalent to only about 10 mm Hg, but in such species the blood pressure (and presumably capillary pressure) is also lower, so that a similar balance is maintained.

Water and solute molecules up to the size of inulin (M.W. 5500, molecular radius, 12 to 15 Å) readily cross the capillary wall but larger molecules such as serum albumin (M. W. 67000, molecular radius 36 Å) do so only with great difficulty. This explains why fluid movement is influenced by the colloid osmotic pressure of the surrounding fluids, as small molecules with their relatively unrestricted movements do not produce an osmotic pressure in such a system. From physicochemical data based on differences of the movements of fluids and solutes by diffusion and hydrodynamic flow, PAPPENHEIMER (1953) calculated that if (as appears likely) fluid movements take place through water-filled 'pores', then their radius would be about 30 Å. This figure corresponds rather well with the limitations of solute movement which are dictated by their molecular size. Such 'pores' have not been seen under the electron microscope. It is considered, that if they exist, they most likely are in the intercellular regions between the endothelial cells. Lipid-so-

18

luble materials cross the capillary boundary very rapidly and, as this property is also exhibited by the plasma membrane, it probably takes place through the cells themselves.

Changes in the filtration coefficient of capillaries do not seem to occur in normal circumstances (but note changes in renal glomerular filtration in lower vertebrates), but their permeability may be altered under extreme conditions associated with mechanical, bacterial, and chemical injury. Conceivably, capillary filtration could be affected by: (1) alteration in capillary pressure, (2) changes in the permeability of the wall to proteins and (3) changes in concentrations of proteins in the body fluids. The latter may occur in severe starvation and malnutrition resulting from a dietary protein deficiency, and result in a characteristic accumulation of fluid in intercellular spaces (oedema fluid). Infections and other injury may increase the permeability of the capillaries to protein, thus disturbing the balance of colloid osmotic pressure. Changes in the capillary hydrostatic pressure may result from changes in the tone of the smooth muscle surrounding the arterioles and venules at either end of the capillaries; thus noradrenaline constricts the arterioles and lowers the capillary pressure.

It is doubtful that hormones normally influence the permeability of capillaries. RENKIN and ZAUN (1955) were unable to show any such changes in perfused tissue from adrenalectomized rats.

c) Skin

Skin is a generic term used to describe the barrier separating most (but note the area of the gills in many species) of the external surface of the animal from the outside environment. It has conservative and protective functions related to the osmoregulation of animals. Such roles are fulfilled in various ways associated with its considerable structural diversity.

In its simplest form the skin consists of several layers of simple epithelial cells overlaying the dermis (or corium). Such simplicity is probably never achieved, as some degree of cellular modification is invariably present. This includes calcified and keratinized structures such s scales, feathers and hairs, which apart from having architectural and decorative significance, may affect the exchanges of water and solutes across this barrier. Glands, if present, produce a variety of secretions, including the mucins that occur commonly in the fishes and the Amphibia and the sweat of mammals. Among the Amphibia the skin has the unique ability to take up sodium from very dilute external solutions, a process which, along with water uptake, may be regulated by the activity of hormones.

There is considerable diversity in the rate of cutaneous water transfer among the vertebrates (Table 1.3), and in the importance of such exchanges relative to the total water metabolism of the animal. Such differences may have adaptive significance; evaporative water losses from the skin of the lizard, *Sauromalus obesus,* and the tortoise, *Gopherus agassizii,* which live in hot desert areas, are one-tenth to one-twentieth that of the lizard, *Iguana iguana,* which lives in tropical forests (Table 1.3). Such cutaneous water losses at 25 ° C make up about two-thirds of the total evaporative loss in such reptiles. Osmotic water transfer across the skin

Table 1.3 *Water exchange across the skin of various vertebrates*

	Fresh water	$\mu l/cm^2\,h$ Sea-water	Evaporation
Agnatha			
Lampetra fluviatilis[1] (River lamprey)	2.5		
Amphibia[2]			
Anura			
Xenopus laevis (Clawed toad)	8		
Rana catesbeiana (Bullfrog)	13	57[*]	
Bufo marinus (Toad)	28		
Urodela			
Necturus maculosus (Mudpuppy)	3		
Amphiuma means (Congo eel)	4		
Ambystoma tigrinum (Tiger salamander)	12		
Reptilia			
Crocodilia			
Caiman sclerops[3]	1		1.2
Chelonia			
Pseudemys scripta[4] (Slider turtle)	0.2	0.4	0.5
Trionyx spinifer[4] (Softshell turtle)	1.1	1.6	—
Gopherus agassizii[5] (Desert tortoise)	—	—	0.06
Lacertilia[6]			
Iguana iguana			0.2
Sauromalus obesus (Chuckawalla)			0.05

References: [1]Bentley (1962 a); [2]Bentley (1969 b); [3]Bentley and Schmidt-Nielsen (1965; [4]Bentley and Schmidt-Nielsen (1970). [5]Schmidt-Nielsen and Bentley (1966); [6]Bentley and Schmidt-Nielsen (1966).
[*]*R. pipiens;* Bentley and Schmidt-Nielsen unpublished observations.

of various species also differs (Table 1.3), being less in aquatic Amphibia, such as the South African clawed toad, *Xenopus laevis,* and the mudpuppy, *Necturus maculosus,* than in more terrestrial species like the toad, *Bufo marinus,* or the tiger salamander, *Ambystoma tigrinum.*

The sweat glands in the skin of many mammals play a role in thermoregulation, the rate of secretion increasing in response to increased body temperature. Solutes, especially sodium chloride, are simultaneously secreted, the amounts being variable and subject to physiological regulation that is at least partly under hormonal control.

d) Water Loss from the Tetrapod Respiratory Tract

Water loss takes place from the tetrapod respiratory tract as a result of evaporation that accompanies both respiratory gas exchange and, in some birds and mammals, a regulatory thermolysis.

The amount of water lost depends on the amount of air inspired and on the difference between the water content of this and the expired air. The latter is saturated with water vapour, usually near the animal's body temperature.

The evaporative water loss, that accompanies gas exchange, depends on the oxygen requirement and the efficiency of extraction of this gas from the inspired air. In mammals this is usually reduced from its normal atmospheric concentration of 21% to 13 or 14%, but in reptiles it may diminish only slightly, to 20%, or to as low as 5% (calc. from BENTLEY and SCHMIDT-NIELSEN, 1966; SCHMIDT-NIELSEN, CRAWFORD, and BENTLEY, 1966). Thus, a high rate of oxygen consumption will result in an increased minute volume and evaporative water loss, but the magnitude of the latter will depend on the amount of oxygen that can be extracted from the inspired air. The rate of oxygen consumption is related to the body temperature and in cold blooded vertebrates this may normally show considerable variation. In addition, homoiothermy places a heavy burden on the water metabolism of the animal, especially at temperatures above the thermoneutral level of about 30° C. Large amounts of water may then be evaporated in order to maintain body temperature within a physiologically acceptable range.

Endocrines can influence general metabolism and so may indirectly influence its water metabolism, due to reflected changes in respiratory gas exchange.

e) Gills

Gills are primarily tissues for respiratory gas exchanges; among the vertebrates, they are present in the 'fishes' and larval and neotenous Amphibia. The permeability of the gills to oxygen and carbon dioxide is accompanied by a diffusivity to water and some solutes. This will affect the osmotic equilibrium of the animal in various ways depending on the properties of the gills, the species and the difference between the composition of the animals' body fluids and the external solution. In addition to transfers down physico-chemical gradients, the gills of some species have an ability to transport sodium and chloride actively against such gradients in a manner that contributes to the animals' osmoregulation.

i. *Diffusion and osmosis across gills*. Freshwater fishes are hyperosmotic to their environment so that water will tend to move osmotically into the animal. A gradient identical in direction, but smaller in magnitude, also exists between sea-water and

21

marine chondrichthyeans and probably myxinoid agnathans. Precise measurements of the osmotic permeability of the gills are difficult, as their complete isolation from the rest of the body surface is onerous *in vivo*. Nevertheless, experiments with perfused and isolated gills indicate that they are permeable to water.

KROGH in 1939 made a careful assessment of the evidence available, and concluded that in freshwater teleost fish a major part of the osmotic water uptake occurs through the gill epithelium. More recent experiments on the gills of the eel (BELLAMY, 1961; MOTAIS *et al.*, 1969) indicate that they can readily take up water osmotically from hypoosmotic solutions. Furthermore, experiments on the perfused gills of a marine chondrichthyean, *Squalus acanthias,* show that they take up water osmotically, but are 10 to 15 times less permeable than membranes like the anuran skin and urinary bladder (BOYLAN, 1967). The overall contribution of the gills to osmotic water uptake by fish probably varies due to differences both in the permeability of the gills of various species and also in the gill surface area in relation to the rest of the body surface. In this latter respect, the gill area in some species has been calculated to be 60 times as great as that of the skin (see PARRY, 1966). Calculations made from experiments on isolated skin and the total water uptake of the agnathan, *Lampetra fluviatilis*, indicate that the gills and the skin have a similar osmotic permeability; about 2.5 μl/cm^{2}h (osmotic gradient 200 mOsm) (BENTLEY, 1962a), compared to 3 to 30μl/cm^{2}h in amphibian skin. The contribution of the gills of fishes to their osmotic water uptake, is a substantial one.

Osmotic water *loss* across the gills is also difficult to determine concisely but SCHLIEPER (see KROGH, 1939) suggested that the perfused eel gill bathed with sea-water has a very low permeability to water. MOTAIS *et al.* (1969), using a cannulated gill preparation in eels found it to be less osmotically permeable in sea-water than fresh water. Substantial amounts of water nevertheless leave the fish by this route.

Both BELLAMY (1961) and KAMIYA (1967) concluded from experiments on isolated gills of the eel, that sodium transfer by diffusion takes place readily in either direction; earlier experiments on intact teleost fish are consistent with this (KROGH, 1939). Sodium and urea diffuse across the perfused gills of the marine chondrichthyean, *Squalus acanthias*, but the permeability is very restricted compared to amphibian membranes (BOYLAN, 1967). Recent experiments on intact fish support these conclusions. BURGER and TOSTESON (1966) measured the influx of sodium through the anterior end of the spiny dogfish, *Squalus;* their results indicate that sodium is accumulated by diffusion through this channel. The rate of sodium influx in the dogfish, *Scyliorhinus caniculus,* is similar to *Squalus* (MAETZ and LAHLOU, 1966) but 30 to 50 times less than in marine teleosts (MOTAIS and MAETZ, 1965).

In summary: it appears that water and sodium chloride move by diffusion across the gills of fish and that this occurs far more rapidly in marine teleosts than chondrichthyeans.

ii. *Active transport of sodium and chloride. Uptake.* Sodium and chloride may be accumulated against their concentration gradients by the gills of the agnathan *Lampetra fluviatilis* (WILKGREN, 1953), by a variety of freshwater teleosts, and possibly by the freshwater chondrichthyean, *Pristis microdon* (KROGH, 1939). In the goldfish, *Carassius auratus,* the transport of sodium and chloride may be independent, and takes place in exchange for ammonium and bicarbonate ions respectively

(GARCIA ROMEU and MAETZ, 1964; MAETZ and GARCIA ROMEU, 1964). It is not clear what proportions of the total ion transfers are coupled in this manner. Sodium transfer across the gills of freshwater chondrichthyeans has not been investigated.

Output. KEYS in 1931 showed that the perfused gills of the eel could secrete chloride into an external solution of sea-water. Active sodium extrusion has also been demonstrated in the isolated gills of these eels (BELLAMY, 1961). Experiments on intact teleost fish also show that this commonly occurs in marine species. An active secretion of sodium or chloride has not been shown among the Chondrichthyes though, as shown by MAETZ and LAHLOU (1966) and suggested by BURGER and TOSTESON (1966), considerable efflux of sodium and chloride takes place from the gills of dogfish. Whether or not this involves active transport is unknown. Extrarenal salt excretion through the gills of the marine myxinoid agnathans is not considered likely (McFARLAND and MUNZ, 1965; MORRIS, 1965) but has not been adequately investigated.

The detailed mechanisms of active ion transport across the gill epithelia of fish are not known, but an overall common denominator of this process in vertebrates, the enzyme Na-K activated ATPase, has been found in the gills of the killifish, *Fundulus heteroclitus* (EPSTEIN et al., 1967) and eels (UTIDA et al., 1966; MOTAIS, 1970). The activity of this enzyme in these fish was also shown to increase when they were adapted to sea-water, a situation in which they may be expected to secrete more sodium. The cellular site for ion transport in fish gills was suggested by KEYS and WILLMER (1932) to be a specialized acidophil cell, termed a 'chloride secreting cell', which is rich in mitochondria (MORRIS, 1957). The evidence for such a site is equivocal and contentious (see PARRY, 1966), but nevertheless may be correct.

The role of the gills in osmoregulation of larval and aquatic Amphibia is not known. In the mudpuppy, *Necturus maculosus,* the external gills are not an important avenue for ammonia excretion (FANELLI and GOLDSTEIN, 1964). The tadpoles of the marine frog, *Rana cancrivora,* excrete salt extrarenally, and circumstantial evidence suggests that the gills may be involved (GORDON and TUCKER, 1965).

f) Gut

The gut is the somewhat fastidious barrier between the dietary food and water and the body fluids. Additional endogenous fluids are continually secreted into the gut as bile, the secretions of salivary glands, exocrine pancreas and the glands in the wall of the alimentary tract. Intake is largely related to 'hunger' and 'appetite'. The secretions are also largely related to the digestive requirements of the animal. The volume of such secretions is considerable; a man in one day secretes into his digestive tract 1600 to 9700 ml. of water, 650 to 1040 m-moles of sodium and 50 to 82 m-moles of potassium (KRUHOFFER and THAYSEN, 1960). The gut regulates the composition of the body fluids with the aid of thirst and 'salt' appetites while also conserving water and solutes by reabsorbing the secretions. In reptiles, birds and mammals the gut represents the only avenue for accumulation of water and salts, but in the Amphibia and fishes these may also be taken up through the skin and gills.

Morphologically the gut may be a simple tubular structure, as in the Agnatha, or it may be anatomically divided into functional regions such as the stomach and small and large intestine (or colon). A cloaca into which the urinogenital ducts also open may be present. The small intestine, colon and possibly the cloaca are the segments from which the water and salts are mainly absorbed.

Drinking, associated with 'thirst', is a common animal habit but is not universally apparent in all vertebrates. Most reptiles, birds and mammals drink, though even in these groups, as in marine mammals and some desert species, there are exceptions. Animals that eat very succulent food may also not require water from drinking. The Amphibia take up water osmotically through their skin and even when dried out do not drink, though they may do so when placed in hyperosmotic salt solutions (KROGH, 1939; BENTLEY and SCHMIDT-NIELSEN, unpublished observations). Freshwater teleost fish and marine chondrichthyeans do not drink, but marine teleosts are dependent on this custom and die of dehydration if prevented from doing so. The sea-water that these teleosts drink is hyperosmotic to their body fluids, but they excrete most of the salt extrarenally by way of the gills (H. SMITH, 1930b; 1931; KEYS, 1931).

The volume of sea-water imbibed varies, depending on such factors as body size and gill surface; in the eel HOMER SMITH found drinking to be about 40 ml/kg body weight in a day, while in the flounder, *Platichthys flesus,* it is about 240 ml/kg in the same period (MOTAIS and MAETZ, 1965). The latter investigators found that the sodium uptake through the gut of the flounder amounted to about 144 m-equiv/kg day, which represents less than one-quarter of the total sodium excreted through the gills in that time. The marine myxinoids, *Eptatretus stouti* and *Myxine glutinosa,* also swallow substantial amounts of sea-water, but this does not appear to be appreciably absorbed (McFARLAND and MUNZ, 1965; MORRIS, 1965).

Ingested water and solutes are principally absorbed from the intestinal region of the gut. Water and many solutes may move down their diffusion gradients in either direction across the intestinal wall but sodium, and probably chloride, can be transported actively across the intestine from the lumen to the blood. Water can also move against an osmotic concentration gradient but this is linked to the transport of sodium and probably results from local osmotic and hydrostatic gradients set up in the gut wall (CURRAN, 1965). While a detailed knowledge of such processes has been mainly obtained on mammals, sodium transport has been indicated in the intestine of the Amphibia(USSING and ANDERSEN,1955), reptiles (BAILLIEN and SCHOFFENIELS, 1961) and marine and freshwater teleosts (HOUSE and GREEN, 1963; M. SMITH, 1964).

While in most vertebrates sodium absorption from the intestine is essential for the maintenance of a positive sodium balance, in the marine teleosts it is primarily related to the animals' water balance. HOMER SMITH (1930b) found that while monovalent ions, like sodium and chloride, were readily absorbed from the intestine of marine teleosts, considerable amounts of magnesium, calcium and sulphate remained and were excreted in the faeces.

Differences between the permeability of the intestines of freshwater and marine teleosts to water and sodium have been observed. The rates of passive water movement are greater in eels from sea-water than those adapted to fresh water (SHARRATT *et al.,* 1964b; UTIDA, ISONO, and HIRANO, 1967). The rate of sodium transfer across the isolated intestine of the freshwater teleost, *Carassius auratus*, (M. SMITH, 1964) is much less than in the marine teleost, *Cottus scorpius* (HOUSE and GREEN, 1963). It is also very interesting that the intestine of Japanese eels adapted to sea-water has a much higher concentration of Na-K activated ATPase than the same species which has been living in fresh water (OIDE, 1967). Hormones may be concerned with such changes.

The cloaca of the reptile, *Caiman sclerops*, (BENTLEY and SCHMIDT-NIELSEN, 1965) and the domestic fowl (SKADHAUGE, 1967) is concerned with active transport of sodium from the luminal side of the gut to the blood, but its role in osmoregulation is still somewhat enigmatical. In domestic fowl only 4% of the urinary sodium is reabsorbed from the gut of sodium-loaded birds, but this proportion may be greater in normal animals. The sodium concentration of the cloacal fluid of the caiman is reduced by about 50% after storage for three hours in this region, suggesting a relatively greater role for the cloaca in this species. KNUT SCHMIDT-NIELSEN and his group (SCHMIDT-NIELSEN *et al.*, 1963) have suggested that the cloaca and the nasal 'salt' gland in some reptiles and birds may act together to conserve water. Thus sodium and accompanying water may be reabsorbed from the cloaca, and the salt subsequently excreted as a very concentrated solution by the nasal gland.

Fluid reabsorption (water and small solutes and ions) may occur from cloacal solutions which have a similar crystalloid osmotic pressure to the blood plasma. Such water absorption is not necessarily linked to active sodium transport. In a series of very ingenious experiments MURRISH and SCHMIDT-NIELSEN (1970) have shown that differences in the colloid osmotic pressure across the cloaca of the desert iguana, *Dipsosaurus dorsalis*, are sufficient to account for fluid reabsorption. They demonstrated the presence of such gradients (measured intracloacally as an equivalent hydrostatic pressure) and these could be increased by dehydrating the lizards or abolished by placing protein solutions in the cloaca. The magnitude of the gradients in colloid osmotic pressure corresponds to the forces holding water in cloacal pellets from the lizards.

Potassium is an important ion in osmoregulation and is absorbed from the alimentary tract. Little is known about the mechanism by which this takes place and it is not clear whether it represents another example of active transport in this tissue (USSING, 1960).

The role of the gut in osmoregulation is not as dramatically apparent as that of the kidney and 'salt' glands (see later), but nevertheless usually is the site for the initial processes involved in accumulation of water and salt.

g) *Urinary Bladder*

The urine in many species is stored for a period of time before being finally voided. Some fishes have a urinary bladder which morphologically represents an expansion of the mesonephric ducts. The typical tetrapod urinary bladder originates as a ventral diverticulum of the cloaca. Such a bladder is present in most species with the exception of the birds and some reptiles, like the snakes and crocodiles. Exchanges of water and solutes can take place across the bladder, but in mammals there is no evidence to suggest that such transfer plays a significant role in osmoregulation. In reptiles and the Amphibia, however, the urinary bladder may act as a useful storage organ for water, and further conservation of urinary sodium may take place by active transport from the urine back into the blood (see BENTLEY, 1966a).

CHARLES DARWIN in his account of the voyage of the 'Beagle' (1839) had the following comments to make about the bladders of frogs and the tortoises of the Galapagos.

"I believe it is well ascertained, that the bladder of the frog acts as a reservoir for the moisture necessary to its existence; such seems to be the case with the tortoise. For some time after a visit to the springs, their urinary bladders are distended with fluid, which is said to decrease gradually in volume and to become less pure. The inhabitants (men) when walking in the lower district, and overcome with thirst, often take advantage of this circumstance, and drink the contents of the bladder if full; in one I saw killed, the fluid was quite limpid, and had only a very slightly bitter taste."

Australian aborigines, who inhabit the arid interior regions of the continent, on occasions drink water stored in the urinary bladders of desert frogs. The volume of fluid held in the bladders of such amphibians may be equivalent to as much as 50% of the normal body weight of species such as *Cyclorana platycephalus* and *Notaden nichollsi*, while in a turtle, *Malaclemys centrata*, I have found urine amounting to 20% of the body weight stored in the bladder.

h) Salt Glands

Extrarenal salt excretion plays a major role in the osmoregulation of some species of reptiles, birds and fishes. Apart from some epithelial cells in the gills of fish, certain glandular tissues also may have this function. The role of such tissues in birds and reptiles was first described by KNUT SCHMIDT-NIELSEN and his collaborators and has been admirably reviewed by him (SCHMIDT-NIELSEN, 1960 and 1965).

'Salt' glands are so named because of their remarkable ability to secrete solutions containing high concentrations of sodium, potassium and chloride. The volumes of such secretions may be large. SCHMIDT-NIELSEN, for instance, has shown that the salt gland of the herring gull secretes at about twice the rate of the human kidney on a unit body weight basis and 20 times the rate in relation to the relative weights of the glandular tissues. The secretions are hyperosmotic to the body fluids and in most instances are even more concentrated than sea-water. They thus afford a channel for ion excretion which is economical to accompanying obligatory water loss, and may even allow the extraction of osmotically-free water from imbibed sea-water (Table 1.4).

Several distinct types of glands may act as 'salt' glands. In the head, these may open into the nasal cavity, termed nasal glands, or they may be modified orbital, Harderian or lachrymal glands. Nasal salt glands have been described among the birds and reptiles while orbital glands have this function in marine turtles. In the chondrichthyean, *Squalus acanthias*, (the spiny dogfish), the rectal gland, which opens into the distal part of the gut, functions as a salt gland (BURGER and HESS,

Table 1.4 *Sodium and potassium concentrations in secretions of vertebrate salt glands*

	m-equiv/l	
	Sodium	Potassium
Sea water	450	12
Chondrichthyes		
Squalus acanthias[1] (rectal gland) (Spiny dogfish)	540	7
Reptilia		
Malaclemys terrapin[2] (Diamondback terrapin)	616—784	
Caretta caretta[2] (Loggerhead turtle)	732—878	18—31
Iguana iguana	340—894	316—862
Dipsosaurus dorsalis[3] (American desert lizard)	494	1387
Uromastyx aegyptus[3] (North African desert lizard)	639	1398
Aves		
Larus argentatus (Herring gull)	718	24
Oceanodroma leucorhoa[4] (Leach's petrel)	900—1100	—

References: [1]BURGER and HESS (1960); [2]SCHMIDT-NIELSEN and FÄNGE (1958); [3]SCHMIDT-NIELSEN *et al.* (1963); [4]SCHMIDT-NIELSEN (1960).

1960). Histologically salt glands are tubular, usually with peripheral channels radiating from central canals into the various lobes that are present. The secretory cells of the tubules have densely packed mitochondria and are rather similar in appearance to the mammalian renal tubular cells. These tissues contain high concentrations of the enzyme Na-K activated ATPase, which undoubtedly plays an important role in their ability to secrete ions.

The initial stimulus for secretion is an osmotic change; this is indicated by the ability of injected hyperosmotic solutions of sodium chloride or sucrose to increase the rate of flow from the glands. In the dogfish 'volume' receptors may be involved (BURGER, 1962, 1965). In the birds and reptiles (but not in the spiny dogfish) the parasympathetic nerve supply has an intermediary action as shown by the increased secretion that follows stimulation of this nerve supply to the gland, or the injection of acetylcholine. The adrenal hormones may also play a role in birds and reptiles, as adrenalectomy and injections of such steroids, may respectively reduce or increase the rate of secretion. These topics will be dealt with in detail later.

i) The Kidney

Kidneys are present in all vertebrates; they play a characteristic and central role in osmoregulation. While it is doubtful that any species could survive for long in the absence of this organ, its relative overall importance differs. This is due, as we have seen, to the presence of other tissues such as 'salt' glands and gills that may also contribute to the regulation of solute metabolism. DICKER (1970) has recently given a complete and lucid description of the function of mammalian kidneys.

α) *Functions.* The kidney is the only organ in vertebrates that can secrete large volumes of water containing solutes at concentrations less than 1% of their total level in the body fluids. It thus has a role as an organ for excretion of water, while at the same time conserving solute. In mammals, the kidney is principally responsible for regulatory secretion of excess mono- and divalent ions. However, among the birds, reptiles and Chondrichthyes, the 'salt' gland also contributes to regulation of sodium and potassium, while in marine teleost fishes the gills are the principal site for expulsion of sodium and chloride. The kidney nevertheless remains, in all species, the main avenue for excretion of divalent ions, like Ca^{2+}, SO_4^{2+} and HPO_4^{2+}.

Solutes that are physiologically undesirable and arise as by-products of metabolism, may be excreted through the kidney, and to some extent through the gut, skin and gills. These include: by-products of nitrogen metabolism such as urea, uric acid and ammonia; creatinine, creatine and the remnants of steroid and protein hormones. A variety of organic molecules may, by fair means or foul, gain access to the animal from its external environment; in a valetudinary species like man this is particularly prevalent, taking the form of various drugs and potions.

Control of the concentration of free hydrogen ion (pH) and the buffering capacity of the body fluids is due to both carbon dioxide exchange by the respiratory organs and the renal regulation of bicarbonate and phosphate concentrations. This is a complex process involving directly the passage of hydrogen ions into the urine, and more indirectly (but vitally) the secretion of ammonia by the renal tubular cells.

The kidney thus has functions as an organ for excretion, conservation and regulation of the constituents of the animal.

β) Structure. The vertebrate kidney is composed of a large number of units called *nephrons.* These are tubular structures which have a tuft of capillaries (the *glomerulus*) invaginated into a capsule (BOWMANS capsule) at their anterior end. The total number of these units differ; in mammals the range is from about 25 000 in a mouse to 15 million in an elephant (H. SMITH, 1951). The usual pattern is one glomerulus to each tubule, though in the Agnatha (GERARD, 1954) a large glomerulus may communicate with several tubules while in some marine fishes, such as the goosefish, *Lophius,* and the toadfish, *Opsanus,* glomeruli are absent. The size of glomeruli also shows considerable variation, and they may be cigar-shaped structures 1 400 μ long in the lamprey (Agnatha) (BENTLEY and FOLLETT, 1963) or, as seen more usually in most other species, oval tufts 60 to 200 μ in diameter (H. SMITH, 1951).

The renal tubules are segmented into units of varying length and diameter. In the Agnatha a prominent ciliated neck segment is present at the junction of the

BOWMAN's capsule, but this is reduced in size, or absent, in other groups of vertebrates (MARSHALL, 1934). The initial segment (proximal tubule) is lined with epithelial cells which have a brush border. In birds and mammals this is followed by a thin section (thin loop of HENLE) which is lined with flattened epithelial cells. In other vertebrates, a small 'intermediate' segment is often present (MARSHALL, 1934) but not invariably so, as shown among reptiles (BENTLEY, 1959b) and the Amphibia (DAWSON, 1951). The posterior portion of the nephron (*distal tubule*) runs into the urinary *collecting ducts*, which proceed to empty into the ureter. The total length of the nephron varies considerably, both within a single kidney and, characteristically, in different species (SPERBER, 1944).

In birds and mammals the loops of HENLE run vertically from the periphery of the kidney (*cortex*), down through the *medulla* towards the renal *papilla*, where

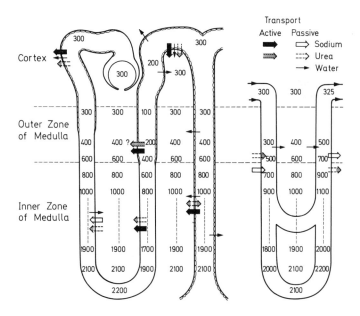

Fig. 1.3 Diagram depicting the countercurrent mechanism as it is believed to operate in a nephron with a long loop and in the vasa recta. The numbers represent hypothetical osmolality values. No quantitative significance is to be attached to the number of arrows and only net movements are indicated. As in the case with the vascular loops; all loops of Henle do not reach the tip of the papilla and hence the fluid in them does not become as concentrated as that of the final urine, but only as concentrated as the medullary interstitial fluid at the same level.

One may propose numerous variations of the countercurrent mechanism which will explain the facts already established, e. g. active sodium transport out of the ascending and into the descending limb of the loop of HENLE with recirculation of sodium and little loss of water from the loop; the possibility that the thin limbs of the loop of HENLE function as a countercurrent diffusion exchanger, etc. Until definitive evidence becomes available, we prefer the hypothesis illustrated above. (From GOTTSCHALK and MYLLE, 1959).

they make a 'hairpin' bend and follow a parallel course back to the cortex, merging into the distal tubule and the collecting ducts. Certain of the renal capillaries (*vasa recta*) follow a similar pattern, and the two types of vessels together function as a *'countercurrent multiplier'* system, to create a hyperosmotic concentration gradient which reaches a maximum at the 'hairpin bend'. Such a system allows birds and mammals to form urine that is hyperosmotic to the plasma, and was first described by HARGITAY and KUHN (1951). An excellent description is given by GOTTSCHALK and MYLLE (1959) (see Fig. 1.3). In 1944 SPERBER noted that mammals with long loops of HENLE and the elongated renal papillae associated with this, concentrated their urine more than those with shorter loops. This is due to the greater osmotic gradients which more elongated 'countercurrent' systems can form.

The blood supply to the glomerulus is arterial, entering and leaving the capillary tuft as afferent and efferent glomerular arterioles. It then breaks down into a second network of peritubular capillaries and vasa recta, which supply the tubular tissue. A renal portal system is present in birds, reptiles, Amphibia and most fishes (it is not for instance present in lampreys). Venous blood from the posterior regions of the body, passes into the renal portal vein from whence it may either supply the tubules (but not the glomeruli), or be shunted by a more direct route into the efferent renal vein (H. SMITH, 1951).

γ) Physiology. The formation of urine involves the ultrafiltration of the plasma through the glomerulus, reabsorption of solutes and water from the lumen of the tubule, as well as secretion of materials by the tubular cells into the tubular fluid. These processes have been studied by a variety of techniques, the most elegant involving collection and analysis of fluid from different regions of the nephron by means of micropipettes.

a) Glomerular filtration. Plasma is filtered in the glomerulus in the same manner as in other capillaries. The filtrate is relatively free of proteins, but otherwise has a similar composition to the plasma. Such ultrafiltration is caused by hydrostatic forces in the arterial system, which are opposed in the nephron by the osmotic pressure of the proteins (equivalent to 25 mm Hg in mammals but usually about half this in other groups), the external tissue pressure on the tubule (10 mm Hg), along with the back pressure of the system.

The glomerular filtration rate (GFR) is thought to be best reflected by the renal clearance of *inulin.* This is calculated as ml/min plasma =

$$\frac{\text{concentration in urine} \times \text{urine volume ml/min}}{\text{conc. in plasma}}$$

Inulin is a polysaccharide that is readily filtered across the glomerulus; it is not subsequently reabsorbed from the lumen and cannot enter by tubular secretion. It thus ideally should reflect the GFR. Creatinine clearance (endogenous or exogenous) is also often used as a measure of GFR, but in some species it may be secreted by the tubules, so that the results should be interpreted with caution. In many mammals, including man, the GFR is very stable and any changes that occur must normally be within the 5 to 10% error inherent in the measurement of inulin clearance. In many species, expecially in non-mammalian ones, the GFR may show considerable variation under physiological conditions.

30

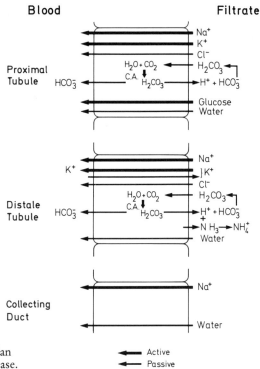

Fig. 1.4. Summary of the principal transfers of water and solutes which take place across the mammalian renal tubule. CA = carbonic anhydrase.

Changes in GFR could result from:

(i) *Dilution of plasma proteins.* By decreasing the osmotic pressure, which opposes the hydrostatic effects in the glomerulus, filtration would be expected to increase. Such changes, although of theoretical importance, may be too small to demonstrate unequivocally (see O'CONNOR, 1962).

(ii) *Haemodynamic effects.* The renal circulation is remarkably stable in the face of large changes in arterial pressure, pharmacological doses of vasoactive drugs like adrenaline, and other circumstances that may be considered physiologically extreme (H. SMITH, 1951). Both the afferent and efferent glomerular arterioles have been considered as possible ways of changing the hydrostatic pressure in the glomerulus, but no acceptable definitive evidence as to whether this normally occurs is available.

(iii) *Glomerular intermittency.* Changes in the *numbers* of functioning glomerule would be expected to alter the overall GFR, without changing the rate of filtration into individual tubules. Such intermittent glomerular activity has been demonstrated in species of birds, reptiles, amphibians, fishes and possibly even in the rabbit. The experimental conditions for demonstrating such effects are sometimes rather exceptional, but not always so (for instance DANTZLER and B. SCHMIDT-NIELSEN, 1966). The concept of glomerular intermittency is most attractive in species that have a renal portal blood supply to the tubular tissue, as this assures its continued metabolic integrity. While it is more usually considered likely that the

31

rates of glomerular filtration affect tubular absorption to some extent, a group in Copenhagen have put forth evidence showing the reverse is so (KRUHOFFER, 1960). Reabsorption of fluid in the tubule may increase the GFR by reducing tubular back-pressure, thus facilitating the hydrostatic pressure in the glomerulus. The magnitude and physiological importance of such effects have not been agreed upon.

b) *Renal Tubular Function.* The glomerular filtrate is progressively modified as it moves along the renal tubule. These changes involve active and passive reabsorption, tubular secretion of solutes, as well as the osmotic transfer of water. The precise details of such processes, both with respect to their regional location in the nephron and interactions with each other, are continually being reconsidered so that the present summary is not a final one. The reader should consult H. SMITH (1951), KRUHOFFER (1960), and MALNIC, KLOSE, and GIEBISCH (1966 a and b) for a detailed description of the available evidence, which is summarized in Fig. 1.4.

Sodium is actively transported, from the glomerular filtrate back into the blood, from most regions of the nephron as well as the collecting ducts. The exception appears to be the descending loop of HENLE in mammals, which is relatively impermeable to sodium and water. Active potassium transfer takes place from the luminal fluid in both the proximal and distal tubules.

Hydrogen ions are secreted from the renal tubular cells into the filtrate in both the proximal and distal tubule. These ions are derived from carbonic acid, which is formed from carbon dioxide and water under the influence of the enzyme carbonic anhydrase. The ability to secrete H^+ is limited at a urinary pH of about 4.5 to 4.8, but these ions may be neutralized in the distal tubule by NH_3 (to form NH^+_4). The ammonia arises in the tubular cells by deamination of amino acids such as glutamine.

The principal anions present in the filtrate are chloride and bicarbonate, and these move out of the tubule down their electro-chemical gradients. The movement of HCO^-_3 is facilitated by its interaction with secreted H^+ to form CO_2 and water. Bicarbonate, equivalent to the H^+, is then transferred from inside the tubular cell to the blood.

Potassium excretion under certain circumstances may exceed the rate at which it is filtered across the glomerulus, indicating its tubular secretion. Such secretion takes place in the distal tubule, and down an electro-chemical gradient, which partly results from the active sodium transport out of the tubule.

The quantity of the various other solutes in the tubular fluid is also modified; calcium, magnesium, sulphate, phosphate and glucose are reabsorbed, but in a-glomerular fish such ions are presumably also secreted. In birds and reptiles uric acid is secreted across the tubular epithelium. In mammals about 40 to 70% of the filtered urea diffuses out of the tubules, but in species that utilize urea for their osmotic equilibrium with sea-water, little may be lost (1%), as seen in the frog, *Rana cancrivora* (SCHMIDT-NIELSEN and LEE, 1962), while in the marine Chondrichthyes urea may even be conserved by active reabsorption (H. SMITH, 1936).

In mammals and amphibians water moves osmotically out of the proximal tubule, so that isoosmoticity with the bathing fluids exists. In the more distal regions of the nephron and collecting ducts, such osmotic equilibration may or may not occur. In the tetrapod vertebrates, such adjustments are largely under the control of the neurohypophysial hormones from the pituitary gland.

The magnitude of the changes that occur in the renal tubule vary in the few species that have been studied sufficiently. In the mammals, about 85% of the filtered solutes and water are reabsorbed in the proximal tubule, while in amphibians this figure is probably nearer 40%. Under normal physiological conditions in mammals, more than 99% of the material filtered across the glomerulus is retained by the animal, though under other circumstances, necessitating substantial excretion, this may be reduced. Thus in freshwater fishes, which must excrete the large volumes of water gained osmotically from their environment, only 30 to 40% of filtered water is reabsorbed.

δ) *Regulation of Kidney Function.* Factors that may affect kidney function have been described in an excellent monograph by O'CONNOR (1962). These are:

(a) Composition of the plasma including the concentration of plasma proteins and solutes such as, glucose, urea, potassium and certain anions.

(b) Arterial pressure.

(c) Renal nerves.

(d) Hormones; including adrenaline, adrenocortical and interrenal steroids and neurohypophysial peptides.

The influences of (a) and (b) on the GFR have already been discussed, while the roles of adrenaline and the renal nerves are not currently considered to be physiologically significant.

The concentration of the plasma solutes has a measure of autoregulatory control within the kidney. The quantities filtered across the glomerulus will be related to their plasma levels, while their reabsorption will be affected by the amounts presented to the tubular epithelium. The latter process may involve either diffusion or active transport. The regulation of plasma urea levels is a good example of such autoregulation; filtered urea is subject to back-diffusion across the nephron wall, so that in mammals about 50% of the amount filtered is excreted. Elevation of the plasma concentration will increase the quantity filtered and a new equilibrium state with the blood will ensue, resulting in a greater rate of excretion. If for some reason the rate of filtration is reduced, an equilibrium with an elevated plasma urea level results. In the case of glucose, the active reabsorption across the tubule wall has a distinct maximum rate. If the rate of delivery exceeds this value, the excess filtered glucose is excreted. Such a process is, of course, not primarily responsible for the regulation of blood glucose levels, but in certain circumstances may contribute to it. The levels of potassium and various anions in the plasma may also directly influence their rates of renal excretion.

The peptide hormones from the neurohypophysial region of the pituitary gland increase osmotic water reabsorption from the distal segment of the renal tubule and the collecting ducts in various tetrapod vertebrates; they can also be shown to decrease the GFR in certain tetrapods and to increase it in some fishes. The precise mechanism for these latter effects is unknown, but they are probably vascular. The adrenocortical and interrenal steroid hormones increase sodium absorption and potassium secretion in the nephron of some tetrapods, and under some circumstances can increase the GFR. The renal actions of such steroids in the amphibians and fishes are not clear.

6. Nitrogen Metabolism and Excretion

The major nitrogenous compounds excreted by vertebrates are ammonia, urea, uric acid, trimethylamine oxide and small quantities of nitrogen containing compounds including free amino acids. Trimethylamine oxide is present in certain marine fishes but its status as a true metabolite is in doubt; it probably is obtained exogenously from the diet (BALDWIN, 1963). Ammonia, urea and uric acid may be derived from the deamination of amino acids, a process preparatory to their subsequent metabolism or conversion to glucose or fat. Uric acid, along with other nitrogenous compounds, including allantoin and urea, may also be formed during the metabolism of nucleic acids (purine metabolism) but the quantities are usually minor compared to those from deamination.

The nitrogenous end products from deamination of amino acids vary in different species, and are closely related to the amounts of water that are normally available. Ammonia is the primary end-product of deamination, but owing to its high toxicity cannot be accumulated in the body. An ammonia concentration of 0.03 mM in the blood is fatal in rabbits. Adequate amounts of water must be available for ammonia excretion, so that it only predominates as a major end-product of amino acid metabolism in certain aquatic species, which are said to be *ammoniotelic*. Ammoniotelic species include teleost fishes, certain aquatic Amphibia, like the mudpuppy, *Necturus macolosus*, and the toad, *Xenopus laevis*, as well as some crocodilian and chelonian reptiles (see BALDWIN, 1963). Ammonia is lost by diffusion across the gills of fishes and the skin of *Necturus* (FANELLI and GOLDSTEIN, 1964) while in *Xenopus* and ammoniotelic reptiles it is excreted in a copious urine.

'The conversion of ammonia to other products is an indispensable adaptation to limitation of the availability of water' (BALDWIN, 1963). It has been calculated (H. SMITH, 1951) that 1 g of nitrogen requires 300 to 500 ml of water for its excretion by an ammoniotelic species like the alligator. In birds and reptiles, which convert the ammonia to uric acid, only about 10 ml of water would be required for such excretion, while if transformed to urea 50 ml would be needed in man and 10 ml in the desert rat, *Dipodomys*. Aquatic amphibians may be ammoniotelic, but terrestrial species form urea. Reptiles vary in this respect; some aquatic species are ammoniotelic while terrestrial species excrete urea, uric acid or both. Such categories are not mutually exclusive; mammals and birds, for instance, all excrete some ammonia.

Apart from reflecting the need for reduced urinary water loss, the formation of urea has a more positive aspect in some species, in which it can be utilized as a solute that contributes to the animals' osmotic equilibrium with its environment. In the marine Chondrichthyes (H. SMITH, 1936), the coelacanth, *Latimeria* (PICKFORD and GRANT, 1967) and the marine frog, *Rana cancrivora* (GORDON, SCHMIDT-NIELSEN, and KELLY, 1961), urea may normally be present in the body fluids at concentrations approaching 500 m-Osmole/l. This contributes to the osmotic constitution of the body fluids, and helps to maintain them at a level that is hyperosmotic to sea-water. The African lungfish, *Protopterus aethiopicus*, aestivates in mud during periods of drought, and accumulates urea in concentrations as high as 500 m-Osmole/l (H. SMITH, 1930a). This is excreted when water becomes available. The desert toad, *Scaphiopus couchi*, also aestivates in the earth during dry periods and stores urea (up to a level of 300 m-Osmole/l) for excretion during the rainy season (McCLANAHAN, 1967). Such tolerance to the presence of urea reflects its high solubility, uniform distribution in the body fluids and low toxicity.

Vertebrates may change modes of nitrogen excretion in different environmental situations. The lungfish, *Protopterus*, is ureotelic only while it is aestivating; in its usual aquatic surrounding it is ammoniotelic. The same sort of transition takes place in *Xenopus*. Such changes have been related to changes in the liver enzymes of the ornithine-urea cycle (GOLDSTEIN, 1968) and these may possibly be controlled by adrenocortical hormones (SCHIMKE, 1963).

7. Osmoregulation and the Origin of Vertebrates

Vertebrates originated in early Palaeozoic times, probably in the early part of the Ordovician age about 400 million years ago. While it is agreed that this took place in aqueous media, there is a sharp dichotomy of opinion as to whether it was in fresh water or the sea. An account of this disagreement is germane to our subject as the evidence is not only palaeontological but draws upon physiological information about the osmoregulation of contemporary species.

In the post-Darwinian period to 1900 it was generally assumed that vertebrates originated in the sea. At that time the oldest fossil remains of the fishes were considered to be present in deposits of Devonian age, which were thought to be of marine origin. However, CHAMBERLIN in 1900 and BARREL in 1916 challenged this interpretation, and showed that some of these deposits had been laid down in fresh water. It is now accepted that Devonian vertebrate remains are found in deposits of *either* marine or freshwater origin (see ROBERTSON, 1957). The earliest vertebrates were jawless fishes called ostracoderms, which have been classified with contemporary agnathans, and have been identified in Ordovician-Silurian deposits which precede the Devonian remains. The question as to whether these fossils are marine or freshwater in origin is still hotly contested (ROMER, 1955 and 1967; ROBERTSON, 1957). The arguments revolve largely around questions as to whether the presence of terrestrial plants and marine invertebrates reflects the true origin of the deposits, or whether they were washed into the area of the beds from another, contrasting, osmotic enviroment. HOMER SMITH (1961) considered the Scottish verdict 'not proven' appropriate to the situation.

The contemporary biological information, relevant to a marine or freshwater origin for the vertebrates, provides an interesting intellectual exercise relevant to our present topic of osmoregulation, though it should be remembered that such evidence is indirect and only circumstantial.

It is usually agreed that vertebrates arose from a marine chordate ancestor, possibly a free-swimming tunicate larva (ROMER, 1967). Three contemporary groups of chordates, the Hemichordata, Cephalochordata and Urochordata are exclusively marine in habitat. The problem is, whether the transition that gave rise to the fourth group, the Vertebrata, took place in the sea or fresh water.

The sea, in early Palaeozoic times, had a composition similar to that which it has today, so that the osmotic problems of the prehistoric animals would have been similar to those of contemporary species. The precise nature of such problems, however, depends on the osmotic concentration of the body fluids and we have no direct information about this in the fossil species. Contemporary lower chordates are essentially isoosmotic with their marine environment, whereas marine vertebrates may be either slightly hyperosmotic, as in the Chondrichthyes or hypoosmotic, as in the Osteiichthyes. The ostracoderms are classified with the Agnatha, in which one contemporary group, the Myxinoidea, is iso-or slightly hyperosmotic to its marine environment, while the other, the Petromyzontoidea, is hypoosmotic to sea-water. It is uncertain which, if either, of these groups represents the primary vertebrate condition. If, as suggested by ROMER and HOMER SMITH, the evolution to a vertebrate took place in fresh water, then osmotic uptake in a myxinoid type would be much greater than in a petromyzontid, though it is easy to envisage the presence of compensatory factors, such as a relatively impermeable integument. If, on the other hand, the evolution occurred in the sea then, as ROBERTSON suggests, the myxinoid pattern 'may well be the primary one', the advantages and osmotic simplicity of this system making it the most logical choice. Nevertheless, despite popular opinion to the contrary, we cannot assume that nature is logical.

Whatever the primary osmotic pattern may have been in the first vertebrate, if the transition occurred in fresh water, novel osmotic stresses not apparent in its marine antecedents would arise. The animal would tend to gain water by osmosis and lose salts by diffusion. This could be compensated for by reducing the permeability of the animals' surface to water and solutes and by excreting the excess water. Such mechanisms exist in nature and presumably would have been apparent early in vertebrate evolution, though whether such changes took place in a marine ancestor and constitute preadaptation for life in hypoosmotic environments (ROBERTSON, 1957), or occurred under the selective pressures in a hypoosmotic environment (H. SMITH, 1961), is not yet apparent.

Early ostracoderms had characteristically heavy dermal armour. ROMER considers this to have evolved as a protection against attacks by eurypterids, which were arthropods contemporary with the ostracoderms. HOMER SMITH on the other hand considers this armour to be an adaptation to fresh water, resulting in reduced osmotic water uptake. ROBERTSON rejects both of these views and cites evidence suggesting that ostracoderms fed on eurypterids rather than the reverse. If the early ostracoderms were marine, as considered by ROBERTSON, then the dermal armour may have constituted a preadaptation to a subsequent move into fresh water. A comparable situation can be demonstrated in two species of turtles, the red-eared turtle, *Pseudemys scripta,* which has the usual chelonian armour, and the softshell turtle, *Trionyx spinifer,* which lacks these hard plates on its carapace. KNUT SCHMIDT-NIELSEN and I (BENTLEY and SCHMIDT-NIELSEN, 1970) have compared the osmotic permeability of these two species in fresh water and in a saline equivalent in concentration to the sea. Water exchange in either solution was four to five times as great in the softshell turtle as in *Pseudemys.* While the hard dermal plates in turtles probably serve the purpose of protection against predators, they could also act as a preadaptation to life in a different osmotic medium, as could have occurred if marine ostradoderms moved into fresh water.

The principal contemporary biological evidence used to support a freshwater origin for vertebrates, is the comparative anatomy and physiology of the vertebrate kidney. As we have seen, the renal glomerulus is a structure through which large amounts of fluid may be filtered by utilizing the hydrostatic pressure of the blood vascular system. Glomerular development and filtration is more pronounced in vertebrate species living in fresh water than in the sea. Such a filtration mechanism, when combined with a tubular system for reabsorption of essential solutes, constitutes an admirable mechanism for excreting excess water. MARSHALL and SMITH (1930) concluded that the glomerulus, which is present in most species of all the major vertebrate groups including the myxinoids, arose as an evolutionary adaptation to life in fresh water. It was assumed that its presence in marine species, such as the myxinoids, is due to their secondary movement into the sea and is not a primary situation. ROBERTSON's interpretation of the glomerulus differs strongly from that of MARSHALL and SMITH; he believes that such a mechanism must have been present before such animals could have lived in fresh water, and points out that such filtration-reabsorption mechanisms exist in many marine invertebrates whose ancestors have never been associated with fresh water. Such filtration mechanisms, when followed by selective reabsorption, act as regulators for excretion of ions in primarily marine species. Studies on contemporary myxinoids (the Atlantic and Pacific hagfishes) suggest that in this group, the nephron is concerned with regulation of divalent ions rather than water excretion and sodium conservation (MUNZ and McFARLAND, 1965). These authors agree with ROBERTSON, that the hagfish kidney reflects a primary marine origin for the myxinoids, and suggests that the Vertebrata originated in the sea, providing of course that cyclostome origins were monophyletic. The recent discovery of a fossil lamprey, *Mayomyzon*, 200 to 300 million years old, which lacks all myxinoid characters, has led BARDACK and ZANGERL (1968) to suggest that the origin of the cyclostomes may indeed be diphyletic as originally suggested by STENSIO. Evidence, based on the characteristics and relationships of plasma proteins in contemporary cyclostomes and other modern vertebrates, supports the concept of a petromyzontoid-type of ancestor (MANWELL, 1963). If this is so, and the petromyzontoids and myxinoids represent two phyletic branches of cyclostomes, then osmoregulatory information about the myxinoids would not be relevant to the speculations on the environment of the main line of early vertebrates. There are too many "buts" and "ifs" to draw any firm conclusions. Evidence of vertebrate origins based on contemporary physiological and serological information cannot be conclusively extrapolated back for 400 million years. The interpretation of such data, as reflecting a primary or secondary type of vertebrate osmoregulatory pattern, must remain an opinion, no matter how well considered.

Chapter 2

The Vertebrate Endocrine System

Communication among cells in multicellular animals is necessary for their overall coordination and survival. Individual cells lying adjacent to one another make contact, either directly or with the aid of ions and metabolites, across the relatively narrow fluid-filled extracellular spaces. Interdependent cells in large animals may, however, be separated by distances of several metres, so that if appropriate coordination is to take place, special methods must be used. Such 'long distance' cell communication takes place in two main ways; through nerve cells and by the secretions of endocrine glands. Nervous communication is characteristically rather faster than endocrine, and attains its specific actions by utilizing discrete tissue pathways to particular organs. Although endocrine communication is slower, once initiated, its actions often may be more prolonged. It is thus peculiarly suited to integration of metabolic, rather than the mechanical and sensory, processes which are more the special province of the nerves. The pathway for endocrine communication is the extracellular fluid, so that endocrine products can ultimately come in contact with many cells of the body. This results in a problem of specificity in communication, that is principally solved by utilizing a host of different molecules, each with special affinities for the physico-chemical characteristics of certain cells. Thus it is not surprising that more different endocrine hormones have been identified than neurotransmitter substances.

Neural and endocrine coordination systems are interdependent and interrelate to each other. The correct function of all cells in the body, including the nerves and endocrine glands, depends on adequate supplies of suitable metabolic substrates, and an optimal osmotic and ionic environment; properties that can be related to the actions of both nerves and endocrine glands. The two systems exchange information, and influence each other's activity, particularly in the region of the hypothalamus and pituitary gland and in the adrenal medulla. Nerve impulses can affect the rates of secretion of endocrine glands, while the hormones may influence neural processes as in behaviour.

Individual cells are usually self contained autoregulated units which, providing they are situated in an optimal environmental solution, can maintain life. Cells in tissue culture media thus may carry through their life cycle in perpetuity. Cells can, however, usually only autoregulate under a rather limited range of ionic and osmotic concentrations, while a constant supply of metabolic substrates, in suitable form and at an adequate concentration, are also necessary. The endocrine-glands play a predominant role in regulating such internal environmental conditions.

The pattern of endocrine coordination is in accord with the classical *servo-control system* (Fig. 2.1.). The main elements of the relevant servo-system are: its sensitivity to change from a 'set point' value as conceived by the 'misalignment detector' which stimulates a 'controller' to feed 'controlled power' to a 'motor'. The 'motor' has an output which can correct the misalignment, a change which then 'feeds back' to inhibit the 'controller'. The endocrine system by analogy is sensitive to changes in osmotic, ionic, metabolic and substrate levels and these can affect

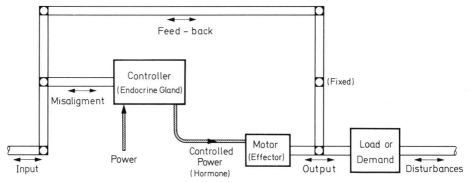

Fig. 2.1 Schematic block diagram of a Servo-system. The double-headed arrows indicate that the various links in the system move to and fro. From BAYLISS (1960). The analogues in the endocrine system have been inserted in parentheses.

an endocrine gland (controller) by acting on its receptors (misalignment detector). Hormones (controlled power) are then released and act on the effector organ (motor). This initiates an action, such as release (output) of a metabolic substrate or another hormone or the accumulation or excretion of water and ions, which by correcting the physiological imbalance inhibits further secretion from the endocrine gland. The 'set point' value of the 'misalignment detector' may be changed under certain environmental and metabolic circumstances, providing the system with a measure of adaptability to changing conditions and allowing the initiation of periodic events, like a breeding season.

1. Endocrine Function. Criteria, Methods and Difficulties

An endocrine gland releases its secretion(s) into blood, which carries it through the circulatory system to its 'target' or effector organ, where it 'triggers' a distinctive physiological response. This response is directed towards achieving metabolic homeostasis in the animal, or integrating an essential activity, like reproduction. The identification of an endocrine pathway in an animal thus requires information that is conjointly derived from at least three anatomical sites: the putative secretory tissue, the blood and the 'target' organs. Initial 'clues' about an endocrine function may be derived from any of these three loci. Thus pathological hypertrophy or atrophy of a putative gland may be associated with characteristic physiological changes, or the extracts from such a tissue may exhibit certain biological actions. Body fluids, like the blood and urine, may have biological activity in certain physiological circumstances. Effector organs may exhibit changes in structure and activity, suggestive of a trophic material. Endocrine function, once initially suspected can then be further investigated until unequivocal evidence of its presence or absence is obtained. Subsequent observations can be extended so as to acquire knowledge of the exact chemical nature of the hormone, its precise manner of formation, release and metabolism, and the mechanisms of its action at the effector site.

a) Criteria

Criteria for endocrine function are usually based on the following;

1. Extirpation of the putative endocrine tissue *should result in characteristic physiological deficiencies,* which can be corrected by administering extracts made from such tissues of other animals. The possibility that a tissue may only assume physiological importance in certain situations should be kept in mind. Prolactin promotes the survival of teleost fish in fresh water, but is apparently unnecessary when they are in salt water. A high salt diet can often considerably ameliorate the effects of adrenalectomy, while diabetes insipidus, due to the absence of the neurohypophysis, is not apparent if the adenohypophysis is also removed.

2. The *biological activity* of the putative gland should also be present *in the circulation*, under appropriate physiological or environmental conditions.

3. The suspected *effector tissue should respond to extracts from the blood and putative gland,* and exhibit a sensitivity consistent with the concentrations present, especially those in the circulation. The reactions of the endocrine gland and the hormone(s), including the factors initiating release of the secretion and the nature of the effector response, should be consistent when they are viewed as an integrated homeostatic mechanism. The response of the supposed effector organ should be directed towards correcting the conditions that brought about the release of the hormone, and such physiological compensation should then reduce ('feed-back') subsequent secretion.

b) Endocrine Extirpation – Surgical and Chemical

Surgical skills play a large role in establishing the endocrine significance and function of a tissue. Such procedures include removal of the putative gland and the alteration of its vascular and neural connections, either directly *in situ*, or by the manoeuvre of transplanting it elsewhere. The basic surgical aim is to perform such tasks in a manner that leaves a minimum of residual tissue, and with as little physiological disruption and inconvenience to the animal as possible. It is often difficult to separate the adverse effects of a surgical procedure *per se*, from the specific endocrine changes that are the experimental aim. Nonspecific surgical trauma and discomfort to the animal can be reduced by the acquisition of the relevant skill by an investigator, as well as by selecting the most appropriate route and method for approaching the tissue in question. Such procedures vary considerably in different species, depending on the anatomical site of the gland and the suitability and sensitivity of the animal to surgery. There are special problems associated with surgery in fishes, as compared to terrestrial tetrapods, due principally to differences in the modes of respiration, but also to their generally different morphological arrangement. Endocrine tissues like the thyroid and interrenals are more diffuse and not as compactly arranged in the fishes as in tetrapods, but even among the latter considerable differences exist. In the reptiles and the Amphibia, adrenocortical tissue is not as compact or discrete as in mammals. Osseous differences make removal of the pituitary a far more difficult operation in the guinea pig than in the rat. Such variations make it necessary to modify surgical methods in different species in order to do the job and to minimize surgical stress. Special pre-operative and post-operative treatment may assist these aims. Thus, antithyroid drugs can be given prior to surgery, in order to reduce the possibility of an 'explosive' release of thyroid hormone during manipulation of this gland. Adrenocortical hormones may be injected during adrenalectomy, in order to alleviate any acute requirements for such hormones as a result of the operation.

Special problems associated with surgical extirpation include the difficulty of separating closely associated endocrine glands, like the adrenocortical and chromaffin tissue, the thyroids and parathyroids, and the adenohypophysis and neurohypophysis. In addition most endocrine tissues produce more than one hormone which surgically cannot be dissociated. Some of these difficulties may be partly overcome by subsequently administering some of the hormones known to be missing. When several secretions are involved this may take the form of serially successive hormone substitutions, in order to find the particular secretion which is responsible for an observed deficiency. Thus the teleost fish, *Fundulus heteroclitus*, living in fresh water dies after removal of its pituitary, but survives if prolactin is injected (PICKFORD and PHILLIPS, 1959), administration of other pituitary hormones being ineffective in this respect. Another problem associated with attempts at endocrine extirpation involve the question of how completely the tissue has been removed, and whether or not any regeneration has occurred; careful histological postmortem examination may be necessary in order to rule out such possibilities.

Chemical extirpation of putative endocrine tissues can be performed by the use of substances that inhibit the formation and release of the endocrine secretions, or block the effector sites. The formation of thyroid hormones can be prevented by administering radioactive ^{131}I which can destroy the tissue, or by giving drugs, like thiourea and propylthiouracil, that prevent hormone synthesis. Adrenocortical steroid formation can be inhibited by metyrapone, while insulin and glucagon production are blocked respectively by, alloxan and synthalin A or cobalt chloride. Such substances, however, may not be very specific, and can have widespread toxic effects.

The release of many hormones into the circulation can also be prevented; ethyl alcohol inhibits the release of neurohypophysial antidiuretic hormone, while gonadal and adrenocortical steroids reduce the release of their appropriate adenohypophysial trophic hormones by a negative feed-back action. Subsequent to its release, the action of a hormone may be prevented by antagonisms at receptor-effector sites; spirolactones, like aldactone, competitively inhibit the action of aldosterone on the kidney and anuran urinary bladder, while structural analogues of the neurohypophysial hormones may also block the action of the parent hormone. The possibilities of immunologically inhibiting endocrine function with specific antibodies are only just beginning to be investigated.

c) Administration of tissue extracts

Tissue extracts often have well defined biological effects *in vivo* and *in vitro,* and such observations may play an important role in assessing the putative endocrine role of tissues. Such decoctions may be administered as an attempted replacement procedure following extirpation of the alleged gland, their actions may be observed in intact animals or they may be tested on isolated preparations like the uterus and anuran skin and urinary bladder.

The initial problem, after obtaining sufficient of the desired tissue, is how to extract it; simple saline homogenates are usually tried out in the first instance. These decoctions may be subsequently modified in attempts to separate any active material from the multitude of other substances that are usually present. If no activity is found in this way, alternate extraction procedures may be tried, as the substance may be poorly soluble in aqueous solution, or even fail to survive such an apparently simple treatment. Thus extraction in such solvents as alcohols and acetone may be attempted.

There are many difficulties in assessing the significance of biological evidence obtained from the effects of such extracts. Lack of a demonstrable effect is inconclusive, as the extraction procedure may be inadequate, the amount of material stored in the tissue may be insufficient to have a stimulatory action, or the biological parameters being examined may be inappropriate. Early attempts to decoct insulin from pancreas were unsuccessful, due to its inactivation by the trypsin which is present in the associated exocrine tissue. It was subsequently found that the enzyme's action could be prevented if ethyl alcohol was present, a procedure which leaves the insulin intact. The storage of corticosteroids is poor, so that large numbers of glands originally had to be used in order to show their biological actions. A suitable choice of the route for administration can assist in instances when only small amounts of material are available; direct injection into a blood vessel supplying the supposed 'target' organ, or the creation of regional storage depots may provide adequate local concentrations. The condition of an animal may be unsuitable for the demonstration of an hormonal effect; thus the neurohypophysial antidiuretic hormone was originally shown to have a diuretic effect when tested on anaesthetized mammals and it was only when hydrated unanaesthetized animals were used that the antidiuretic action became apparent. This problem arose as a result of the high endogenous circulating levels of ADH that are present in animals exposed to many anaesthetics. Apart from physically removing the source of glandular secretion, high endogenous levels of hormones can often be reduced by providing conditions calculated to reduce their release; for instance administration of water reduces circulating ADH levels, and placing toads in hypotonic sodium chloride solutions for several days lowers their plasma aldosterone levels.

The biological action of a tissue extract can reflect its physiological role, but such evidence alone is not conclusive, and at the most is only suggestive. Pharmacological effects of such extracts are more common than any physiological actions. Mammalian antidiuretic hormone has a well established physiological role in aiding reabsorption of water from tetrapod kidney tubules, but it may also raise the blood pressure of anaesthetized mammals, elevate blood glucose levels, increase renal sodium excretion, initiate release of adenohypophysial, adrenocortical and thyroid hormones and increase the motility of the gut. These latter effects are most probably all pharmacological and reflect the actions of quantities of hormone that do not normally appear in the circulation. It should also be remembered that after removal from a moribund animal, tissues are often subjected to a multitude of chemical insults; boiling, freezing, thawing, extraction in various solvents and the like. Substances that are not normally secreted may be decocted, while novel materials may appear as a result of tissue degradation.

d) Histological Procedures

Histological and cytological methods, using the light and electron microscopes, play an integral part in the study of endocrinology. Such techniques provide information about the morphology of endocrine glands and their vascular and nerve connections that is relevant to their function. The cytological appearance of the gland assists in the identification of homologus tissues in other species, or even in different parts of the same animal. In addition such precedures aid the precise localization of the cells in which the hormones originate and are stored, and also may provide an index of activity.

Endocrine tissues and their secretions can be stained in a multitude of ways that facilitate observations of their structure. Some of these procedures are histochemical, being directed towards the staining of specific substances like lipids, glycoproteins and cholesterol. Many staining routines are, however, more arbitrary, their success being judged by their ability to differentiate between various cells and to make secretory products visible. A vast number of dyes, which may give comparable results in homologous tissues from a wide range of species, can be used for such purposes. In many instances, however, a particular staining technique may be peculiarly adapted to a particular species or group of animals.

Differential staining of the various cells in the pituitary gland provides a good example of the types of procedures used. In the adenohypophysis a broad distinction can be made between cells which are stained by acid (acidophils) and basic (basophils) dyes. The basophils can be stained by the periodic acid – SCHIFF (PAS) method, while the acidophils are exposed by a modified MALLORY stain or HERLANT's tetrachrome. These two groups of cells can each be further subdivided, by differential staining with other dyes, to distinguish five or six different types of cells. In the neurohypophysis, secretory material can be seen in nerve cells after staining the tissue with GOMORI chrome-alum haematoxylin or paraldehyde-fuchsin. In adrenocortical tissue more precise histochemical procedures can be used, as this contains lipid droplets, cholesterol and ascorbic acid, all of which may be associated with its endocrine activity. The lipid droplets can be seen after staining with fat soluble SUDAN dyes, and specific methods are also available for the other materials. The thyroid hormones are associated with the colloidal material stored in the follicles of the gland, and this stains with periodic acid – SCHIFF reagent. Suitable histological procedures have been described for all endocrine tissues in a large variety of species.

Once the types of cells present in an endocrine tissue have been histologically characterized by their situation, shape, size and reaction to dyes, an attempt can be made to identify the role of each in the production and storage of the different hormones that may be present. This can be quite a problem, even if only one or two hormones are involved. Thus, the pituicytes present in the neurohypophysis were for a long time thought to be the site of formation of the antidiuretic hormone, though it is now clear this hormone originates in the nerve cells that extend into this tissue. BARGMANN, using the chrome-alum haematoxylin stain, was able to identify secretory material in the nerve terminals and axons of the neural lobe, and was able to trace such material back along the nerve fibre to its site of origin in the hypothalamus. The distribution and quantity of this material paralleled that of antidiuretic hormone, and so initiated our contemporary views on the cellular sites for the formation and storage of this hormone. The problem may be somewhat greater, when, as in the instance of the adenohypophysis, six or seven hormones may originate in proximity to each other. Cytological changes in each cell type can be observed under a variety of natural

43

and experimental conditions when the production and secretion of a particular hormone changes. In this way an assessment of the role of each type of cell in the secretion of a certain hormone can be made. An interesting new approach to this problem has been described by EMMART, PICKFORD and WILHELMI (1966) using prolactin antibodies to which are attached a fluorescent material, fluorescein isothiocyanate. Such fluorescent antibodies were allowed to react with frozen sections of the pituitary of the teleost, *Fundulus heteroclitus,* while the subsequent serial sections were stained by classical methods. The fluorescent antigen-antibody material was found to be localized near the *eta* cells which are thought to be the site of origin of prolactin.

Direct measurements of the concentration of the hormones in endocrine glands and in the circulating blood afford the most satisfactory way of assessing the level of endocrine activity. Such procedures are not always feasible, and a histological examination of the secretory tissue may provide some information about this. Observations that can provide criteria of activity include; the size and number of the secretory cells, their mitotic activity, the appearance of the cytoplasm, nuclei, nucleoli, mitochondria, GOLGI apparatus, granules and stainable secretory products that are present. The correlation of such changes with direct measurements of the quantities of hormone present facilitates the physiological interpretation of such cytological observations. Histologically the activity of the thyroid gland can be judged from measurements of its secretory cells, the greater their height, the greater their activity, while the density of the neurosecretory material present in the neurohypophysis can be assessed and related to the function of the tissue. Endocrine activity may also be judged, in some instances, from the histological condition of their target tissues; reproductive organs like the uterus, testis and ovary afford the clearest examples of this. Adequate interpretation of histological observation on endocrine function can be difficult, especially with regard to the related hormone levels in the circulation. It thus may be difficult to judge, whether accumulation of secretory material in a tissue reflects increased production, or decreased release of the hormone into the blood, alternatives that may indicate contrasting physiological conditions.

e) Identification and Measurement of Hormones

These procedures provide a veritable host of problems that have resulted in a multitude of methods. Many of the difficulties arise from the necessity to identify and measure the active materials, not only in the glandular tissues, but also in body fluids like blood and urine where the concentrations may be very low (of the order 10^{-12}M). This results in a two-fold problem; firstly how to handle and measure so little material, and secondly how to differentiate it from other substances of similar chemical nature (like proteins) which are present at a 10^6 to 10^9 higher concentration. A battery of biological, physico-chemical and immunological methods are used.

Initial endocrinological investigations usually utilize biological methods for identification and measurement of the material present. In the first instance, a hormonal activity is usually identified by its effects on an animal or tissue and these afford the criteria for its presence. If the hormone has several demonstrable actions so much the better, as this provides a characteristic profile that can be used to aid in distinguishing it from other substances. More precise assessment of the amounts of material present can be made by measuring the magnitude of its effects on a particular biological system(s), such methods of measurement being called bioassays. Two recently identified candidates for hormone status in fishes are paralactin (a homologue of tetrapod prolactin), and a principle that can be extracted from the caudal urophysis. Paralactin (and mammalian prolactin) increase the ability of hypophysectomized fish (kept in fresh water) to maintain adequate plasma sodium levels, and this function has been used as a basis of measuring its concentration in the pituitary (ENSOR and BALL, 1968a). Plasma sodium concentrations in such fish are used as an index of the amount of hormone suspected to be in an injected extract. Ovine prolactin is used as a 'standard' preparation for comparison with the 'unknown'. In the instance of the urophysis, tissue extracts increase water transfer across the urinary bladder of toads and also contract the trout's urinary bladder (LACANILAO, 1969; LEDERIS, 1969), and these responses may not only be used to help identify the material but could also provide a basis for its bioassay.

Analogous biological methods have been (and in some cases are still) used for initially measuring quantities of many hormones. Thus, the potency of adrenocortical extracts were originally measured by their ability to promote survival of adrenalectomized mammals, alter renal sodium excretion or promote the formation of liver glycogen. Thyroid extract activity has been estimated by its ability to increase metamorphosis of tadpoles, and decrease the suvival time of mice in closed jars (due to increased rate of oxygen usage). Changes in blood-glucose levels and glycogen content of rat diaphragm muscle can be used to gauge insulin concentration, while corticotrophin can be measured by its ability to decrease the amount of ascorbic acid in the adrenal cortex. Amounts of neurohypophysial hormones are still assessed by measurement of their ability to produce effects such as antidiuresis, increased blood pressure, contraction of the uterus or promotion of 'milk let down' in mammals. Other assays of neurohypophysial hormones utilize biological responses in such diverse groups as birds, reptiles, and amphibians.

Adequate quantitative biological expression of a hormonal activity requires the adoption of a 'standard', with which the strength of an 'unknown' extract can be compared. If the precise chemical structure of the hormone is known, and it is available in a pure form, the effect of a known amount (the 'standard') can be compared with the 'unknown,' and the result expressed in terms of the weight or molecular concentration. If, on the other hand, the exact chemical nature of the hormone is unknown, or if it is unavailable in a sufficiently pure chemical form, an arbitrary '*biological standard*' can be made. This usually consists of a suitably prepared and preserved quantity of the endocrine tissue, which is assigned a potency, usually in terms of units (U.) of a certain biological activity per unit weight. Thus, the international standard for neurohypophysial hormones consists of dried and powdered beef neurohypophyses, stored under refrigeration in the presence of a dessicant. Each milligram of this powder has been assigned 2 I.U. (international units) of vasopressor-antidiuretic activity, as measured by its ability to increase the blood pressure or produce an antidiuresis in rats, and 2 I.U. of oxytocic activity, as shown by its ability to contract the rat uterus. Pure neurohypophysial peptides have been made and one milligram of vasopressin (ADH) is equivalent to 450 I.U. activity, while one mg of oxytocin has 400 I.U. A 'standard' need not have an international status, though this helps comparison of results in different laboratories, but can be made up for present and future reference by a particular investigator.

Identification of a hormone can be made by measuring its various activities and properties. Such 'profiles' or 'fingerprints', may utilize various biological actions, physico-chemical properties and immunological behaviour. A *pharmacological profile* of some of the many activities of neurohypophysial hormones is shown in Table 2.1. Many biological actions may be shared by several hormones, but the ratios of the potencies of the different hormones, with respect to each of their activities, vary so that when several such criteria are used, almost positive identification of a substance is possible. In addition to pharmacological criteria, chemical and enzymological evidence can also be used to aid identification of the neurohypophysial peptides; incubation with thioglycollate breaks the disulphide bridge in the hormone and abolishes its actions, while tyrosinase and trypsin also destroy the activities of some such peptides. Chromatographic behaviour can also be used to aid identification of hormones, though in some instances, as with oxytocin, mesotocin and isotocin, they can behave similarly.

Despite the fact that the chemical structure of many polypeptide hormones is known, their chemical determination is not feasible. This is partly due to their presence in low concentrations, especially compared to other protein materials that are also present. Separation of the various components would be essential before chemical determination would be even possible, while the tedious nature and inevitable losses make such procedures routinely inconceivable. Thus, apart from immunological procedures, biological methods have been retained as the most convenient, sensitive, specific and accurate ways for the determination of the polypeptide hormones. In the instance of the neurohypophysial hormones, bioassay procedures are still superior to immunoassay methods.

Identification of *steroid hormones* in tissues is often hampered by the small amounts of material that may be present. In order to increase the concentrations, suspected steroidogenic tissues are often incubated *in vitro* prior to extraction and identification procedures. These methods can be refined further by addition of exogenous steroid substrates,

Table 2.1 *Partial Pharmacological Profile of Some Neurohypophysial Hormones (U/mg)*

Hormone	Rat antidiuretic	Rat uterus oxytocic with Mg 2+	Rat uterus oxytocic without Mg 2+	Mammary gland milk let down	Frog bladder Water transfer	Frog bladder Sodium transport
Arg[8] — vasopressin (ADH)	400	20	34	70	26	33
Lys[8] — vasopressin (ADH — Pigs)	250	5	5	60	5	5
Arg[8] — oxytocin (Vasotocin)	186	139	264	100	10 000	1100
Oxytocin	5	450	450	450	450	450
Ile[8] — oxytocin (mesotocin)	1	289	375	328	—	507
Ser[4] — Ile[8] — oxytocin (ichthyotocin, isotocin).	1	150	420	300	—	5

Other assay preparations include frog and toad skin, toad bladder, hen and turtle oviduct, decrease in blood pressure of chicken, increase in blood pressure of rat, antidiuresis in frogs and chickens. From: MOREL and JARD (1968); FITZPATRICK and BENTLEY (1968) such as progesterone, that are usually isotopically labelled to facilitate identification of any products that may be formed. Such procedures, utilizing added substrates, can be used to demonstrate the presence of certain steroidogenic enzymes, but do not indicate whether or not the products are normally formed or released.

Although in the past steroid hormones were principally assayed and identified by biological means, these are becoming less frequent as sophisticated chemical and radioisotopic methods become available. Steroid hormones can be separated from other biological substances, and from each other by extraction and partition in suitable solvents such as chloroform, methylene chloride or ethyl acetate, followed by such procedures as gel filtration and chromatography. They can be measured by colourimetric methods such as that utilizing the yellow colour developed with phenylhydrazine (PORTER-SILBER reaction for cortisol and corticosterone), or by using their ability to fluoresce after treatment with ethanolic-sulphuric acid, sodium hydroxide or potassium terbutoxide. Steroid hormones can also be measured by gas chromatography and radioisotope methods, such as the double – isotope dilution method, which is particularly sensitive for hormones, like aldosterone, that are present at low concentrations. Isotopes of the steroids can be utilized in many quantitative procedures, and especially to assess losses of material that inevitably occur during their purification and preparation for assay. *Identification of steroid hormones* is generally based on their physico-chemical behaviour, this includes such things as:

1. Chromatographic mobilities in different solvent systems as compared with parallel standards which, in order to assist their location, can be isotopically labelled. Changes in chromatographic behaviour following acetylation of the steroids can be used as an additional criterium for identification.

2. Characteristic colour reactions such as take place with PORTER-SILBER reagent and tetrazolium blue.

3. Fluorescence, either natural or after treatment with such agents as ethanolic sulphuric acid or sodium hydroxide.

4. Measurement of ultra-violet absorption spectra in sulphuric acid, ethanol or methanol and infra-red spectra in chloroform.

A consensual definitive statement regarding the problems and desired criteria for identification and measurement of steroid hormones has recently been made (BROOKS *et al.*, 1970).

f) Immunoassay

Immunoassay procedures, especially those using isotopically labelled hormones, have in recent years allowed great strides to be made in the measurement of various protein hormones. Specific antibodies can be made to most peptide hormones, though in some instances when the molecule is small, like the neurohypophysial hormones, it is first conjugated to a larger molecule like serum albumin. Peptide hormones are labelled with ^{131}I or ^{125}I, and these react with the specific antibody to form a complex. When unlabelled hormone, either as a 'standard' or 'unknown', is added, it competes with the labelled hormone for the antibody so that the ratio of 'bound' to 'free' labelled hormone decreases. The 'bound' and 'free' components are separated by absorption onto materials like charcoal, cellulose or a second antibody, and the radioactivity of each counted. The methods are not free of problems and the presence of interfering substances can necessitate special separation procedures. The use of immunological assays has been largely confined to mammalian hormones, though they have been used to indicate the degrees of structural similarity of peptide hormones from different vertebrates. Their potential use, as a tool for investigating the different peptide hormones throughout the vertebrate series, is considerable.

2. The Pituitary Gland

The pituitary gland plays a central role in the maintenance and regulation of the secretions of three other endocrine glands, produces several additional hormones that act directly on effector tissues, and it also serves as the major link connecting the nervous and endocrine systems. Information concerning conditions in the external and internal environments is transmitted from the neural elements of the brain to the pituitary, where it is translated into relevant endocrine language which is then dispatched to the appropriate effector tissues. The external neural information comes from such things as light, temperature, the presence of the opposite sex and otherwise stressful or potenially harmful situations. Internal information about the general homeostatic state of the animal, such as water, solute and hormone levels in the body fluids, is also transmitted through the brain.

a) Morphology

Phyletically the pituitary gland can first be distinguished in the cyclostome fishes but is absent in protochordates like *Amphioxus*. The gland arises from two types of tissue; ectoderm from the anterior region of the head (hypophysis) and neural tissue (infundibulum) arising as a downgrowth from the forebrain. The glandular parts of these tissues remain in a juxtaposition with the hypothalamic region of the brain just behind the optic chiasma. In the Agnatha the different regions of the pituitary are separate, while in other vertebrates they are usually more tightly associated, but still retain their separate endocrine integrity and connections with the brain. The ectodermal portion of the pituitary differentiates into three main regions; the pars distalis, the pars intermedia and the pars tuberalis, which when grouped together make up the *adenohypophysis*. The neural tissue forms the *neurohypophysis*, which in its most complex condition consists of the median eminence, infundibular stem and pars nervosa (neural lobe). The neurohypophysis and pars intermedia are sometimes grouped together and called the posterior lobe of the pituitary (PLP) while the rest is then termed the anterior lobe (ALP). A diagrammatic representation of the amniote pituitary gland is given in Fig. 2.2.

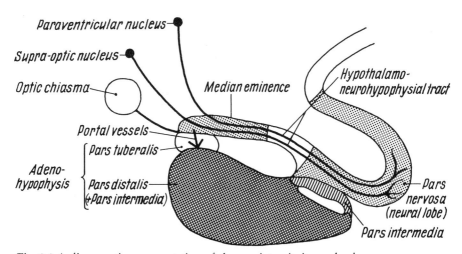

Fig. 2.2 A diagramatic representation of the amniote pituitary gland.

Communication between the pituitary gland and the brain takes place along nerve fibres to the neurohypophysis (*hypothalamic-neurohypophysial tract*) and pars intermedia, or through blood vessels to the pars distalis. The hypothalamic-neurohypophysial tract (which also sends some branches to the median eminence) originates in amphibians and fishes in the praeoptic nucleus, situated in the anterior region of the hypothalamus, while in the higher tetrapods (amniotes) these fibres arise from two areas, the paraventricular and supraoptic nuclei. Blood vessels connect neurons, which converge on the hypothalamus (in the region of the median eminence of lungfishes and tetrapods), with the pars distalis. These capillaries may form distinct portal vessels, as in some fishes and most tetrapods, or a less well defined vascular plexus like in many teleosts. In the fishes the arrangement of the vascular link varies considerably in different species.

The precise anatomical disposition of the tissues of the pituitary gland varies in the different vertebrate classes. This is summarized in the now classic illustration of the late J.D. GREEN (Fig. 2.3). One of the more interesting phyletic developments of the tetrapod neurohypophysis is the appearance of a more distinct median eminence, and its expansion posteriorly to form the pars nervosa (neural lobe). WINGSTRAND (1966) has suggested that this may be correlated with adaptation to a terrestrial way of life. It is rather interesting that such a change can also be seen in the lungfishes, which are often considered to be closely associated with the evolution of the tetrapods. FOLLETT (1963) summarized information about the hormonal content of the vertebrate neurohypophysis and found its concentration (moles/kg body wt) to be greater in those species with a well developed pars nervosa.

b) Secretion of the Pituitary Gland

α) *Pars Distalis.* Eight hormones with distinct actions have so far been found to originate in the pars distalis. These (see Table 2.2) include the *gonadotrophic hormones* (FSH and LH or ICSH), which influence breeding cycles, by promoting the maturation of the germ cells in the ovaries and testes, as well as the production of the steroid sex hormones. These hormones are sometimes indirectly associated with a change in the animals' osmotic circumstances. Thus fish, like the salmon and lamprey, migrate from the sea into rivers in order to breed while eels move from rivers back into the sea for the same reason. Periodic breeding migration of birds, mammals and reptiles does not result in such dramatic changes in the osmotic situation but, nevertheless, may alter the availability of water. Thus humpback whales migrating from polar to equatorial waters to breed do not feed, and, as indicated by their urine composition, do not drink sea-water (BENTLEY, 1963), depending on their metabolic tissue reserves for water during these months. Terrestrial birds making transoceanic or transdesert flights must endure similar privations. It is doubtful if such hormones have any direct effects on osmoregulation. The *thyrotrophic hormones* (TSH) and *corticotrophic hormones* (ACTH) influence the general integrity, growth and secretion of the thyroid and adrenocortical tissues respectively, the former may influence the osmotic balance of some fishes while the latter assists the regulation of sodium and potassium metabolism in many groups of ver-

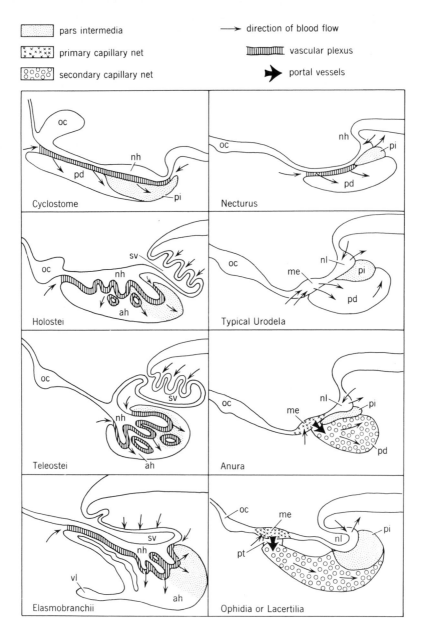

Fig. 2.3 Vascular relationships between the brain and pituitary in cold-blooded vertebrates, according to J.D. GREEN, illustrating the trend toward evolution of a hypophysial portal system. Abbreviations: *ah* Adenohypophysis; *me* Median eminence; *nh* Neurohypophysis; *nl* Pars nervosa; *oc* Optic chiasma; *pd* Pars distalis; *pi* Pars intermedia; *pt* Pars tuberalis; *sv* Saccus vasculosus; *vl* Ventral lobe. (From GORBMAN and BERN, 1962 as modified from GREEN, 1951).

tebrates. *Growth* and *lipotrophic hormones* help regulate the overall general protein, carbohydrate and fat metabolism of the tissues but evidence for a direct role in osmoregulation is lacking. *Prolactin* promotes the secretion of progesterone by the ovaries of rats and mice, but in other mammals has a more general role in instituting the secretion of milk by the mammary glands. In non-mammalian vertebrates the physiological role of prolactin is not clear, but when injected it may influence a variety of processes including, growth and water and salt metabolism of some urodele Amphibia and euryhaline fishes.

β) Pars Intermedia. The intermediate lobe of the pituitary gland secretes a hormone called *melanocyte stimulating hormone* (MSH) or *intermedin* which in the amphibians and some fishes may influence melanophores that are concerned with colour change (WARING, 1963). In higher vertebrates its role is unknown.

γ) Neurohypophysis. Arginine vasotocin is present in representatives of all vertebrate groups from the Agnathan fishes to the birds (see SAWYER, 1968), and is usually accompanied by a second hormone whose structure varies characteristically in the different taxonomic classes. In the Agnatha, and probably in the chondrostean fishes, like the sturgeons, vasotocin alone is present. The mammals secrete *vasopressin* (antidiuretic hormone, ADH) and *oxytocin.* In tetrapods vasotocin and vasopressin influence water metabolism by acting on the kidneys and certain other tissues like amphibian skin and urinary bladder. The role of such hormones in the fishes is far from clear. The second hormone in mammals, oxytocin, assists in the initiation of 'milk let down' from the mammary glands and may aid the contraction of uterine smooth muscle during parturition. However, in non-mammalian species the role of the analogous hormones are undetermined. Neurohypophysial hormones are stored in highest concentrations in the pars nervosa (neural lobe), but can also be identified in neurohypophysial tissue extending to the region of the praeoptic or supraoptic and paraventricular nuclei in the hypothalamus.

In mammals the median eminence is the site of storage for at least seven further molecular species that can initiate or inhibit the release of six different hormones from the pars distalis (see Table 2.2). These substances may be released in response to neural stimuli from the brain, pass through a portal circulation from the median eminence to the pars distalis and bring about the specific discharge of a particular hormone into the systemic circulation (see MCCANN, 1968). The non-mammalian vertebrates have been somewhat neglected with respect to these transmitters but there is evidence that at least one, the corticotrophic releasing factor (CRF), is present in the Amphibia (JORGENSEN and LARSEN, 1967). However JORGENSEN and LARSEN consider that it is doubtful that such a control system exists generally in the fishes, as the required anatomical portal circulation only occurs sporadically in diverse species, where it may be preadaptive, rather than of contemporary physiological significance.

51

Table 2.2 *Secretions of the Pituitary Gland and its Associated Hypothalamic Regions*

Adenohypophysis	Secretion	Abrev.	Hypothalamic — Median-eminence releasing factors	Physiological target organs and effectors
Pars distalis	Follicle-stimulating hormone	FSH	FSHRF	Ovaries and testes
	Luteinizing hormone or Interstitial cell stimulating hormone	LH ICSH	LRF	Ovaries and testes
	Thyrotrophic hormone	TSH	TRF	Thyroid
	Corticotrophic hormone (adrenocorticotrophic hormone)	ACTH	CRF	Adrenocortical tissue
	Lipotrophic hormone		?	(adipose tissue?)
	Prolactin (Luteotrophic hormone)	LTH	PIF (Prolactin inhibiting factor)	Mammary glands, osmoregulation, gills, etc.
	Growth hormone (somatotrophic hormone)	GH	GHRF, GHIF	Tissue metabolism (endocrine pancreas??)
Pars Intermedia	Melanocyte stimulating hormone (intermedin)	MSH	(Nerve transmission)	Melanophores
Pars Nervosa	Arginine-vasotocin (vasotocin) Arginine (or Lysine) vasopressin Oxytocin, etc.	AVT ADH	(Nerve transmission)	Kidney, amphibian skin, urinary bladder, mammary gland, uterine smooth muscle.

RF = releasing factor
IF = inhibiting factor

c) The Chemical Nature of the Pituitary Hormones

The pituitary hormones are proteins or glycoproteins, ranging in molecular weight from about 45 000 down to 1 000 (polypeptides). Some of the higher values may result from molecular associations, and have been recently revised with the aid of improved techniques for their isolation. The amino acid sequence has been determined for the smaller molecules including the neurohypophysial hormones, MSH, corticotrophin as well as growth hormone. Considerable differences often exist in the precise amino acid arrangements in homologous hormones from various species, and these somewhat alter their pharmacological, immunological and physiological properties. Such differences in activity are especially apparent when equivalent hormones from widely divergent species are cross-tested; thus YVES FONTAINE (1964) has pointed out that adenohypophysial thyrotrophic gonadotrophic and growth hormones from fishes, are ineffective when tested on mammals, though the equivalent hormones of mammals act in the fishes. Corticotrophins, which have a smaller molecular weight, can be shown to act in either direction. Similarly in the instance of the neurohypophysial hormones, mammalian arginine vasopressin (antidiuretic hormone) has an antidiuretic effect in the frog, *Rana esculenta* but is 50 times less active than the homologous amphibian hormone arginine vasotocin from which it only differs by a single amino acid substitution. (HELLER and BENTLEY, 1965).

α) Adenohypophysis. The gonadotrophins (FSH and LH) and the thyrotrophic hormone are glycoproteins with molecular weights of about 30 000. Prolactin and growth hormone are proteins with molecular weights of about 20 000 to 24 000. The amino acid sequence of human growth hormone has been elucidated (LI, LIU, and DIXON, 1966) and it consists of 54 amino acids. Corticotrophin is a smaller molecule (M. W. about 4 000) and the sequence of its 39 amino acids has been described (and shown to differ) in several mammalian species (see EVANS, SPARKS, and DIXON, 1966). The precise structures of melanocyte-stimulating hormones (MSH) from several mammals have also been described. In pigs there are two such active polypeptides, β-MSH with 18 to 22 amino acids and α-MSH with 13. Corticotrophin and MSH exhibit a cross-over in their biological activities; MSH has some corticotrophic actions (only 1.5% as strong), while corticotrophin can expand amphibian melanophores. Before their structures were known, some confusion existed as to whether they really occurred in nature as distinct entities (see WARING 1963). Knowledge of the amino acid sequences of each shows that the 13 amino acids of α-MSH are identical with the N-terminal tridecapeptide portion of corticotrophin (see I. HARRIS, 1966), so that the crossover in biological activity reflects similarities of their structures. As we shall see, other examples of analogous biological activity occur between different hormones, and these should be considered when assessing the physiological significance of a particular hormone in a species. In addition the occurrence of such catholic endocrine actions may be relevant to assessing the evolution of the physiological role of different hormones; a contemporary pharmacological action could reflect a past physiological role.

   Histochemical techniques have shown the presence of six different types of cells in the pars distalis, and these have each been assigned to one of the hormones present. The cell types can be divided into two major groups, acidophils and basophils, depending upon reactivity of their constituent granules with acidic or basic dyes. HYMER and McSHAN (1963) have separated the basophilic and acidophilic granules of the adenohypophysis by ultrafiltration, the acidophil material sedimenting more rapidly. The basophil granules were found to contain glycoprotein hormones, FSH, LH, and TSH, while the acidophilic granules contained prolactin and growth hormone. Corticotrophin was contained in the small-particle fraction. The further identification of the cell types with particular hormones, is based on observed histological changes associated with altered endocrine conditions, such as thyroidectomy or gonadectomy, and correlation of this with the hormone content of the gland. Analogous types of cells have been identified in all the major vertebrate groups. These have the generic suffix 'troph' so that the cells secreting thyrotrophic hormone are called thyrotrophs and so on.

The precise chemical structure of an adenohypophysial hormone from the non-mammalian vertebrates has not yet been described, presumably reflecting the difficulty of the procedures involved and human preference for the mammals. However, the considerable divergency in biological activity of the non-mammalian and mammalian hormones suggests that such knowledge may be particularly interesting with respect to the relationship of the structure and biological activity of such molecules.

β) Neurohypophysis. The hormones of the neurohypophysis are polypeptides with a molecular weight of about 1 000. Like the adenohypophysial hormones, there are differences in structure among the various vertebrate species. This is combined with a certain conservatism, as illustrated by vasotocin, which, with its distinct amino acid sequence, is represented in all the major non-mammalian groups.

The precise structures of the releasing and inhibiting factors found in the median eminence have not been determined, but it is known that they are polypeptides with a relatively low molecular weight. Amino acid analysis of corticotrophic releasing factors (CRF) suggests that they have similarities to both vasopressin and MSH (SCHALLY, SAFFRAN, and ZIMMERMAN, 1958; SCHALLY and GUILLEMIN 1963), and that this accounts for the early difficulties in distinguishing, and separating the activities of these peptides. The polypeptides of the neurohypophysial regions thus have a cross-over in their biological activities, that sometimes makes it difficult to relate each precisely to a distinct physiological action.

The structures of the mammalian neurohypophysial hormones, arginine-vasopressin (lysine-vasopressin in the pig family) and oxytocin were determined by Du VIGNEAUD and his collaborators in 1953 (Du VIGNEAUD *et al.*, 1953a and 1953b) and they were subsequently made by chemical synthesis. Structural analogues of these hormones were also synthesized, and one of these, arginine-vasotocin, was subsequently shown to be identical with the amphibian 'water balance principle', which was pharmacologically identified in lower vertebrates by HELLER 18 years previously (HELLER, 1941; PICKERING and HELLER, 1959; SAWYER, MUNSICK, and VAN DYKE, 1959). This created the interesting precedent of a hormone being synthesized in the laboratory before it was known to exist in nature. The neurohypophysial peptides consist of a ring of five amino acids joined by a disulphide bridge, contributed by two half-cystine residues, to which is attached a side chain of three amino acids. The amino acid composition and sequence of the peptides varies slightly in different species, but this results in considerable changes in their biological activity. So far eight such polypeptides have been chemically characterized in the vertebrates; usually two per species (though one or three may sometimes be present in a single gland). The structures of these hormones are shown in Fig. 2.4; arginine-vasopressin (ADH) and oxytocin which are present in mammals, and arginine-vasotocin, which has been identified in species from all the non-mammalian groups. Vasotocin has the ring sequence of oxytocin with the side chain of arginine-vasopressin, and when injected into mammals has some of the properties of each hormone; it contracts the uterus, brings about 'milk let down'

Amino acid sequences of the known active natural neurohypophysial principles*

A. VASOPRESSOR-ANTIDIURETIC PEPTIDES
1. Arginine vasopressin

Cys-Tyr-Phe-Gln-Asn-Cys-Pro-Arg-Gly-NH$_2$
1   2   3   4   5   6   7   8   9

2. Lysine vasopressin

Cys-Tyr-Phe-Gln-Asn-Cys-Pro-Lys-Gly-NH$_2$

3. Arginine vasotocin

Cys-Tyr-Ile-Gln-Asn-Cys-Pro-Arg-Gly-NH$_2$

B. NEUTRAL OXYTOCIN-LIKE PEPTIDES
1. Oxytocin

Cys-Tyr-Ile-Gln-Asn-Cys-Pro-Leu-Gly-NH$_2$

2. Mesotocin (8-isoleucine oxytocin)

Cys-Tyr-Ile-Gln-Asn-Cys-Pro-Ile-Gly-NH$_2$

3. Isotocin (4-serine, 8-isoleucine oxytocin)

Cys-Tyr-Ile-Ser-Asn-Cys-Pro-Ile-Gly-NH$_2$

4. Glumitocin (4-serine, 8-glutamine oxytocin)

Cys-Tyr-Ile-Ser-Asn-Cys-Pro-Gln-Gly-NH$_2$

* Amino acids differing from those in the corresponding positions in the arginine vasopressin molecule are underlined.

Fig. 2.4 Amino acid sequence of the known active natural neurohypophysial principles which are distributed in different groups of the vertebrates. (From SAWYER, 1967).

and produces an antidiuresis. Conversely the mammalian hormones when injected into lower vertebrates may act like vasotocin, and promote retention of water in the amphibians, reptiles and birds.

The neurohypophysial peptide hormones are stored in greatest concentration in the median eminence and pars nervosa (or its equivalent region in fishes), in proximity to the blood vessels supplying these regions. Contrary to earlier suggestions, the peptides do not originate in these regions but travel down the nerve axons of the hypothalamo-neurohypophysial tract after their formation in the regions of the praeoptic or supraoptic and paraventricular nuclei. They are merely stored at the termini of the axons. The elucidation of this role of the nerve cells in the formation and storage of the neurohypophysial hormones arose from careful histological observations of this tissue under a variety of physiological circumstances. Certain nerve cells, widely distributed among the invertebrates, as well as in the neurohypophysial region of vertebrates and caudal parts of the spinal cord of some fishes (urophysis), have been shown to contain granules, that can be stained by methods such as the GOMORI chrome-alum haematoxylin procedure. The quantities and distribution of this stainable material can be related to the physiological circumstances in the animal, and in the instance of the neurohypophysis with the hormones present. The formation in nerve cells of histologically stainable material that may be secreted into the blood stream, is termed 'neurosecretion' and was developed by ERNST and BERTA SCHARRER (1945). BARGMANN (1949) related the presence of GOMORI stainable substances in the hypothalamo-neurohypophysial region, to that of the neurohypophysial hormones. He was able to trace its formation to the region of the supraoptic nucleus, and to follow its movement down the axons to storage sites at the nerve termini in the pars nervosa (see BARGMANN, 1968). The hormones are present in membrane-bounded granules attached to a protein called neurophysin (M. W. 14 000) (GINSBURG and IRELAND, 1966) and it is these granules that constitute the stainable material. The neurosecretory cells, like ordinary nerve cells, can transmit impulses that are involved in the release of the hormones, but differ in that they do not enter into synaptic contact with other nerves, but release their products directly into the circulation (for a review see BERN and KNOWLES, 1966). The hormones of the median eminence are, no doubt, also of neurosecretory origin, but the neurons here are not readily stainable, except in the birds (FARNER and OKSCHE, 1962).

d) Release and Metabolism of Pituitary Hormones

The levels of the adenohypophysial and neurohypophysial hormones in the blood are influenced by a variety of specific stimuli, that can be related to the particular roles of the secretions in physiological homeostasis. In addition, release may also be initiated by many stimuli (non-specific) that bear no apparent relationship to physiological adjustments by the animal (for a recent review see GINSBURG, 1968). Thus, the physiological significance of a relationship between a stimulus and the release of a hormone must be interpreted with some caution. The release of pituitary hormones may be affected by changes in internal conditions such as temperature, hydrostatic pressure, and the concentrations of solutes and metabolites

56

in the body fluids (interoceptive stimuli). External physico-chemical changes in the environment may also influence the release of hormones: these include temperature, light, noxious chemicals, injury, anoxia and the proximity of other animals with all their associated implications, friendly or otherwise. These stimuli may be translated by the nervous system into fear, frustration, anxiety, rage and pain (the more extreme being termed 'stress'), or in certain instances, as when the opposite sex is involved, (usually) desire and pleasure. Such responses may involve the release of endocrine secretions.

Physiological levels of hormones appear in the blood for varying periods of time, dependent on their rate of release and their metabolic fate. The destruction and removal (clearance) from the circulation, depends to some extent on all tissues of the body, but in many instances the blood, liver and kidney are especially concerned. The survival of hormones in the circulation may vary from a few minutes, as with the neurohypophysial hormones, to several hours for thyroxine. The balance between release and destruction determines the circulating concentration, and could influence the physiological role of the hormone. Thus, the neurohypophysial hormone oxytocin is destroyed by an enzyme that appears in the blood of women during pregnancy, a time when the uterine-contracting ability of the peptide is undesirable.

(α) *Adenohypophysis*. Corticotrophic and growth hormones may be released in response to 'stress', while thyrotrophic hormone secretion is inhibited by such stimuli. The normal release of gonadotrophic hormones may also be upset by unusual exteroceptive stimuli, and this may alter the normal processes of the reproductive cycle. The implantation of fertilized eggs in mice can be prevented by exposure of the female to the odour of a strange male mouse (PARKES and BRUCE, 1961), while the emotionally caused vagaries of the human menstrual cycle have plagued most of us.

The normal physiological stimuli that control the secretion of corticotrophin, are the circulating levels of the steroid hormones secreted by its target organ, the adrenocortical tissue. High circulating levels of the adrenocortical steroids inhibit, or reduce, release of corticotrophin, while low concentrations facilitate its discharge. Such a control mechanism is an example of the *'negative feed-back' system*. The pituitary gonadotrophic-gonad and thyrotrophic-thyroid axes also exert mutual control in this manner. The receptor site for the feed-back inhibition by the steroid hormones is in the median eminence region of the hypothalamus, though it is still considered possible that a direct effect may also occur in the adenohypophysis itself. Thyroxine influences TSH discharge by a direct action on the anterior pituitary. Stimuli (apart from stress) may also impinge on these functional axes. In the instance of the gonadotrophic hormones, this may rest in the obscurity of a hypothalamic 'biological clock,' which can be influenced by such factors as the external temperature, light (day length), rainfall and the nutritional state of the animal.

The mechanisms that control the release of corticotrophin are of a dual nature; the effects of 'stress' take place rapidly within minutes and are little, if at all, affected by the steroid feed-back system, while the changes initiated by steroid levels only take place after many days. Prolactin is released in response to 'suckling,' a mammalian prerogative, but in teleost fishes it apparently takes place when the animals are transferred from salt water to fresh water (ENSOR and BALL, 1968). These latter observations remind us that mechanisms for the physiological release of hormones may differ among the vertebrates, in a manner commensurate with their differing roles.

(β) Neurohypophysis. The neurohypophysis has direct neural connections with the brain, and release of its hormones may be initiated through these by a variety of stimuli, many of which are not specifically related to the physiological responses. Thus, neurohypophysial secretion may be initiated by a variety of noxious physical and chemical stimuli, pain and even emotion. Stimuli for release in normal physiological circumstances have been studied almost exclusively in mammals, but these have been extrapolated to the non-mammalian groups. As in the instance of the adenohypophysis care must be taken to distinguish between such secretion in response to non-specific stimuli, and those that may be initiated in relation to precise homeostatic requirements and adjustments.

Osmotic stimuli play a major role in regulating the release of the antidiuretic hormone vasopressin in mammals, and probably also have this effect in other tetrapods. Unequivocal evidence for this was provided by the classical experiments of VERNEY (1947) with dogs. VERNEY showed that injections of hypertonic solutions of sodium chloride into the common carotid artery, resulted in a reduction of the urine flow of hydrated bitches. An increase of only 2% in the osmotic concentration of the blood produced a 90% reduction in urine flow. Injections of glucose also resulted in such responses, but urea was ineffective, presumably reflecting its permeant nature which fails to initiate an osmotic change. JEWELL and VERNEY (1957) repeated these experiments and tied off various branches of the common carotid artery in an attempt to localize the region of the brain that responds to these osmotic changes. It was found that the blood supply to the anterior hypothalamus was essential; this region includes the supraoptic nucleus, and it has been suggested that small vesicular cells near this may function as the osmoreceptors. Osmotic modulation of such receptors then may be transmitted cholinergically to the neurons of the supraoptico-hypophysial tract, and result in increased, or decreased, release of vasopressin from the nerve terminals. Osmotic stimuli such as dehydration and immersion in hypertonic saline solutions have been shown to deplete the neurohypophysis of its stored peptides in some birds, amphibians, fishes and mammals, and this probably reflects release of the hormones. Vasotocin has also been measured in the circulation of frogs and toads, following such osmotic excitation (BENTLEY, 1969a).

In mammals, vasopressin is released in response to a variety of non-osmotic stimuli such as haemorrhage, exposure to an increased environmental temperature, suckling, administration of anaesthetics, and following a large variety of major and minor noxious stimuli. A depression of hormone release follows exposure to cold, a rising blood alcohol level, and some tranquilizing drugs like chlorpromazine. Some of these stimuli are also effective in frogs and toads. The second neurohypophysial peptide present in mammals is oxytocin; commensurate with its probable physiological roles in initiating 'milk let down' from the mammary gland and assisting uterine contractions during parturition, it is secreted in response to suckling and stimulation of the female genitalia. Vasopressin is simultaneously released. Most non-mammalian species also possess two such neuro-

58

hypophysial peptides, but if there are separate stimuli for their release, these are unknown.

Neurohypophysial peptides are rapidly removed from the blood in mammals, sustained circulatory levels presumably being maintained as a result of continuous secretion. Injected vasopressin or oxytocin have half-lives of less than 5 minutes in rats, dogs and rabbits. This is largely the result of their enzymatic inactivation in the liver and kidneys, accompanied by a small (about 10%) excretion in the urine. The only observations on such clearances in non-mammalian species are those of HASAN and HELLER (1968) in chickens and toads, *Bufo marinus*, in which injected vasopressin, oxytocin and vasotocin were found to disappear three to twenty times as slowly as in the mammals. Such variations could provide a basis for differences in the role and function of the neurohypophysis in different groups of vertebrates. Sustained circulating hormone levels may be better suited to mediate functions that have a prolonged hormonal requirement.

3. The Thyroid Gland

Despite exhaustive studies extending over the last 30 and more years, the thyroid gland has not been shown to have a direct role in the integration of osmotic homeostasis in vertebrates. It does, however, influence water and salt metabolism indirectly in many ways and it may potentiate the actions of hormones that are directly involved in osmoregulation.

a) Structure and Secretions

Embryologically the thyroid gland is formed as a median downgrowth from the floor of the pharynx, and this reflects its anatomical site in the region of the 'neck,' and its affinities to the pharyngeal endostyle in lower chordates. Phyletically, the thyroid gland has the longest continuous history of any vertebrate endocrine organ, as homologous tissue and thyroid hormone have been identified in the endostyle of the ammocoete larva of the lamprey, and in protochordates like *Amphioxus* and the ascidians. In vertebrates the thyroid gland contains hollow follicles surrounded be secretory cells; colloidal material containing the stored hormones is accumulated within these follicles. The tissue is surrounded by a discrete capsule in the tetrapods, but in the fishes it is more diffuse and often extends along the region of the ventral aorta.

The thyroid secretions are unique among the vertebrate hormones, in that they contain an inorganic element, iodine, as an essential part of their structure. Synthetic analogues containing bromine and isopropyl moieties, instead of iodine, may have considerable biological activity (TAYLOR, TU, BARKER, and JORGENSEN, 1967) but these do not appear to exist in nature. The iodine occurs in the aromatic ring of a tyrosine molecule (Table 2.3), and four such compounds have been identified in thyroid tissue; monoiodotyrosine, diiodotyrosine, tetraiodothyronine (thyroxine) and triiodothyronine. The latter two substances are the secreted hormones. Thyroxine is the predominant secretion, and differs from triiodothyronine in the

Table 2.3 *Chemical structures of some vertebrate hormones*

Corticosteroids

Cortisol   Corticosterone   Aldosterone

Sex Steroids

Oestradiol   Progesterone   Testosterone

Catecholamines   HO—⟨ ⟩—CHOH·CH₂·NH·CH₃   HO—⟨ ⟩—CHOH·CH₂NH₂

Adrenaline   Noradrenaline

Thyroid Hormones   HO—⟨ ⟩—O—⟨ ⟩—CH₂·CH·COOH   Thyroxine

Triodothyronine

Angiotensin II (equine)   H. Asp-Arg-Val-Tyr-Ileu-His-Pro-Phe. OH

duration and potency of its biological actions, which are measured in terms of days for thyroxine and hours for triiodothyronine. These four iodinated tyrosine compounds have been identified in thyroid tissue throughout the vertebrates, including the ammocoete of the cyclostome, *Petromyzon planeri* (ROCHE, SALVATORE, and COVELLI, 1961).

Iodinated derivatives of tyrosine, with thyroid activity, occur widely in animals, including the protochordates and invertebrates. This is not surprising as such compounds can be made in the laboratory by simply incubating iodine with tyrosine containing proteins, such as casein. The process of formation in vertebrates consists initially of accumulation of iodide from the nutrients in the gut or by absorption through the gills in fishes (LELOUP, 1966). The thyroid gland preferentially traps this iodide and oxidizes it to iodine which then interacts with tyrosine to form mono-and di-iodotyrosine, two of the latter molecules condensing to form thyroxine, with the expulsion of an alanine side chain. Triiodothryonine may be formed by the removal of one of the iodine atoms from thyroxine or, as considered most likely, by the separate condensation of mono-and di-iodotyrosines. The hormones are stored in combination with a large glycoprotein molecule, thyroglobulin, which make up the colloid in the follicles, and are released from this by the action of a protease.

The activity of the thyroid gland may be judged in several ways. The histological appearance varies, the secretory cells being tall and columnar when it is in an active state, and low and squamous when inactive. The quantity of protein-bound iodine and the uptake of radioiodine (^{131}I) is greatest in active glands. The significance of such changes, however, must be interpreted with considerable caution, as they do not necessarily reflect circulating blood levels of the hormones. The environmental iodine levels thus may profoundly affect such values; limited availability will result in the appearance of a highly active gland, yet the animal may be functionally hypothyroid. The pharmacological standardization of therapeutically used thyroid gland preparations can be carried out by measurement of the protein-bound iodine in the gland. A batch of such material once released on the American market was found to be almost devoid of biological activity, but it is not clear whether this was due to a biological inconsistency, or a human one of 'salting' the material with iodine.

b) Function

The uniform presence of thyroid tissues and secretions throughout the vertebrates, along with the persistence of the controlling pituitary thyrotrophin-thyroid axis, strongly suggests that it has a physiological role in all vertebrates. This has been well characterized in the mammals, birds and larval amphibians, but despite extensive investigations an unequivocal function has not been established in the rest of the vertebrates.

In the homoiotherms the thyroid hormones have a *calorigenic* function and stimulate metabolism and oxygen consumption. In the poikilotherms, a similar role has been claimed in some species under certain conditions, but most evidence strongly indicates that this is not a characteristic action in such animals.

The thyroid is also necessary for optimal growth and differentiation in some species, but again, this has only been consistently shown in the homoiotherms. Similar claims have been made among the poikilotherms, especially the fishes, but a consensus has not been attained. The thyroid secretions are necessary for the metamorphic transformation of larval Amphibia to the adult form, the *morphogenetic effect*, but while evidence for a comparable role in fishes has been sought, it has not been forthcoming.

Evidence derived principally from homoiotherms, suggests that the thyroid secretions have a diffuse action on all the cells in the body and, by a process which is not yet clear, facilitate their optimal function, growth and differentiation. In the poikilotherms changes of thyroid activity have been associated with alterations in their physiological and environmental situation including the reproductive cycle, migration, temperature, life in salt and fresh water and movements between such osmotic media. Thus, it is often felt that thyroid hormones have a widespread role in the life functions of such animals. Nevertheless, fish may get along perfectly well after their thyroid has been removed (MATTY, 1957).

c) Relationship to Osmoregulation

While there is a lack of evidence concerning a direct role of the thyroid hormones in osmoregulation, there are many examples of their physiological interactions. The osmotic situation of an animal may affect thyroid function by influencing the amounts of iodide that are available; less iodine is available to fish in fresh water than to those living in the sea (GORBMAN, 1963). Goitrogenic substances that inhibit the formation of the thyroid hormones occur in nature, and these include excesses of sodium chloride. Mice given sodium chloride solutions to ingest suffer from hypothyroidism, as renal excretion of the excess salt also results in loss of iodide (ISLER, LEBLOND, and AXELRAD, 1958). Thus care must be taken to distinguish between those effects of environmental osmotic changes on thyroid activity that are due to altered iodine metabolism, and a supposed additional osmoregulatory requirement for the hormone.

The calorigenic effects of thyroid hormones in homoiotherms will increase the respiratory water loss that accompanies the increased respiratory gas exchange. Evaporative water losses from the skin may also be increased, especially if there is a resulting need for thermoregulation.

Inhibition of thyroid function in rats results in an increased exchange of sodium chloride and water. This is due to a failure of the kidney to conserve salt adequately; the renal tubules do not reabsorb as much sodium, their sensitivity to aldosterone is reduced about ten-fold, and the production of this steroid hormone is depressed (FREGLY and TAYLOR, 1964). The effects of the antidiuretic hormone are also reduced. In ducks, thyroidectomy results in a delay in the onset of secretion from the nasal salt gland following the administration of sodium chloride solutions (ENSOR, THOMAS, and PHILLIPS, 1970). This probably reflects a delay in the absorption of the saline from the gut. It is not known whether such actions extend to poikilotherms, but it has been suggested that seasonal changes in the sensitivity of frogs to neurohypophysial hormones, may be mediated by the thyroid gland (HELLER, 1930).

The 'maturation effect' of thyroid hormones in the Amphibia profoundly affects their water and salt metabolism. The metamorphosed adult animals usually assume a terrestrial, as opposed to an aquatic manner of life, and this is associated with the development of a number of relevant physiological processes. Tadpoles of the bullfrog, *Rana catesbeiana*, do not appear to transport sodium actively across their skin until after the fore and hind limbs appear, a process that can be accelerated by exposure to thyroxine (TAYLOR and BARKER, 1965). Amphibian tadpoles also respond poorly, if at all, to the neurohypophysial hormones, but the response is facilitated after metamorphosis (HOWES, 1940).

The migration of fishes is accompanied by increased activity of the thyroid gland (BARRINGTON, 1963). Such migration often takes the animals into a contrasting osmotic milieu, such as fresh water from the sea, or in the opposite

62

direction. It is not clear if such changes in thyroid function influence osmoregulation, but OLIVEREAU (1948) has demonstrated increased glandular activity when teleost fishes are moved from the sea to fresh water, and HICKMAN (1959) has noted a similar change when the starry flounder is moved from fresh water into solutions of increased salinity. Treatment of fish with thyroid and antithyroid preparations also may influence their salinity 'tolerances' and 'preferences,' but the sites of action, and normal physiological significance of these observations are not clear (M. FONTAINE, 1956). HOAR (1959) has suggested that the thyroid may aid fishes in adapting to altered environmental situations and stresses, including osmotic changes, a concept, which in my opinion, makes good sense. We should, however, reserve judgement as to whether such changes have adaptational significance, remembering that in other endocrine glands, like the adrenal cortex and neurohypophysis, changes that appear to be irrelevant physiologically can be initiated by non-specific stresses.

4. The Parathyroid Glands and Ultimobranchial Bodies

These endocrine tissues are concerned with the regulation of calcium metabolism. Embryologically they originate from the region of the branchial (gill) pouches; the four parathyroid glands are derived from the third and fourth pair of pouches, and the ultimobranchial bodies from tissue near the last pair. The parathyroid glands make their initial phyletic appearance in the amphibians and are absent in the fishes, while the ultimobranchial bodies are present in all vertebrates, with the apparent exception of the cyclostomes. The parathyroid glands lie on either side of the thyroid gland, while in non-mammalian vertebrates the ultimobranchial bodies lie in the pericardial region. In mammals the latter tissues are present within the thyroid tissue, where they make up the parafollicular or 'C' (for calcitonin) cells.

The secretion of the parathyroid glands is a polypeptide made up of 83 amino acids and is called parathormone (GREEP, 1963). The ultimobranchial hormone (thyrocalcitonin or calcitonin) of mammals was identified by COPP et al. in 1962 and has since also been found in representatives of the rest of the tetrapods and in the chondrichthyean and bony fishes (COPP, 1968). Mammalian calcitonin is a polypeptide, which consists of 32 amino acids (MacINTYRE, 1968).

Parathormone elevates the plasma-calcium concentration, while calcitonin lowers it. They thus have an opposing action, which is probably physiological, and is predominantly mediated by their effects on the rate of calcium reabsorption from bone. Thus, parathormone facilitates resorption of calcium from bone while calcitonin reduces it. In addition there may also be actions on other tissues, including the kidney.

a) Relation to Osmoregulation

Calcium is an important mineral constituent of the body, both with respect to structure and to its essential role in many biochemical reactions. It is necessary for the adequate functioning of all cells, but is not usually considered to be an os-

63

motic component of the body, as it makes up less than 2% of the total solute in the body fluids. Parathormone and calcitonin, thus, are only indirectly concerned with osmoregulation, in that the structural integrity of the membranes of the body and the release and actions of hormones, such as those from the neurohypophysis, are partly dependent on calcium. In mammals, elevated plasma calcium levels, as occur in hyperparathyroidism, result in increased renal water loss (FOURMAN, 1963). This seems to be due to inhibition of the effects of vasopressin (ADH) on water reabsorption from the renal tubule.

5. 'Islets of Langerhans' of the Pancreas

The endocrine secretions of the pancreas are formed in groups of tissue called the 'Islets of Langerhans'. At least two histologically distinct types of cells can be identified in these tissues; the α-cells which secrete glucagon and the β-cells insulin. α-cells have not been identified in the cyclostomes or urodeles. The hormones are polypeptides that affect the metabolism of glucose and also that of fat and protein.

Insulin consists of two peptide chains one of which, the A-chain, contains 21 amino acids while the other, the B-chain, has 30 amino acids. These chains are connected by two disulphide bridges. The amino acid sequence has been completely or partly determined for several mammals, the chicken and several teleost fishes (L. SMITH, 1966). Of the 59 amino acid positions in the molecule, substitutions have so far been found at 29 different places in the insulins of different species. Mammalian insulins usually differ from each other by substitutions at 2 or 3 positions, though guinea pig insulin has 17 changes as compared to the pig or human hormone. Chicken insulin has 6 amino acid substitutions, compared to the latter, while teleost fishes have as many as 17 such alterations. The various amino acid changes result in different immunological behaviour, but apparently have surprisingly little effect on biological activity, even when the insulins are tested in contrasting species (L. SMITH, 1966). Physiologically, the most obvious action of insulin is to lower the blood-glucose concentration. This effect is seen throughout the vertebrates from the cyclostome fishes to the mammals (see BENTLEY and FOLLETT, 1965), though the reptiles and birds are less responsive than other species. The effect of insulin on glucose principally reflects its action in facilitating conversion of glucose to glycogen and fat. It also has important effects in increasing protein synthesis, and reducing the mobilization of fat from adipose tissue, at least in some species.

Glucagon has been identified in species of mammals, birds and bony fishes (BERTHET, 1963). Mammalian (pig) glucagon contains 29 amino acids. This hormone elevates the blood glucose concentration in representative species from most vertebrate groups, but the urodele Amphibia and the cyclostomes (BENTLEY and FOLLETT, 1965) are insensitive to the mammalian hormone. Glucagon increases the blood-glucose levels by promoting liver glycogenolysis.

a) Relation to Osmoregulation

Insulin and glucagon are not directly involved in the regulation of water and salt metabolism. Insufficient insulin results in an elevated blood-glucose level, which if it exceeds the reabsorptive capacity of the renal tubule is excreted in the urine, with the simultaneous osmotic loss of extra water. This results in the abnormal urinary water losses and thirst that are associated with the disease of diabetes mellitus. When diabetes mellitus is induced experimentally in laboratory rats their water consumption increases four-fold while the plasma concentrations rise by 24 m-osmole/l (YOUNG, 1969). The vasopressin content of the neurohypophysis of such rats declines by one-third and this seems to be due to an increased rate of release of this peptide. Such deficiencies can be corrected by injecting insulin. Such effects are not the principal metabolic inconvenience of the disease, except possibly if water is not freely available. The sand rat, *Psammomys obesus,* which lives in the deserts of North Africa and the Middle East, develops diabetes mellitus when kept under certain laboratory conditions (SCHMIDT-NIELSEN, HAINES, and HACKEL, 1964). This is associated with an increased caloric intake when the sand rats are fed on rat checkers instead of fresh vegetables. It is doubtful that this condition occurs in the natural desert environment, but death could perhaps be facilitated by dehydration.

6. Steroid Hormones

The steroid hormones are made up chemically of three fused cyclohexane rings, and one cyclopentane, which together constitute the steroid nucleus (Fig. 2.5). Modifications of this nucleus, including the addition of methyl, hydroxyl and aldehyde groups, confer the various physiological activities that are associated with the

Fig. 2.5 Steroid nucleus showing conventional numbering of carbon atoms.

steroid hormones. Such chemical entities have been identified throughout the vertebrates, and in many instances have a well-defined hormone action. Unlike the polypeptide hormones, the steroid hormones may be structurally identical throughout a large phyletic range of animals. It is also interesting that such steroids are not confined to the vertebrates, but also occur in invertebrates and even in plants, so that they are not unique biochemical entities of the endocrine system, but have been utilized for this purpose in certain animals. Indeed, the structure does not even represent the sole key for unlocking their physiological metabolic effects. A group of synthetic compounds called stilbenes, have biological actions similar to the female sex hormone oestrogen, but are not steroids. Another substance with an action like oestradiol, genistein, occurs in subterranean clover, and this, by upsetting the normal levels in foraging sheep, brings about infertility in the ewes.

A variety of steroids have been identified in the blood and certain tissues of vertebrates, and some of these are specific hormones. Others probably represent products and by-products in the metabolic synthesis and degradation of the hormones, and also result from the chemical insults perpetrated on the tissues by the investigator. The vertebrate steroid hormones can be conveniently classified into three main groups chemically, or four physiologically. Chemically this is based on the number of carbon atoms ($C_{21,}$ C_{19} and C_{18}) and physiologically (*in italics*) on the general spectrum of their action and origin. 1. C_{21} steroids. These include; the *adrenocortical* (or *interrenal*) *steroids*, principally cortisol (hydrocortisone), corticosterone and aldosterone, and the *progestogens*, including progesterone, which are secreted mainly by the ovary. 2. C_{19} steroids include: the *androgens* like testosterone from the gonads of the male animal. 3. C_{18} steroids: the *oestrogens*, like oestradiol, which are secreted by the ovary. This list is not exhaustive, but is relevant to our subject. The chemical structures of the principal examples from each group are given in Table 2.3. It is interesting to see that relatively small changes in the chemical groups of the steroid nucleus confer widely different physiological activities on the molecules. It is, nevertheless, not surprising that each substance is not completely specific in its actions, a small mutual cross-over in biological activity often being apparent.

7. The Gonads

Oestrogens, progestogens and androgens have been identified in various species throughout the vertebrate series (see NANDI, 1967), though among the lower vertebrates demonstrations of their presence in the circulation are sparse.

Oestrogens are secreted by the follicular tissue of the ovary, and have also been found in the mammalian placenta, the testis and adrenocortical tissue. In the female animal the oestrogens integrate the processes of growth, differentiation, maturation and maintenance of the organs and tissues concerned with reproduction. That ethereal entity, female 'behaviour,' is also largely due to the actions of the oestrogens. Oestradiol is the most potent of the naturally occurring oestrogens, and is usually considered to be the principal and 'parent' hormone. Other oestrogens also widely found, include oestriol and oestrone which are probably derived from the 'parent' compound as a result of its metabolic transformation.

Progestogens, of which the principal natural example is progesterone, are formed by the ovary of a wide variety of vertebrates, and have also been found in adrenocortical tissue and in the mammalian placenta. Progesterone is necessary for the maintenance of pregnancy and the growth of the secretory alveoli of the mammary glands. Its function in the lower vertebrates is unknown, but it may be concerned with gestation in viviparous and oviviviparous species of the Reptilia, the Amphibia and the Chondrichthyes. Progesterone may be a metabolic intermediate in the biogenesis of all steroid homones (GRANT, 1962), so that its presence in tissues that form steroid hormones need not imply a direct endocrine function.

66

Androgens are secreted by the interstitial cells of the vertebrate testis, and have also been identified in the ovary, adrenal cortex and the mammalian placenta. In the male animal they are necessary for the normal growth, differentiation and maintenance of the reproductive organs and other sexual characters. Small amounts of androgens are necessary for the optimal function of the female reproductive system, just as oestrogens are necessary in the male. Testosterone is the principal and most potent naturally occurring androgen, but others such as androstenedione have a wide distribution.

The secretions of the gonadal steroid hormones are regulated by the gonadotrophic hormones of the adenohypophysis (FSH and LH or ICSH), and the respective circulating levels of these two groups of hormones are controlled by the negative feed-back system described previously.

a) Osmoregulatory Significance of the Gonadal Steroids

The actions of the hormones of the ovary and testis on water and salt metabolism are indirect and somewhat conjectural. Osmoregulation in gonadectomized animals appears to be adequate. The added needs for water and salt during reproduction and care of the young are obvious. These requirements appear to be adequately met by the usual osmoregulatory mechanisms, though some added adjustments may be necessary.

Many species migrate to certain areas to breed, and in many fishes this may involve movement from the sea into rivers or in the opposite direction. It is doubtful that the gonadal hormones play a direct role in initiating such migrations, as sexually immature animals may also make the journey and gonadal development usually does not accelerate till arrival at or near the breeding ground (BARRINGTON, 1963). Oestrogens and androgens, respectively, facilitate and inhibit the development of the adrenocortical tissue, an action that is due to changes in the release of the corticotrophic hormone (see VAN OORDT, 1963). Potentially they could thus influence osmoregulation through the adrenal cortex. Such effects may account for the hypertrophy of the adrenocortical tissue during sexual maturation of the migrating salmon (ROBERTSON and WEXLER, 1959). It should be remembered however, that the role of the adrenocortical tissue in osmoregulation of these fish is uncertain.

The similarities in the chemical structures of the gonadal steroids, with those secreted by the adrenal cortex, suggest that some interactions of their effects may occur. Progesterone, thus, can facilitate sodium excretion in man (LANDAU et al., 1955), an effect that may be due to its antagonism to aldosterone in the kidney tubules. CRABBE (1963) has shown such an antagonism in vitro; high concentrations of progesterone reduce the increases in sodium transport produced by aldosterone in the toad bladder. Oestrogens and androgens have also been shown to produce salt retention by mammals (see KRUHOFFER, 1960), but their sites of action are not clear. It is unknown whether such effects of gonadal steroids occur normally, or even in abnormal circumstances. It has been suggested that fluid retention in women prior to menstruation may be due to altered sex steroid levels. The increased levels of circulating aldosterone seen during pregnancy, could be the result of its antagonism by the high concentrations of progesterone that occur in this condition. Such effects may increase the need for osmoregulation, but do not constitute a part of the process.

8. The Adrenal (Suprarenal) Glands

The adrenal glands in mammals, birds and reptiles are usually discrete paired structures lying adjacent to the kidneys. In the Amphibia the adrenal tissues are attached to the ventral surface of the kidney, either in a continuous layer as in the Anura, or in scattered patches as in the Urodela. In the tetrapods the adrenal gland consists of two anatomically, embryologically and physiologically distinct regions; the adrenocortical (or interrenal) and chromaffin tissues. In mammals the chromaffin tissue makes up an internal medullary region, while the interrenal cells surround this to form the cortex of the gland (hence adrenocortical tissue). In non-mammalian tetrapods these two tissues are mingled with each other, so that there is no

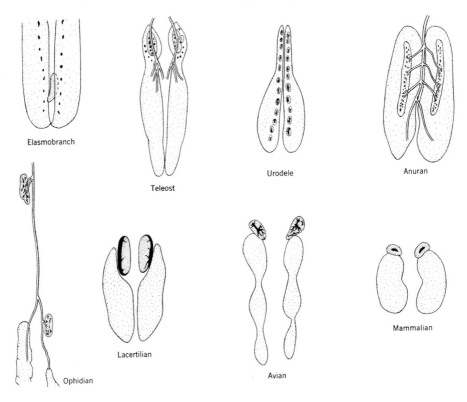

Fig. 2.6 Types of vertebrate adrenals and their association with the kidneys, in ventral view. Solid black indicates chromaffin tissue, dense stippling indicates interrenal tissue, and light stippling indicates kidney. Teleost kidneys show postcardinal venous drainage; anuran and ophidian kidneys show postcaval venous drainage. Teleost interrenal and chromaffin tissues are embedded in "head" kidney. Adrenals are shown as if in section. (From GORBMAN and BERN, 1962).

68

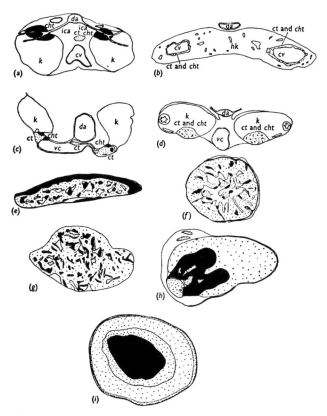

Fig. 2.7 Diagrams of cross-sections of cortical and chromaffin tissue from representatives of the different classes of gnathastomata.

a Dog-fish (elasmobranch)   *b* Perch (teleost)
c Ichthyophis (amphibian)   *d* Frog
e Lizard   *f* Crocodilia
g Pigeon   *h* Echidna (prototherian mammal)
i Rat (eutherian mammal)

Stippling Cortical tissue; *black* Chromaffin tissue; *cht* Chromaffin tissue; *ct* Cortical tissue; *cv* Cardinal vein; *da* Dorsal aorta; *hk* Head-kidney; *ica* Intercostal artery; *k* Kidney; *vc* Vena cava. (From CHESTER JONES, 1957)

distinct cortical and medullary zone. In the fishes, homologous tissues are present in all groups, but these usually (but not always in the bony fishes) are present as separate structures. In the Chondrichthyes, chromaffin tissue exists as paired segmentally arranged units, situated along the borders of the kidneys and Wolffian ducts. The adrenocortical cells in the Chondrichthyes usually coalesce to form a pair of distinct glands lying between the kidneys hence their name: interrenals. In the Osteichthyes (except the lungfishes which have a Urodele-like arrangement) the chromaffin tissue may extend in clumps along the cardinal veins where it is sometimes embedded in pieces of adrenocortical tissue. The cyclostome chromaffin cells are spread throughout the body, lying along the blood vessels, while the adrenocortical tissue exists as a series of bodies embedded in the posterior cardinal veins. The distribution of the interrenal and chromaffin tissues in the major vertebrate groups is shown in Fig. 2.6 and 2.7.

Anatomically, there appears to be an evolutionary trend towards a closer association of the chromaffin and adrenocortical tissue as one ascends the phyletic scale, but the biological significance of this is unknown. For detailed descriptions of the anatomy of the adrenal glands in vertebrates the monographs by HARTMAN and BROWNELL (1949) and CHESTER JONES (1957) should be consulted.

a) Chromaffin Tissue

This tissue is named because of its brown staining reaction with bichromate or chromic acid. Tissue identified in this manner is not only associated with the adrenal endocrine secretions, but also exists in other parts of the body, particularly in the ganglia of the sympathetic nervous system. Indeed the adrenal medulla is anatomically homologous to the sympathetic nerve ganglia, but discharges its products directly into the circulation, instead of towards a post-ganglionic nerve fibre. Chromaffin tissue secretes two principal hormones: adrenaline (or epinephrine) and noradrenaline (or norepinephrine). These, along with dopamine (which may act as their metabolic intermediate), are termed catecholamines. Adrenaline acts mainly as an endocrine hormone, while noradrenaline may have the role of a hormone, a neural transmitter, or an intermediate in the formation of adrenaline. The proportions of these two hormones in the endocrine chromaffin tissue vary considerably in different vertebrate groups, and even within the same species. Among the mammals, the percentage of the combined total of the hormones existing as noradrenaline, ranges from 86% in whales to 2% in rabbits, while in frogs and toads it is about 50%, and in the chicken and dogfish, *Squalus acanthias*, 70%. Histological evidence suggests that each hormone is stored in a different type of cell. The structure of adrenaline is shown in Table 2.3. It is formed in the chromaffin tissue from accumulated tyrosine, which is oxidized to dopa which, in turn, is decarboxylated to form dopamine. Dopamine is taken into granules in the cells where it is stored or converted by β-hydroxylation to noradrenaline, which, in turn, may undergo N-methylation to form adrenaline. The latter two products can be released from the granules in response to nerve stimuli conveyed through the splanchnic nerves that supply the gland. Reflex release can be initiated from other parts of the nervous system such as the sciatic nerve and the vagus, the centre for control being situated in the brain. The neural release of the catecholamine hormones is initiated by a variety of stressful, uncomfortable and unpleasant conditions including emotion, fear, pain, excesses of temperature, anoxia, a drop in the blood pressure, a low blood sugar level and physiological experiments. The adrenal chromaffin tissue thus affords, in addition to the pituitary, another site where nerves and endocrines may interact. Denervation does not result in degeneration of the chromaffin tissue. In addition to responding to neural stimuli, it is thought to discharge small amounts of its products continually into the circulation.

Animals survive adequately after ablation of the endocrine chromaffin tissue, even though the secretions in certain circumstances facilitate physiological adap-

70

tation to altered environmental conditions. Adrenaline and noradrenaline have widespread physiological (and pharmacological) actions, each hormone having some distinctively different actions and potencies from the other. Adrenaline increases the heart rate, constricts the blood vessels of the skin and visceral regions and dilates vessels in the skeletal muscles, so that the blood pressure is changed little. Noradrenaline, in contrast, constricts the blood vessels of the muscles in addition to those in the other regions, so that the blood pressure rises, and there is a reflex slowing of the heart. These hormones also dilate the pupils and bronchi, and relax the smooth muscle of the gut and urinary bladder, adrenaline being the more potent in these respects. Metabolic processes, like glycolysis and mobilization of fatty acids in the tissues, along with peripheral utilization of substrates and oxygen are increased, adrenaline again being more active than noradrenaline. Some effects of adrenaline and noradrenaline are mediated through the formation of a nucleotide, adenosine, 3',5'-phosphate (cyclic AMP), which is also an intermediary in the actions of a number of other hormones, including those from the neurohypophysis.

Relationship to osmoregulation. The catecholamines are not of direct physiological importance in osmoregulatory processes, but they may exert indirect effects in certain environmental, experimental and pharmacological circumstances. By influencing oxygen consumption, and changing the regional distribution of blood in the skin, kidneys and gills the catecholamines could conceivably influence evaporation and diffusion of water, as well as electrolyte transfers. Such effects are probably too transient to have chronic effects on the osmoregulation but could have an influence over short periods of time.

The actions of catecholamine hormones on kidney function have been explored repeatedly. Emotional stress may result in a transient antidiuresis in mammals, as a result of the release of adrenaline, which at the same time inhibits release of ADH (O'CONNOR and VERNEY, 1945). Injections and infusions of adrenaline may reduce renal sodium excretion (O'CONNOR, 1962). In large doses it facilitates the release of ADH and corticotrophin, which in turn can influence water and sodium losses. The amounts of adrenaline required to initiate the above effects are in considerable excess of those that accelerate the heart rate, so that it seems unlikely they are normally of physiological importance. In fishes, adrenaline dilates the branchial vessels (KEYS and BATEMAN, 1932) which can influence the exchange of electrolytes (STEEN and KRUYSSE, 1964). In the anuran amphibians, injected adrenaline also alters water and sodium exchange; water uptake in toads is increased (ELLIOTT, 1968), and in the isolated frog skin, water and sodium uptake is increased (JARD, BASTIDE, and MOREL, 1968; BASTIDE and JARD, 1968). These effects are similar to those of the neurohypophysial hormones, and it would appear that these analogous actions are both mediated by the formation of cyclic AMP. It is usually considered that such osmotic effects are normally mediated by neurohypophysial hormone, but it is possible that an interaction with adrenaline could facilitate its action. On the other hand injections of catecholamines may reduce the action of ADH on the mammalian kidney and produce a diuresis (see for instance FISHER, 1968). Such an antagonism can be demonstrated *in vitro* on the isolated toad urinary bladder and appears to be mediated by α-adrenergic effects of the catecholamines (HANDLER, BENSINGER, and ORLOFF, 1968; BENTLEY, 1971 c). The formation of cyclic AMP is a β-adrenergic response. High concentrations of adrenaline increase the rate of chloride loss from isolated frog skin, an effect that is due to stimulation of the secretory glands present (USSING, 1960). It seems unlikely that the catecholamines normally influence water and sodium exchanges, except in acute circumstances when their actions could contribute to the confusion of an investigator who is carrying out a distressing experimental procedure. However, the possibility of interactions with the effects of other hormones, particularly those of the neurohypophysis, should be considered.

b) The Adrenal Cortex and Interrenals

The adrenal cortex of marsupial and placental mammals usually contains three types of cells arranged in successive layers, or zones. These layers, in succession from the external capsule of the gland, are the *zona glomerulosa*, the *zona fasciculata* and the *zona reticularis*. The cells in these areas can be distinguished by their cytological appearance, shape and arrangement. Cells in the zona glomerulosa are round, not very prominent, rich in mitochondria, but poor in lipid droplets. The zona fasciculata is the largest of the zones; its cells are arranged in columns, and they are rich in lipid droplets, but poor in mitochondria. The innermost zona reticularis, is made up of flattened 'compact' cells that contain little lipid and numerous mitochondria. The various cortical zones and cells-types are not distinguishable in all mammals; the zona glomerulosa cannot be seen in certain mice, lemurs and monkeys. Analogous types of cells have not been identified in non-mammalian vertebrates.

The presence of different types of cells in the mammalian adrenal cortex suggests that they may be related to some of the endocrine functions of the gland. Originally it was thought that they represented a centripetal migration of cells, from the region of their formation to an area (zona fasciculata) where they produced their secretions, followed by final degeneration in the zona reticularis. However, it is now clear that the various zones represent areas where different steroids may be principally made and stored. Hypophysectomy greatly reduces the glandular secretions of cortisol and corticosterone, but has only a modest effect on the secretion of aldosterone. This is correlated morphologically with maintenance of the zona glomerulosa, and degeneration of the other two groups of cells. The adrenogenital syndrome in man is due to an excessive production of androgenic steroids by the adrenal cortex, and this is associated with hyperplasia of the zona reticularis. This evidence suggests that the zona glomerulosa is primarily concerned with the production of aldosterone, the zona reticularis with androgenic steroids, and the zona fasciculata with the secretion of cortisol and corticosterone. It is doubtful that such partition of labour is entirely discrete, the zona glomerulosa for instance seems to be able to secrete corticosterone. Attempts have been made to identify and measure the levels of the different biogenic enzymes and steroids present in the various zones, a process that is facilitated in large adrenal glands. RACE and WU (1961) dissected out the adrenocortical zones of the sperm whale, *Physeter catadon*, and their analyses generally support the above relationships. There is, nevertheless, still a considerable crossover among occurrences

72

of the steroids in the different zones. Whether such distribution of labour exists in production of different steroids in the adrenocortical cells of non-mammalian vertebrates is unknown.

α) *Biogenesis of the Corticosteroids.* Adrenocortical hormones are formed principally as the result of a series of hydroxylations to the steroid nucleus of cholesterol. It is considered possible that some steroid hormones are formed without the mediation of cholesterol, but the details of this are unknown. The adrenal cortex, using

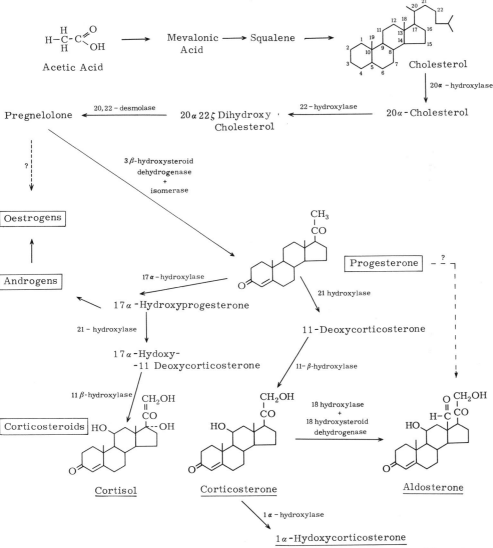

Fig. 2.8 Some of the steps in the biogenesis of corticosteroids and their relationship to the formation of other steroid hormones.

acetate as a substrate (Fig. 2.8), readily makes cholesterol, but it is possible that some of this may also be accumulated from the blood. Cholesterol is subsequently converted, via a series of intermediates, to progesterone, a pathway that seems to be common to all endocrine glands that secrete steroid hormones. In the adrenal cortex, progesterone may be converted to any of the three principal secretory products: cortisol, corticosterone or aldosterone, via two, and possibly three, principal pathways. A series of hydroxylations take place involving the enzymes 17-hydroxylase and 21-hydroxylase, which are associated with the endoplasmic reticulum, an 11 β-hydroxylase and 18-hydroxlase from the mitochondria. Progesterone, when hydroxylated at carbon 17, is converted to 17-hydroxyprogesterone, which is successively further hydroxylated at carbons 21 and 11, to give cortisol. This hormone can be oxidized, at carbon-11, to cortisone. If the progesterone is alternatively hydroxylated at carbon-21, 11-deoxycorticosterone is produced, which is then converted to corticosterone by hydroxylation at carbon-11. Corticosterone can be oxidized to 11-dehydrocorticosterone or, by a series of changes at the carbon-18, be converted to aldosterone. The formation of all aldosterone secreted by the adrenal cortex cannot be accounted for by such a conversion, so it seems likely that it is also made from progesterone in another way. A summary of these reactions is given in Fig. 2.8 but it should be remembered that these processes have not been demonstrated in many species and the information available is usually derived from mammals.

β) *Physiology.* The physiological importance of the adrenals was first described by THOMAS ADDISON in humans with diseased glands. These observations were confirmed by extirpating the gland surgically in other mammals. The part of the adrenal critical to life is the cortex, the chromaffin tissue of the medullary region being dispensable. Acute removal of the adrenals in mammals is usually followed by death in a few days, the associated physiological changes being comparable with those seen in pathological situations. In mammals there are metabolic changes, such as reduction in blood glucose and liver glycogen, a decrease in the level of plasma sodium, and an increase in plasma potassium. Probably largely secondary to such changes, there may be a reduction in extracellular fluid volume, a decrease in blood pressure, reduced renal blood flow and glomerular filtration rate, and a low renal clearance for urea and water. The effects on carbohydrate metabolism reflect reduced gluconeogenesis from protein and fat, as well as changes in peripheral utilization of metabolic substrates. The changes in plasma sodium and potassium levels in mammals are mainly mediated by the kidney, which fails to conserve enough sodium and secrete sufficient potassium.

While most mammals only survive for a few days following adrenalectomy some animals, like rats, can occasionally live for extended periods due to the presence of accessory adrenocortical tissue. The American opossum, *Didelphis virginiana,* often survives for several months following removal of the adrenals but this is not a marsupial character as one of its Australian relatives, the wallaby *Setonix brachyurus,* dies in two to three days after such an operation (BUTTLE, KIRK, and WARING, 1952). Survival of adrenalectomy can often be prolonged, though not indefinitely, by providing the animals with saline solutions to drink, thus moderating the changes in body electrolyte levels.

Adrenalectomy or interrenalectomy in non-mammalian vertebrates is often a difficult procedure, and the changes observed in different species are inconsistent (CHESTER JONES, 1957). Birds, reptiles and amphibians die after removal of the adrenals, but some species, like the frog *Rana temporaria*, may survive for several months if the operation is performed in winter, or die in two or three days if it is done in summer. Sharks and rays (Chondrichthyes) survive about one to three weeks after interrenalectomy (see HARTMAN *et al.*, 1944). CHAN *et al.* (1967a) have performed this formidable operation in eels (Osteichthyes), and their animals survived for at least three or four weeks. The metabolic condition and environmental situation of animals, especially poikilotherms, appears to influence their survival following adrenalectomy.

In non-mammalian vertebrates, the metabolic changes associated with adrenalectomy or interrenalectomy may be similar to those in mammals, or sometimes vary from this pattern. HARTMAN *et al.* (1944) found that interrenalectomy in the skate, resulted in a decreased blood glucose and liver glycogen, but no detectable changes in sodium and potassium metabolism. Adrenalectomy in the desert lizard, *Trachysaurus rugosus*, produces an elevated plasma potassium level but no change in sodium (BENTLEY, 1959b). Removal of the adrenocortical tissue of the eel does not affect plasma potassium levels, but results in a decreased plasma sodium in fish kept in fresh water, and elevated sodium if the animals are in sea-water (CHAN *et al.*, 1967a). It thus appears that the effects of adrenalectomy on electrolyte metabolism are at least partly related to the osmotic situation and relative availability of sodium and potassium and the physiological need for conservation or excretion of these ions.

The three principal adrenocortical steroids (corticosteroids) produced by vertebrates are cortisol, corticosterone and aldosterone. In addition, small amounts of androgens and oestrogens may be secreted by this tissue. The distribution of the three major steroids varies in different vertebrates. Cortisol and corticosterone are present in species throughout the vertebrate series, but their presence and relative abundance varies greatly. Thus the ratios in the plasma levels of cortisol/corticosterone varies from 0.05 in rats, to 4 in the cat and 20 in a monkey, *Macaca mulatta* (BUSH 1953). Aldosterone is widespread among the tetrapods, but rare or absent in the fishes.

The natural and synthetic adrenocortical steroids are, for convenience, often classified into two main groups. 1. *Mineralocorticoids*, which have prominent effects on sodium and potassium metabolism. Aldosterone and deoxycorticosterone are the main natural examples of this, while fluorocortisol is a potent synthetic analogue. 2. *Glucocorticoids*, which have more pronounced actions with respect to the interconversions among carbohydrates, fats and proteins. These include cortisol and corticosterone and synthetic analogues like prednisone and betamethasone. It should be remembered that the dividing line between these groups is not a clear one, and that there may be considerable crossover in their respective activities. Cortisol and corticosterone also exhibit effects on electrolyte metabolism, while on a purely molecular basis, aldosterone exhibits considerable glucocorticoid activity. The ratios of the two major activities vary, the mineralo/gluco-corticoid activity (measured in mammals) is: aldosterone, 100; corticosterone, 0.015; and for cortisol, 0.01; a 10 thousand-fold range.

The sites of glucocorticoid action are probably situated in most tissues of the body, but the liver and muscles are particularly active in this respect. Mineralocorticoid effects are mediated principally by the tetrapod kidney, the skin and urinary bladder in some of the amphibians, and the gills of certain fishes. Sodium and potassium exchanges in such tissues as the gut, the salivary glands, the sweat glands and the various 'salt' glands may also be influenced by corticosteroids. The relative importance of such sites varies in different species depending on their osmotic problems and methods of regulation. Adrenocortical steroids may act on the kidney by increasing tubular reabsorption of sodium from the glomerular filtrate and by promoting tubular secretion of potassium into this fluid. In some instances the corticosteroids may also facilitate the rate of glomerular filtration. Sodium uptake from environmental bathing fluids is increased through the skin of the Amphibia, while urinary sodium loss is reduced by its reabsorption across the wall of the urinary bladder in this group. Sodium absorption across the wall of the gut may also be promoted by corticosteroids, while the sodium/potassium ratio in saliva and sweat can be decreased. Sodium excretion by the nasal 'salt' gland of birds may be facilitated by such steroids. The gills of some fish are also considered to be sites for the regulatory action of endogenous corticosteroids, and they may increase sodium excretion through the gills of marine fish and increase uptake of this ion in freshwater species. These effects will be discussed in later sections.

γ) *Control of Secretion.* The adrenocortical tissue is subject to a number of stimuli that control its functional integrity (DENTON 1965; GANONG, BIGLIERI, and MULROW, 1966), and the synthesis and release of its hormones. The secretion of aldosterone and the glucocorticoid hormones are individually controlled, though this dichotomy is not rigid. Corticotrophin from the adenohypophysis stimulates the secretion of both aldosterone and the glucocorticoids, but only affects the former in the high part of the physiologically effective dose range. Aldosterone secretion can be initiated by the direct influence on the adrenal cortex of a reduced plasma sodium and/or elevated plasma potassium. A more complex mechanism controlling mineralocorticoid secretion is provided by the renin-angiotensin system. Renin is an enzyme secreted into the blood by the juxtaglomerular cells of the kidney in response to changes in renal arterial blood flow and pressure. This enzyme interacts in the plasma with an α-2-globulin called angiotensinogen, to produce a decapeptide called angiotensin. This decapeptide is then converted by a plasma enzyme to an octapeptide, angiotensin II (Table 2.3), which has the ability to constrict peripheral blood vessels and also to initiate aldosterone release. The peripheral vasoconstriction results in elevation of the blood pressure, and in this respect angiotensin II is the most potent substance known, being about 40 times as active as noradrenaline. The possible physiological role of angiotensin as a vasoactive agent is uncertain, but in a number of species it undoubtedly influences secretion of aldosterone. Such a role has been shown in man, the dog, the Virginia opossum and the bullfrog, *Rana catesbeiana* (DAVIS *et al.*, 1967). A note of caution is, however, conveyed in that the evidence for such a role in the rat is equivocal. Secretion of renin by the kidney can be shown to reflect sodium metabolism, this being greatest in sodium-depleted animals. The sodium depletion, and associated reduced blood volume, is thought to promote release of this enzyme, by way of small decreases in the renal blood flow and pressure. It is noteworthy that haemorrhage is also a powerful stimulant for release of aldosterone and renin. Renin and the plasma 'converting' enzyme activate angiotensin II which acts on the zona glomerulosa to initiate release of aldosterone and some corticosterone. In higher concentrations angiotensin II will also release glucocorticoids. These physiological events result in sodium retention, restoration of the blood volume and inhibition of further renin release (Fig. 2.9). Renin is bioassayed by measuring its effects on the blood pressure in mammals and its activity has been identified in non-mammalian vertebrates including kidneys of freshwater and marine teleost fishes, reptiles and amphibians (FRIEDMAN, KAPLAN, and WILLIAMS, 1942; MIZOGAMI et al., 1968; CAPELLI, WESSON, and APONTE, 1970). The effects of angiotensin on corticosteroid levels in fishes has not yet been described but the levels of renin are greater in fish kept in fresh water and this could reflect a role in sodium conservation.

DENTON (1965) has suggested that the mechanism controlling secretion of mineralocorticoids has been subject to evolution, and that the earliest method may be represented by the direct effects of plasma sodium and potassium concentrations on the gland. The renin-angiotensin system could be a more recent and sophisticated product that is sensitive to volume, and hence total sodium levels in the body.

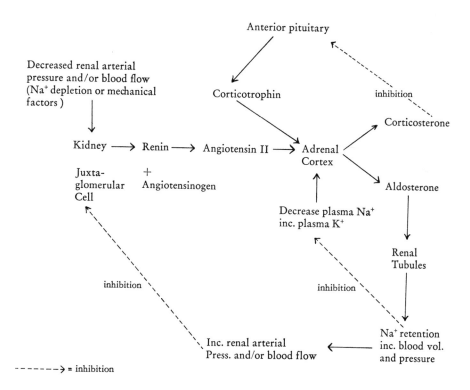

Fig. 2.9 Factors controlling release of aldosterone (adapted from Davis et al. 1967).

The corpuscles of STANNIUS in the fishes have been shown to contain a vasopressor material, which has been tentatively suggested as the equivalent of the latter control system (CHESTER JONES et al., 1966) and this will be discussed in the next section.

9. Putative Endocrine Tissues in Fishes

The fishes have an evolutionary history that has been independent of tetrapods for a long time. Phyletically and physiologically they are a very diverse group, with a vast number of species. It would thus not be surprising to find endocrine tissues amongst the fishes which have no homologues in the tetrapods, or even with other piscine groups. Two such putative endocrine tissues are the corpuscles of STANNIUS and the urophysis.

a) Corpuscles of STANNIUS

These islets of tissue are present in teleostean and holostean bony fishes, but are apparently absent from other piscine groups. The corpuscles of STANNIUS vary in number from two, in most teleosts, to forty or fifty in the holostean, *Amia calva.* Embryologically they arise as evaginations of the pronephric duct and are situated ventrally and posteriorly on the kidney, or they may lie caudally to this organ.

Since the original description of these corpuscles by STANNIUS in 1839, speculation about their function has been considerable. It has generally been thought that they are endocrine glands, but this has yet to be established. RASQUIN (1956) examined the histology of these tissues in the freshwater teleost, *Astyanax mexicanus*, after it had been immersed in 1% sodium chloride solutions, or injected with deoxycorticosterone. Cytological changes in the corpuscles that followed such treatments suggested that they may be involved in osmoregulation. Ablation of the corpuscles of STANNIUS in the eel results in a decreased plasma sodium level, accompanied by an elevated concentration of potassium and calcium (LELOUP-HATEY, 1964; M. FONTAINE, 1964). Such changes were ameliorated by the injection of corpuscle tissue extracts. These results have been confirmed by CHAN et al. (1967a), so that the evidence strongly supports the possibility of an osmoregulatory role for these tissues.

The exact actions of the corpuscles of STANNIUS in osmotic regulation, and the nature of their putative secretions are at present uncertain. Due probably to the proximity of the STANNIUS corpuscles and the adrenocortical tissues in fishes, the possibility that they are both parts of the same steroid secreting system has been explored extensively. It should be noted, however, that the two groups of tissues have distinct embryological origins and morphological sites, even though they do both lie in the caudal regions of a fish. Many attempts have been made to identify steroids in extracts of the corpuscles of STANNIUS, and to show formation and accumulation of such substances, after their *in vitro* incubation in the presence or absence of steroid substrates. The results will be discussed in detail in a later section, but they are variable and contradictory. There are some reports of the presence of steroids (or steroid-like materials?) under certain conditions, but

78

these have not been substantiated (see for instance CHESTER JONES et al.,1965). A fresh approach towards the problems of identifying a secretion has been made by CHESTER JONES and eight of his collaborators (1966); extracts of the corpuscles of STANNIUS from eels, when injected into rats, or eels, increased the blood pressure. Removal of the corpuscles from the eel reduced its blood pressure. CHESTER JONES et al. have indicated certain similarities between this pressor material and renin, and suggest that it may be part of a system homologous to the renin-angiotensin-aldosterone axis of tetrapods. This would account for some of the similarities between ablation of the STANNIUS corpuscles and interrenalectomy in eels, though the effects of the two operations are not completely identical; both procedures result in a decreased plasma sodium level, but only ablation of the corpuscles produces the elevated potassium and calcium concentrations (CHAN et al., 1967 a). Endocrine tissues usually have been found to have several roles and secretions; this could also apply to the corpuscles of STANNIUS.

b) The Caudal Neurosecretory System and the Urophysis

Neurosecretory nerve fibres have been histologically identified in the caudal regions of the spinal cord in a number of fishes. The literature has been reviewed by FRIDBERG and BERN (1968). These neurons (DAHLGREN cells) were described by SPEIDEL (1919) in chondrichthyeans. ENAMI(1955) described a discrete bundle of such cells in the Japanese eel, Anguilla japonica, where they form a distinct lobe of neurosecretory cells abutting onto a bed of capillaries to form a neurohaemal junction. The degree of aggregation of such cells to form a lobate-structure, the urophysis, differs in various species, and achieves its most distinct form in certain teleosts. In the Chondrichthyes the cells are situated in a diffuse area, which in Raja batis, may extend over the area of the last 55 vertebrae. These caudal neurosecretory fibres have only been identified in the Chondrichthyes, teleostean and chondrostean bony fishes.

The neurosecretory cells in the caudal regions of fish, and those of the hypothalamo-neurohypophysial region of the vertebrate brain, are analogous in their general histological properties. However, the caudal cells, in contrast to the cephalic ones, do not produce secretions that can be stained with iron-alum haematoxylin. Both groups of neurosecretory material are thought to be protein-like. The caudal neurosecretory neurons, like typical neurosecretory cells, can transmit nerve impulses (MORITA, ISHIBASHI, and YAMASHITA, 1961).

An endocrine role for the caudal neurosecretory cells remains to be established. Their possible role in osmoregulation has been explored, probably partly as a result of their morphological analogies with the neurohypophysis. Extirpation of the urophysis in some teleosts has failed to reveal evidence for such a function, but the potential regeneration of such tissues makes such available evidence inconclusive (FRIDBERG and BERN, 1968). MAETZ, BOURGUET, and LAHLOUH (1963) found that when extracts of the urophysis were reinjected into the goldfish, Carassius auratus, there was an increased branchial uptake of sodium and an increased urine flow, with a reduced renal loss of sodium. These experiments need extending to other species of fish before any general conclusions can be drawn, but at the moment they afford the strongest available evidence for a urophysial role in piscine osmoregulation. Electrical recordings from caudal neurosecretory cells have shown that they can respond to osmotic changes in the fish. Injection of distilled water into the marine teleost, *Paralichthys dentatus,* increased the rate of discharge of these cells (BENNETT and FOX, 1962), while in the freshwater species, *Tilapia mossambica,* YAGI and BERN (1963) found that there was an increased rate of discharge when the gills were exposed to saline. These effects were further examined in *Tilapia* and it was found that some of the neuro-secretory fibres were stimulated by high sodium concentrations while others responded to sodium-free solutions (YAGI and BERN, 1965). The specificity of such stimuli to the particular neural activity is in question.

The biological activity of urophysial lobe extracts has been examined pharmacologically in order to characterize it, and in an attempt to identify it with other known biological agents. Possible homologies with the neurohypophysial peptides have been explored without success (SAWYER and BERN, 1963), though the extracts do exhibit an ability to increase the permeability of the frog urinary bladder to water and this effect has also been shown in toads (LACANILAO, 1969). Extracts of this tissue also have the ability to produce water retention in toads and elevate blood pressure of eels; the latter effect was not abolished by incubation with sodium thioglycollate which distinguishes it from neurohypophysial peptides (BERN *et al.,* 1967). The active material thus mimics neurohypophysial peptides in some of its actions. The effects on anuran membranes are particularly interesting, as apart from neurohypophysial-type peptides, only the nucleotide cyclic AMP, in high concentrations, is known to change their osmotic permeability. The chemical nature of the materials in the urophysis will thus be of considerable interest.

10. Mechanisms of Hormone Action
(with special reference to osmoregulation)

The responses of 'target' or 'effector' organs to hormones can be classified broadly into four groups:

1. *Trophic* (or tropic) actions on other endocrine glands, such as the effect of gonadotrophins on the gonads and corticotrophin on the adrenal cortex.

2. *Metabolic effects,* such as the formation of proteins and glucose from amino acids, interconversions between fatty acids, glucose and glycogen, and the increased catabolism of substrates which provide energy. This category can also include the interconversions and transfer of mineral reserves such as calcium and phosphate from 'fixed stores', as in the bones, to a more mobile form in the plasma. Hormones that act in this way include insulin, growth hormone, the various gonadal and adrenocortical steroids, as well as the thyroid and parathyroid secretions.

3. *Mechanical responses,* like the contraction and relaxation of the smooth (nonstriated, involuntary) muscle that surrounds the blood vessels, oviducts, uterus and the gut. Adrenaline, oxytocin and vasopressin may initiate such effects.

4. *Permeability changes* in membranes, which result in changes in movements of water and ions across the tubules of secretory glands, like the kidneys, the wall

of the gut, amphibian urinary bladder and skin, and the gills of fish. The neurohypophysial peptides and corticosteroids may initiate such effects.

The precise manner (in terms of cells and molecules) in which any hormone may exert such actions is unknown, but over the last decade enough information has become available to indicate that the process is very complex. The information that is available in several instances has provided a framework of knowledge about these final events in the actions of hormones and has given us plenty of ideas about what to look for at the effector sites. This is particularly applicable to the mechanisms of action of the neurohypophysial and corticosteroid hormones on osmoregulatory processes. The action of a hormone at its effector site involves a series of chemical and metabolic events that culminate in one of the types of responses just listed. The preceding events can be broadly classified as:

1. The approach of the hormone to the vicinity of the effector tissue. This may be influenced by such factors as its concentration, size, shape (steric configuration), electrical charge and polar or non-polar nature.

2. The molecular interaction of the hormone with its 'receptor'. The 'receptor' is a pharmacological concept, which is envisaged as a chemical substance, or group, that complements chemically, sterically and electrically, the structure of the hormone, so that the two can interact in a specific way. Other substances can also sometimes interact with such receptors, but may initiate little or no effect, though they may exclude the hormone from this site and so prevent its action. The receptor may be an enzyme, or part of an enzyme-system, or occupy some strategic structural site in the cell. Little is known about their chemical nature, though in some instances, such as the receptors for neurohypophysial hormones, oestrogens and insulin, it is suspected that a sulphydryl group may be involved. The isolation of specific receptors *in vitro*, is the ultimate aspiration of many pharmacologists and endocrinologists. Receptors are thought to be in the cell membrane, and in specific sites within the cell such as in the nucleus. In the instance of the oestrogens and corticosteroids, the evidence for a nuclear site is strong. Receptor sites at the cell surface have been favoured for other hormones, but the evidence is usually weak and largely intuitive; they may well be located there nevertheless.

3. Secondary reactions follow as a direct result of the hormone-receptor interaction. Such reactions presumably serve the purpose of forming metabolites to be used for 'triggering' the molecular machinery of the ultimate effector, as well as amplifying the small energetic changes associated with the initial event. Our knowledge of such secondary reactions is very incomplete, though it does warn us of considerable complexity. An intermediate associated with such processes is the nucleotide cyclic adenosine-3',5'-monophosphate (cyclic AMP). This is formed in effector tissues as a result of the actions of a number of hormones like corticotrophin, vasopressin, adrenaline and glucagon. Cyclic AMP arises from ATP by the action of adenyl cyclase, an enzyme that is activated in some way by the hormone. Another intermediate is ribonucleic acid (messenger-RNA), which is synthesized at increased rates in the presence of oestrogens and corticosteroids, and mediates cytoplasmic genetic translation for the synthesis of specific proteins.

4. The effector response is finally 'triggered', but how this is done by the mediation of cyclic AMP or the actions of newly synthesized proteins is not always clear. Cyclic AMP can activate phosphorylase, which in metabolic responses like

glycogenolysis, plays an established biochemical role. In other instances, when the result is a change in membrane permeability, the nature of the final steps is unknown, though they are the subject of much speculation.

a) Mechanism of Action of Neurohypophysial Peptides and Aldosterone

The mechanism of action of two groups of hormones, the peptides from the neurohypophysis and steroids from the adrenal cortex, is particularly relevant to our theme of osmoregulation. These hormones can alter the transfer of water, sodium and potassium across epithelial membranes.

Our knowledge of the mechanism of action of the neurohypophysial and corticosteroid hormones is derived principally from *in vitro* studies of the urinary bladder of anuran amphibians. This membrane survives and functions for many hours in physiological-saline solutions (such as RINGER solution).

EWER (1952 a) showed that an African toad can reabsorb water from its urinary bladder at an accelerated rate when dehydrated, or injected with neurohypophysial peptides. Apart from being physiologically useful to many of the Amphibia, this organ furnishes an admirable, and unique, preparation for studying hormonal effects on membrane permeability. The anuran urinary bladder *in vitro* can actively transport sodium from mucosal to serosal side, a function that is accelerated by neurohypophysial hormones (LEAF, ANDERSON, and PAGE, 1959) and aldosterone (CRABBE, 1961 a, b). It is also osmotically permeable to water and this is increased by neurohypophysial peptides (BENTLEY, 1958).

Structurally the anuran urinary bladder is made up of a layer of epithelial cells, supported by a thin serosal membrane, containing connective tissue, blood vessels and interdigitating smooth muscle fibres. The epithelial cells line the inside of the bladder, their outer borders (mucosal side) being bathed *in vivo* by the urine. The epithelial cells are the portion of the tissue that is responsible for the special properties of the membrane with respect to movements of water and sodium, and are the effector sites for the hormones. Three different types of cells make up this layer. The predominant cells present (about 85%) are termed '*granulated epithelial cells*' and are characterized by apical concentrations of small dense granules and a paucity of mitochondria. About 10% of the epithelial cells contain numerous mitochondria and are called '*mitochondria-rich cells*'. In addition to these two cell-types, there are also a few mucous cells present. It is uncertain whether the 'mitochondria rich' and 'granulated cells' have special functions, but they exhibit different osmotic behaviour. 'Mitochondria-rich cells' swell more than 'granulated cells' when exposed to toxic concentrations of the polyene antibiotic amphotericin B, an effect related to their sodium permeability (SALADINO, BENTLEY, and TRUMP, 1969). On the other hand, 'granulated cells', but not 'mitochondria-rich cells', swell when they are exposed to vasopressin, an effect resulting from an increase in their osmotic permeability (DiBONA, CIVAN, and LEAF, 1969). It is likely that the 'mitochondria-rich cells' and 'granulated cells' play a special role in the response of the tissue to hormones, but at this time the evidence is incomplete.

Molecules may conceivably move across the anuran urinary bladder by intercellular and intracellular routes. The magnitude of intercellular transfer is uncertain, but for reasons of metabolic involvement associated with active sodium transport and hormonal stimulation, a substantial movement would appear to take place in this way. This is thought to involve at least two major steps, entry into the epithelial cell across its mucosal (outer) border, and exit through the opposite or lateral borders. LEAF in 1959 showed that the lactate that accumulates in the bladder under anaerobic conditions, passes more readily into the serosal than into the mucosal bathing fluid. This suggests differences in the permeability of the two surfaces of the epithelial cells, and also that the mucosal barrier may have a limiting action on molecular transfers. Net water movement only occurs down its concentration gradient, so there is no need to postulate a special energetic mechanism for its transfer. Sodium, however, moves against a concentration gradient, so that the presence of at least one sodium 'pump' is postulated, and this is usually considered to be present at the serosal boundary of the cell. Sodium entry from the external solution (urine *in vivo*) is widely thought to take place by diffusion; though as this process exhibits saturation kinetics with respect to the external sodium levels, some interaction of sodium with the membrane presumably occurs. Sodium can be transported from external solutions as dilute as 10^{-4}M, but chemical and electrical measurements of the intracellular media do not indicate that a sufficient gradient for sodium diffusion under such conditions is present. There may, however, be a compartmentalization of cell sodium, as suggested in the frog skin by CEREIJIDO and ROTUNNO (1967) and this may result in favourable local gradients for diffusion of sodium into cells. Alternatively there could be a sodium 'pump' at the mucosal barrier.

Such is the current view of the stage on which water, sodium, and the neurohypophysial and adrenocortical hormones play. The anuran urinary bladder and the tetrapod kidney share many physiological properties, including responses to the same hormones, so that information about the amphibian membrane throws light on the corresponding renal mechanisms.

α) Neurohypophysial Hormones. Eight of these hormones have been described in different vertebrates, 8-arginine vasotocin (vasotocin, AVT) has the widest distribution while the vasopressins (ADH) are the most celebrated, probably because of their presence in man and other mammals. These hormones are octapeptides (see Page 55) containing a disulphide bridge, the breaking of which abolishes the specific biological activity. They can increase the permeability of the tetrapod kidney tubule, and amphibian urinary bladder and skin, to water. In some instances they also increase the rates of sodium and urea movement across the amphibian membranes, while effects on the permeability of the gills of fish have also been described. These peptides have an additional ability to constrict, or dilate, peripheral blood vessels, and to contract the smooth muscle of the uterus and gut.

The biological potency of the various natural hormones and their synthetic analogues varies considerably. Thus, when tested for its ability to increase the osmotic permeability of the toad bladder, vasotocin is 200 times as active as the other amphibian neurohypophysial peptides, oxytocin and mesotocin. The antidiuretic effect of vasopressin in rats is 100 times greater than oxytocin. Vasopressin also acts on the toad bladder, but is 100 times less active than vasotocin. These variations in potency are thus associated with contrasting species, and with differences of one or two amino acids in the molecular structure. There would appear to be differences in the structure of the receptors, as well as the hormones, in different species, the magnitude of the response being related to how favourable an interaction between the two is possible. Apart from their concentration this depends on the steric configurations and electrical charges of the hormones and receptors. Once the molecules are suitably aligned and in apposition it is possible that some chemical interaction occurs, which could involve the disulphide bridge of the hormone and sulphydryl groups on the receptor. This suggestion has not gained consensus however; since the hormone only acts when present at the serosal side, the receptors may be present at this surface.

Subsequent to the primary interaction of the hormone and receptor, the enzyme adenyl cyclase is activated and cyclic AMP is formed from ATP. This observation was originally made by ORLOFF and HANDLER (1962, 1967), using the isolated toad urinary bladder, and constitutes a major advance in our ideas about the hormone's action. Cyclic AMP subsequently initiates an increased permeability of the membrane to water and sodium, but the details of how this is brought about are unknown. The nucleotide activates phosphorylase and is destroyed by another enzyme, phosphodiesterase. The process of the hormones action can be blocked at various points; neurohypophysial peptides with little biological activity can reduce the activity of vasotocin by excluding it from its receptor site (MOREL and JARD, 1963), Ca^{2+} and Mn^{2+} block the hormone's effect but not that of cyclic AMP, while Zn^{2+} also blocks the nucleotide (BENTLEY, 1967). Xanthines, like theophylline and caffeine, inhibit phosphodiesterase and so imitate the action of the hormone by preventing the breakdown of endogenous cyclic AMP (ORLOFF and HANDLER, 1962). Though the dual increased permeability to water and to sodium can be demonstrated in the toad bladder (and skin), these effects are physiologically separate, and it seems likely that the nucleotides that mediate each effect are separated. This could be an anatomical sequestration, such as that involving 'granulated' and 'mitochondria-rich' cells.

Water moves more readily down an osmotic gradient across the bladder in the presence of neurohypophysial hormones. This normally takes place from the mucosal to serosal side, but if the gradient is reversed *in vitro*, acceleration can also be shown to occur from the serosal to mucosal surface. Thus it is clear that no special energy input mechanism, to account directly for the water movement, need be postulated. The effect is indeed relatively independent of metabolism, and still can be shown to occur after the tissue has been exposed to cyanide or dinitrophenol. The site of the effector change in the epithelial cells is generally thought to be the cell membrane on the mucosal side of the bladder, though the evidence for this is equivocal. Changes in permeability to water can be measured in two ways; in the presence of an osmotic gradient (osmotic permeability), or in its absence, with the aid of the labelled water molecules, deuterium or tritium (diffusion permeability). Neurohypophysial peptides have a greater effect on the osmotic than diffusion permeability and this has been interpreted to indicate that the hormones increase 'bulk' flow of water through pores. This suggestion would be consistent with an increase in the diameter of such pores (rather than the total pore surface area, which is directly related to movement by diffusion). The evidence is circumstantial, but affords an explanation of the observations. Such changes in pore size

84

may be envisaged as taking place by an alteration in the configuration of structural proteins and/or lipids in the cell membrane but there is no direct evidence about this.

Alternatively such differences in osmotic and diffusion permeability could arise as a result of the presence of 'long pores' (HARRIS, 1960). The molecules must 'line up', in file, so that the labelled ones present cross more slowly and so give a low estimate of the rate of diffusion. HAYS and FRANKI (1970) have shown that when the fluid bathing each side of the toad bladder is stirred rapidly the estimated diffusion permeability in the presence of vasopressin actually increases as much as 10-fold. They have suggested that the anomalous results obtained previously were the result of a masking effect of unstirred layers of water and that these could be synonymous with 'long pores'.

Sodium movement in response to neurohypophysial hormones is conceptually even more difficult than that of water, as active transport is involved. This latter process need not be directly related to the action of the hormones, as merely supplying the 'pump' with more of its ionic substrate could be sufficient to increase its activity. The effector mechanism for mediating the action of neurohypophysial hormones on sodium transport across the bladder is physiologically distinct from that concerned with water movement. This is apparent at all levels of the response. At the primary receptor, various neurohypophysial peptides have relatively different potencies in eliciting each response (BOURGUET and MAETZ, 1961). In the secondary reactions, the osmotic response is blocked by Ca^{2+} and Mn^{2+}, but the increase in sodium transfer is unaffected (BENTLEY, 1960a and 1967; PETERSEN and EDELMAN, 1964). The final effector mechanisms can exist independently, as shown in the skins of the anuran, *Xenopus laevis*, and the urodele, *Triturus alpestris*, which respond by increasing sodium, but not by water transfer, (MAETZ, 1963; BENTLEY and HELLER, 1964). This distinction, between water and sodium effectors, is important in assessing the nature of the change in permeability, and also with respect to the evolution of the physiological actions of these peptides.

Early experimental evidence suggested that the neurohypophysial hormones increase the rate of admission of sodium into the cell, by increasing the permeability of the membrane on the mucosal side. This evidence was based on the measurement of changes in the concentration of radioactively labelled sodium in the bladder ('sodium pool'). An increase in the level of labelled sodium was assumed to reflect an increased rate of admission across the mucosal border into the cell. However, there is now considerable uncertainty as to the distribution of the sodium throughout the bladder tissue, making such evidence equivocal (SHARP *et al.*, 1966). CIVAN and FRAZIER (1968) have impaled toad bladder epithelial cells on microelectrodes, in order to measure the D. C. resistance across the mucosal cell border. They found that after exposure to vasopressin, 98% of the decrease in D.C. resistance that is observed across the whole membrane can be accounted for by the change at the mucosal barrier. This is consistent with the earlier ideas about the site of the hormone's action.

β) *Aldosterone.* Aldosterone, as well as corticosterone and cortisol, increases sodium transport across the toad bladder *in vitro*. Other steroid hormones, like oestradiol and progesterone, lack this activity and may competitively inhibit the corticosteroid's action. Synthetic analogues of corticosteroid hormones, such as fluorocortisol and prednisolone, also increase sodium transport across the bladder. In contrast to the neurohypophysial hormones, whose effects can be seen in less than a minute, the increase in sodium transfer that follows exposure of the serosal side of the bladder to aldosterone has a latent period of about 60 to 90 min. This could be due to a delayed rate of access to receptor sites, or to a prolonged series of 'secondary reactions.' As the delay is independent of the hormone's concentration, CRABBE (1963) suggested the latter explanation.

The receptor sites for aldosterone in the toad bladder appear to be situated intracellularly, probably in the nucleus. EDELMAN, BOGOROCH, and PORTER (1963) demonstrated by radioautography, that ^{3}H-aldosterone was accumulated preferentially to progesterone in the nucleus. The nuclei of these cells have since been separated by centrifugation procedures, and have been shown directly to bind aldosterone in preference to other steroids. The relative abilities of different steroids to displace bound aldosterone were found to correspond roughly with their activities in increasing sodium transport (AUSIELLO and SHARP, 1968). These results are interpreted with some caution, as it is difficult to be sure that the binding sites are those specifically related to the physiological response. FANESTIL and EDELMAN (1966) found comparable binding sites for aldosterone in the nuclei of rat kidneys, and concluded that such sites are probably proteinaceous, as only proteolytic enzymes could increase the rate of release of aldosterone from such loci.

In an effort to unravel the nature of the secondary series of reactions involved in the effect of aldosterone, EDELMAN et al. (1963) and CRABBE and DE WEER (1964) tested the effects of actinomycin D and puromycin on the bladder. Actinomycin D prevents transcription of nuclear DNA to messenger RNA, while puromycin inhibits the subsequent ribosomal protein synthesis. Both these agents inhibited the effects of aldosterone on sodium transport. Aldosterone was also shown to increase the rate of incorporation of ^{3}H-uridine into RNA (PORTER, BOGOROCH, and EDELMAN, 1964) in the toad bladder. The increase in RNA formation is independent of changes in intracellular sodium levels, so that the action of the hormone is probably direct and not mediated, at this stage, by ionic changes (DE WEER and CRABBE, 1968). The results all strongly suggest that aldosterone changes the activity of certain genes according to the theory of JACOB and MONOD (see MONOD, 1966). Thus, aldosterone may interact in the nucleus with a repressor, thereby facilitating the expression of an operator gene to allow the operon to transcribe messenger RNA that initiates protein synthesis.

The final effect of aldosterone, increased sodium transport, appears to be mediated by the action of a protein whose formation is induced by the hormone. Such a protein could conceivably act in several ways: 1. It could increase the action of the 'sodium pump'. This could be done either directly, as by increasing the activity of Na-K activated ATPase or indirectly by furnishing it with additional energy substrate (ATP). 2. Outflux (or back diffusion) of sodium from the serosal to mucosal surface could be reduced with a resulting increase in the net flux. 3. If more sodium were admitted to the cell across its mucosal border, then the 'sodium pump' may be expected to respond by increasing the rate at which it extrudes sodium from the cell.

Numerous measurements of sodium fluxes in the presence of aldosterone indicate that the increased net flux results from facilitated influx from mucosa to serosa, and not a decrease of movement in the opposite direction. The possibility that aldosterone increases the rate of admission of sodium across the mucosal surface into the cell has received serious consideration based originally on measurements of the isotopic sodium 'pool' in the tissue. As mentioned previously, uncertainties about the precise location of this sodium make such results difficult to interpret. However, indirect evidence involving possible analogies between the action of aldosterone and that of amphotericin B (which destroys the selectivity of the mucosal border and admits sodium into the cell) suggests that this mode of action is possible (SHARP and LEAF, 1968). The protein induced by aldosterone could be a permease. EDELMAN and his collaborators do not consider this explanation likely (FIMOGNARI, PORTER, and EDELMAN, 1967; FANESTIL, PORTER, and EDELMAN, 1967) as they found that aldosterone increases the electromotive force of the sodium 'pump' and this is related to the levels of certain substrates in the tricarboxylic acid cycle. Furthermore CUTHBERT and PAINTER (1969) have found that when they abolish sodium transport across the toad bladder by exposing it to ouabain, but maintain the p.d. by passing a current through the membrane, the electrical properties of the membrane are unaltered by aldosterone. Vasopressin on the other hand was still effective under these conditions. CUTHBERT and PAINTER consider that if a permease, which acts on the mucosal membrane of the epithelial cells, were involved, its action, like that of vasopressin, should still be apparent. The proteins induced by aldosterone may then be enzymes that effect energy substrate levels, and eventually facilitate the formation of ATP. The activity of certain enzymes, involved in the tricarboxylic acid cycle, increase in the presence of aldosterone, and this effect is independent of the presence of sodium (KIRSTEN et al., 1968). KIRSTEN et al. do not consider that their results rule out a parallel action of the hormone on sodium entry into the cell, but suggest that the two effects could occur together and be mutually supportive.

Chapter 3

The Mammals

1. General Introduction

About 1050 contemporary genera of mammals are named by WALKER (1964) in his monograph 'Mammals of the World'. Three of these belong to the order Monotremata (egg-laying mammals), ninety to the order Marsupialia (typically undeveloped young kept in pouches), and the remainder are placentals. Monotremes are only known from the Australian region and are represented by the echidna and platypus. This order probably represents a separate line of mammalian evolution, beginning in the Triassic period. The marsupials and placentals probably diverged from a common stock early in the Cretaceous period, about 85 million years ago. The Australian (75 genera) and American (15 genera) marsupials subsequently diverged late in the Cretaceous. The placental mammals today contain about 950 genera, of which 360 are rodents (Rodentia) and 180 are bats (Chiroptera). The other generically numerous groups are the Insectivora (hedgehogs, shrews), 66 genera; Primates (man and monkeys) 62; Cetacea (whales) 35; Carnivora 103; Pinnipedia (seals) 20 and Artiodactlya (pigs, cattle, sheep) 82. The mammals have a wide geographic and osmotic distribution; they are predominantly terrestrial and are even found in the driest desert regions, though some are aquatic and live in rivers and lakes or in the sea. Only one genus of the monotremes (*Ornithorhynchus*, the platypus), and one marsupial (the South American water opossum, *Chironectes*) live in fresh water and these groups have no marine representatives. Most mammalian orders have representatives that live in desert regions.

The ability to osmoregulate in such contrasting environments depends on physiological and behavioural adaptations, many of which are uniquely suited to the particular needs of mammals. In the absence of such specializations most temperate terrestrial species find it difficult to survive in areas where water is more restricted, though there are some exceptions such as the dog and European rabbit which thrive in the desert conditions of Australia. Dispersal of mammals is, nevertheless, relatively restricted by deserts, and especially limited by salt-water barriers (DARLINGTON, 1957). Bats, due largely to their aerial capabilities, have the widest distribution among the mammals reaching isolated oceanic islands, such as the Azores, New Zealand and Hawaii, that are otherwise inaccessable to all but marine species. The rodents also have often managed to cross salt-water barriers (probably on rafts), but if the aid of man can be excluded, this occurs only across the short distances to islands lying adjacent to the continents. Such journeys, whether as aviators or mariners, nevertheless constitute an osmoregulatory feat of some substantial magnitude considering the rather unfavourable conditions experienced by terrestrial mammals in oceanic regions.

2. Osmoregulation

Two evolutionary events have had profound effects on the osmoregulatory problems of tetrapods. The first was the adoption of a terrestrial, instead of an aquatic, habitat. Evaporation of water then made its initial appearance as a vertebrate problem. In addition, there arose the novel necessity to replenish the body water periodically from the sporadic sources available in a terrestrial environment. The second major event to influence tetrapod water metabolism, was the adoption of homoiothermy by the birds and mammals (or possibly an ancestor of these). Potential evaporation from the body surface is increased at the relatively high, and sustained body temperatures of such animals, while the added metabolic need for respiratory gas exchanges also increases evaporative water loss from the respiratory tract. Heat loss accompanying evaporation of water is utilized for cooling in hot environmental conditions, when additional evaporation is facilitated by the secretion of sweat or by panting. Viviparity also increases the needs for water and salts, especially in placental mammals which retain the young *in utero* until they are relatively well developed. The secretion of milk to nourish the young places yet another periodic osmoregulatory burden on the mother, but has obvious advantages for the offspring. The mammals thus have the usual tetrapod problems with respect to their water and salt metabolism, and a few additional ones as part of the price that they must pay for their unique physiology. However, the high metabolic rate of mammals allows them to travel considerable distances for food and water, while their viviparity and ability to suckle their young, foster reproduction in situations in which it could otherwise be difficult, both nutritionally and osmotically. It should be remembered that while the high rate of metabolism in mammals increases their requirements for osmoregulation, such adjustments may be facilitated by the increased energy produced by the cells of such animals.

The relative abundance or scarcity of salts may limit the distribution of mammals. An excess of water or salts in the food can be tolerated by some species, but not others. Marine mammals like whales and seals live in a hyperosmotic salt solution, equivalent in concentration to 3.3% sodium chloride. Many marine mammals eat invertebrates that are isoosmotic to sea-water. Such mammals do not appear to drink sea-water normally, but when feeding on invertebrates take in large quantities of salts, far more than most other mammals could tolerate with the limited quantities of osmotically free water that are available. Most terrestrial mammals cannot survive if provided with drinking water that contains 2 to 3% sodium chloride, but there are exceptions. The North African sand rat, *Psammomys obesus*, normally eats halophytic plants with a salt concentration higher than that of sea-water (SCHMIDT-NIELSEN, 1964a), while the western harvest mouse, *Reithrodontomys r. halicoetes*, of the United States eats similar plants and can drink sea-water (HAINES, 1964). Other species may drink sea-water sporadically, and this has been shown among the Australian marsupials, including the Tammar wallaby, *Macropus eugenii* (KINNEAR, PUROHIT, and MAIN, 1968). In many areas of the world there is a deficiency of sodium in the soil, a condition that appears most often in the interior of continental areas and high mountainous regions. The vegetation reflects the low sodium content of the soil, so that herbivorous animals may have restricted amounts of salt available. Thus the herbage in grazing areas of central Australia may only contain 1 to 3 m-equiv/kg dry weight of sodium, compared to 100 to 350 m-equiv/kg dry weight in coastal areas (DENTON, 1965). The popularity of 'salt licks' among mammals in such regions is well known.

a) Water Loss

α) Cutaneous. Cutaneous water loss takes place through the skin by diffusion, and as a result of secretion by the sweat glands. This water is evaporated from the body surface, resulting in a loss of heat; a process that can be used physiologically for cooling in hot environments. In man the eccrine sweat glands are usually considered to be associated principally with thermal cooling while the apocrine glands, which are associated with hair follicles, respond to other stimuli. Sweating is initiated by exercise in the horse, a response that is stimulated by elevated levels of adrenaline in the blood and one which contributes to heat and water losses (LOVATT EVANS, SMITH, and WEIL-MALHERBE, 1956). The sweat glands can secrete large volumes of fluid; a man working in the desert loses up to 15 litres of water a day in this manner. Sweat glands are typically mammalian structures though their presence, and relative importance for thermal cooling is not the same in all species. MACFARLANE (1964) found that the maximal rate of thermal sweating in merino sheep was 32 g/m$^2 \cdot$h, while the camel can form 260 g/m$^2 \cdot$h (SCHMIDT-NIELSEN *et al.*, 1957). Many mammals such as the rabbit and rat, do not possess sweat glands though water loss can still take place across the skin. Dehydration may result in a decreased rate of sweating. The camel after 24 h without water perspires at 60 to 70% of the rate seen when it is normally hydrated (SCHMIDT-NIELSEN *et al.*, 1957). Thermal cooling response of sweat glands are under the control of the sympathetic nervous system. In man the final transmitter is acetylcholine. However, in bovidae, like domestic ox and various wild species like the oryx, eland, wildebeest and buffalo which live in East Africa, these nerves are adrenergic (FINDLAY and ROBERTSHAW, 1965; ROBERTSHAW and TAYLOR, 1969). The injection of adrenaline in these species as well as in the horse, sheep and camel results in the secretion of sweat. This type of response is generally thought to arise physiologically during exercise rather than in response to heat stress. Noradrenaline is relatively ineffective. Thermal sweating is abolished in the dehydrated oryx, though sweat glands can still respond to injected adrenaline (TAYLOR, 1969). Neurohypophysial antidiuretic hormone does not alter the rate of sweating in man (AMATRUDA and WELT, 1953). It seems likely that reductions in the volume of sweat secretion during dehydration are mediated by nervous, rather than endocrine, mechanisms but changes in the level of circulating adrenal medullary hormones could have an effect.

β) Respiratory. The respiratory tract is a major route of water loss in tetrapods and more especially in the mammals and birds. This loss takes place by evaporation, as an unavoidable consequence of the uptake of oxygen and excretion of carbon dioxide, but this avenue can also be utilized physiologically for thermal cooling by 'panting'. The normal overall obligatory losses of water from the respiratory tract increase or decrease with parallel changes in metabolic rate and oxygen consumption. Normally mammals extract about 30% of the oxygen present in the inspired air, but in certain circumstances this may be altered, with reflected changes in the accompanying water loss. DICK (C. R.) TAYLOR (1969) has shown that an African antelope, the oryx, which lives in desert areas, may reduce its oxygen consumption by 30% when it is dehydrated, and, by breathing more deeply, can extract additional oxygen from the inspired air. The net result is a considerable reduction in water loss. Similarly, SCHMIDT-NIELSEN and his collaborators (1967) have shown that the camel reduces its oxygen consumption when dehydrated. Such a decrease in metabolism is not invariably seen in dehydrated mammals. Thus the East African waterbuck, *Kobus defassa ugandae*, fails to change its rate of oxygen consumption when deprived of water (TAYLOR, SPINAGE, and LYMAN, 1969). Largely as a result of the high rate of accompanying evaporative water loss these antelope withstand water deprivation poorly, even when compared with domestic Hereford cattle. It seems possible that the ability to reduce the metabolic rate under such conditions constitutes a physiological reaction that could adapt the animals more readily to the changed circumstances. The thyroid gland influences the metabolic rate in mammals, but it is not known whether it is concerned with the changes that are observed in dehydrated animals. Such a possibility should be simple to test.

γ) Urinary. Urinary water losses occur either as a homeostatic response like excreting a dietary excess of fluid, or as an unavoidable consequence of the necessity to excrete solutes. Excess water may be taken in, especially by herbivorous species eating large amounts of succulent food. Solutes that must be excreted may represent an excess which is taken in with the food and water, or be metabolic solutes, principally urea in mammals.

The magnitude of the renal water loss varies and depends on the diet, the environmental conditions and the ability of the animal to concentrate its urine. Diets with a high salt content or that are rich in protein, necessitating additional urea excretion, increase urinary water loss. The proportion of the total water loss that is represented by the urine will vary, depending mainly on the environmental conditions that influence the extrarenal loss. In normal circumstances urinary water loss is less than 50% of the total water loss, and may be lower than 10%, especially in hot environments where evaporative loss is high. Mammals and birds, in contrast to other vertebrates, can form a urine which is hyperosmotic to their body fluids – the more concentrated, the greater the saving in water. The maximal ability to concentrate the urine varies considerably in different mammals; the mountain beaver, *Aplodontia rufa,* can secrete urine with a maximum concentration of about 500 m-Osmole/l, (DICKER and EGGLETON, 1964) while the North African sand rat, *Psammomys obesus,* can form urine over 6000 m-osmole/l, (SCHMIDT-NIELSEN, 1964a) and Australian hopping mice, *Notomys alexis,* 9000 m-osmole/l. (MACMILLEN and LEE, 1969). Neurohypophysial hormone normally initiates the formation of a hyperosmotic urine.

δ) Faecal. The faeces of mammals usually contain 50 to 60% water, but this may be as high as 85% in grazing cattle. The faeces of camels with water to drink, contain about 55% water, but after a day without water this level drops to about 45% (SCHMIDT-NIELSEN, 1964a). It is unknown whether such changes involve special regulatory mechanisms mediated by hormones.

b) Salt Loss

α) The Sweat Glands. The secretion of the sweat glands contains dissolved solutes, principally sodium, potassium, chloride and bicarbonate. In mammals that utilize sweat secretion for thermal cooling the losses of these solutes, especially sodium, may be considerable. The sodium concentration of the sweat in man varies from 5 to 60 m-equiv/l, being greatest when the rate of sweat secretion is high (THAYSEN, 1960). During sodium depletion the amount of this solute in the sweat decreases more than 5-fold, while the potassium level rises. Such changes require several days. SCHMIDT-NIELSEN (1964a) suggested that in herbivorous animals, like the donkey, little sodium is lost in sweat and measurements by MACFARLANE (1964) on sheep and camels support this view. The sweat of the camel contains 9.5 m-equiv/l sodium and 40 m-equiv/l potassium. In man a decrease in the Na/K ratio in the sweat has been related to the action of the adrenocortical hormones (CONN et al., 1947), and it seems likely that these steroids may contribute to the acclimatization in the salt levels of the sweat that is seen in other species.

β) The Gut. The gut is the avenue of entrance and exit of salts from the external environment. Salts are taken into the gut with the food and water, while large volumes of fluids, rich in electrolytes, are continually secreted into the lumen of the alimentary tract in order to facilitate digestion. These fluids include saliva, various gastric and intestinal secretions, the bile and the pancreatic juices. In order to maintain adequate salts in the body all of the electrolytes must ultimately be reabsorbed. Inadequate intestinal absorption, which may occur, for instance, as a result of infection, can result in death from water and electrolyte deficiency.

The composition of the secretions derived from the different *salivary glands* varies, and depends on the species and the physiological condition of the animal. Salivary secretion in the sheep has been extensively studied by DENTON and his collaborators (BLAIR WEST et al., 1967). The total volume secreted by this species in a day is 6 to 16 litres, principally by the parotid glands. Normally in a sheep, with adequate dietary sodium, the parotid secretion contains 180 m-equiv sodium/l and 5 m-equiv potassium/l (Na/K is 36). If sheep are deficient in sodium, the levels can change to give a Na/K ratio of 10/170 m-equiv/l, or 0.06. Adrenalectomized animals exhibit little adjustment in their response to sodium depletion. The infusion of aldosterone, however, can restore this deficiency (Fig. 3.1).

The rate of secretion of aldosterone from the adrenal cortex can increase 13-fold in sheep depleted of 300 to 500 m-equiv of sodium, and the levels of this steroid are consistent with the amounts which must be infused in order to change similarly the composition of the saliva. Cortisol and corticosterone can also alter the electrolyte level in the saliva, but the necessary concentrations are large, and probably only of pharmacological significance. Antidiuretic hormone has no effect on salivary secretion.

The site(s) of action of the corticosteroids on the salivary glands is uncertain; they conceivably could act on the primary secretory mechanism in the acinar cells, or alter reabsorption from the ducts. BLAIR WEST et al. (1967) suggest that aldosterone probably acts at both sites.

The biological importance of the action of aldosterone on the electrolyte composition of saliva is more readily apparent in animals that secrete large volumes

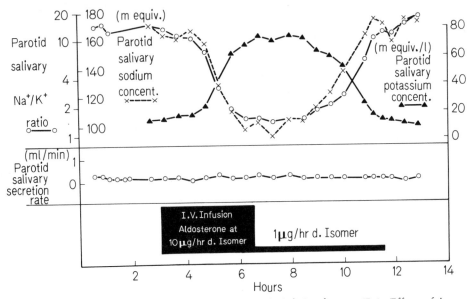

Fig. 3.1 Sheep Zachary (adrenally insufficient, Na$^+$ deficit of 250 mEq). Effect of intravenous infusion of d-aldosterone on parotid salivary Na$^+$ concentration, ×——×; K$^+$ concentration ▲——▲; and parotid salivary Na$^+$:K$^+$, O——O. Parotid secretion rate is shown also. (From BLAIR-WEST et al., 1967).

of saliva. These include many of the Artiodactyla, which digest their herbivorous diet with the aid of microbial fermentation. The volume of the fluid in the digestive tract of these animals is largely derived from the salivary glands, making up 10 to 15% of the body weight, and containing sodium equivalent to half that present in the extracellular fluid. A physiological mechanism to limit such sequestration of sodium could well be vital for survival when the amounts of this ion in the body are depleted, and its dietary availability is limited. In addition some mammals drool saliva from the mouth when the rectal temperature is elevated in hot situations. Such salivation has been observed in several marsupials and placentals, including some rodents, rabbits, cats, and cattle. If this saliva is spread over the fur it may aid in cooling. Considerable losses of electrolytes could occur in this way during heat stress, and these could be somewhat ameliorated by physiological adjustments of the solutes that are present.

Absorption of electrolytes in the diet and reabsorption of those secreted into the gut take place continually across the walls of the intestines. Reduced loss of sodium in the faeces has been observed in mammals that are depleted of sodium or that have a surfeit of aldosterone. This suggests that corticosteroids may be concerned with regulation of sodium absorption from the intestine, but in mammals direct evidence has been difficult to obtain. COFRE and CRABBE (1965), using the isolated colon of the toad, *Bufo marinus*, have demonstrated an accelerated rate of sodium transport from the mucosal to serosal side in the presence of aldosterone. It appears likely that such effects also occur in the mammalian gut, though for technical reasons they are more difficult to demonstrate.

γ) The Kidney. Large volumes of plasma are filtered across the glomerulus; this ultrafiltrate passes into the renal tubule, where most of its constituents are reabsorbed. The quantities of solutes filtered are enormous, and usually more than 99.9% of the sodium is reabsorbed back into the plasma. These processes are relatively greater in mammals than in cold-blooded species. If the diet contains an excess of sodium, or if changes in the composition of the body fluids are necessary, more sodium may be excreted. Most vertebrates can form a hypoosmotic urine which may contain only a few m-equiv/l of sodium. In animals on a normal diet such urinary losses are easily replenished, but in some circumstances even these may be physiologically significant. The formation of large volumes of urine, due to excessive drinking, or to feeding on a very succulent diet, exaggerates such losses. The role of the adrenal cortex in renal sodium conservation and potassium excretion has already been described; aldosterone promotes reabsorption of sodium and secretion of potassium across the wall of the renal tubule. The absence of a functional adrenal cortex results in an excess loss of sodium in the urine and failure to excrete adequate potassium. Changes in the sodium content of the diet are reflected in the sodium content of the urine, and the circulating levels of the adrenocortical hormones, especially aldosterone. Thus, rabbits living in the sodium-poor areas of the plateau of the Australian Snowy Mountains have urinary sodium levels as little as one-thirtieth of those rabbits living in adjacent sodium-replete grasslands (BLAIR WEST *et al.*, 1968). The adrenal glands of the sodium-deficient rabbits are enlarged and the levels of aldosterone and renin in the blood are 3 to 6-fold greater than in the sodium-replete animals. Other hormones that may influence urinary sodium losses are the thyroid and neurohypophysial hormones. The administration of antithyroid drugs increases renal sodium loss, an action related to its decreased tubular reabsorption, and a decline in the rate of production and the efficacy of aldosterone (FREGLY and TAYLOR, 1964). Injections of vasopressin and oxytocin also may increase urinary sodium and chloride losses in a number of species, including the rat, dog, camel and sheep, but not man (see HELLER, 1963; MACFARLANE, 1964). Such effects of the neurohypophysial hormones are often unpredictable, depending on such factors as, the urinary concentration and volume, as well as the sodium content of the diet. In many instances such effects can be related to the vascular actions of such hormones, which can result in alterations of the glomerular filtration rate. A normal physiological role for the thyroid and neurohypophysis in the control of urinary salt excretion is doubtful, but in abnormal circumstances large changes in the circulating levels of such hormones could facilitate loss of sodium.

δ) The Mammary and Lacrymal Glands. The secretions of these glands are an additional potential source of salt loss.

Human milk, immediately post-partum, contains 20 m-equiv sodium/l and 24 m-equiv potassium/l, but after 120 days the concentration of these salts changes to 6 and 16 m-equiv/l respectively, the Na/K ratio drops from 0.8 to 0.4, (see THAYSEN, 1960). The young of marsupials are born in a less developed condition than those of placental mammals, and it is interesting that in the wallaby, *Setonix brachyurus*, the initial sodium level of the milk is higher than in man (BENTLEY and SHIELD, 1962), being 40 m-equiv/l while potassium is 20 m-equiv/l (Na/K ratio is 2). These levels change continually, until after 180 days of lactation the Na/K ratio declines to 0.5. The milk constitutes an additional drain on the salt resources of lactating mammals, and the concentration of the constituents exhibit physiological changes. The mammary glands are embryologically homologous with the sweat glands, which, as we have seen, are subject to regulatory actions by aldosterone. The effects of corticosteroids on the composition of the milk do not appear to have been examined.

The lacrymal glands secrete tears, which contain about 150 m-equiv sodium/l and 15 m-equiv potassium/l (see THAYSEN, 1960). Excessive weeping during a period of sodium deficiency would thus be inadvisable, but in most mammals this situation is an unlikely one. Our interest in these glands is mainly due to the presence of a modified lacrymal gland (orbital 'salt gland') in turtles, which functions as a major avenue for salt excretion (SCHMIDT-NIELSEN, 1960), and which in these reptiles may be influenced by corticosteroid hormones (HOLMES and McBEAN, 1964). In sheep, however, the infusion of aldosterone does not alter the lacrymal secretion (BOTT et al., 1966).

c) Accumulation of Water and Salts

Water and salts are obtained:

1. *In the food.* The water content of plants is often equivalent to more than 95% of their weight, and even a carnivorous diet contains about 70% water. In addition to such directly available fluid, the metabolism of fats, carbohydrates and proteins results in formation of additional water; the yield from oxidation of one gram of fat is 1.1 ml, while the same amount of carbohydrate gives 0.6 ml and protein 0.4 ml. There has been much conjecture about the importance of fat reserves in assisting animals to maintain a positive water balance. The camel's hump was probably the earliest recipient of such speculation, but this has been discounted by KNUT SCHMIDT-NIELSEN. It must be remembered that the oxidation of an energy substrate requires an exchange of gases in the respiratory tract, and this may result in evaporative water loss. If the animals are breathing dry air, oxidation of such substrates will result in a net loss of water, due to saturation of the expired air with water vapour. Inspiration of more humid air will decrease this loss, so that in marine mammals, like the whale, which breathe air already saturated with water vapour, a net gain of water from metabolism can be expected.

As already described the sodium content of plants may vary considerably and in some inland continental and alpine regions the sodium levels may be low. This is the prime determinant of the salt status of all herbivorous animals living in the area, while carnivorous species should obtain adequate sodium from the prey on which they feed. The diet of other mammals may consist predominantly of halophytic plants with a high concentration of salts such as eaten by African sand rats (SCHMIDT-NIELSEN, 1964a) and possibly Australian hopping mice (MacMILLEN and LEE, 1969).

2. *In the drinking water.* Drinking supplies the water requirement beyond that obtained from the food. O'CONNOR and POTTS (1969) have emphasized the predominant importance of this process in regulation of a positive water balance. The urin-

ary volume and concentration of dogs on a standard diet, in contrast to what they drink, does not vary over a wide range of environmental circumstances. This suggests that it is drinking, rather than kidney function, that is responsible for regulating the positive water balance. The continual and persistent action of ADH on the kidney is, however, necessary in order to maintain renal water conservation. Fresh water, containing at the most only a few m-equiv of salts, is usually the beverage of choice. However, as already described, brackish water or even sea-water may be imbibed and tolerated. Prolonged drinking of such solutions can affect the endocrine system. Rabbits living in areas which are deficient in sodium chloride exhibit an hypertrophy of the adrenal cortex (BLAIR WEST et al., 1968). Rats given sodium chloride solutions to drink show a decreased storage of neurohypophysial hormones (JONES and PICKERING, 1969) and a decreased rate of secretion from the thyroid gland (AXELRAD, LEBLOND, and ISLER, 1955).

3. The Monotremes (Egg-Laying Mammals)

The monotremes are confined to Australia and New Guinea where they occupy a variety of contrasting habitats. The platypus is aquatic, living in lakes and rivers, while echidnas live in areas ranging from deserts to tropical rain forests.

Most of our knowledge about the physiology of monotremes is based on observation of echidnas, especially *Tachyglossus aculeatus*. This mammal weighs about 3 kg and in its native habitat is insectivorous, eating ants and termites. The echidna, in common with the platypus, has a body temperature of only 30 °C (MARTIN, 1902) and a metabolic rate about half that expected in a placental mammal of similar size (SCHMIDT-NIELSEN, DAWSON, and CRAWFORD, 1966). The extraction of oxygen from the inspired air is similar to that in other mammals (BENTLEY, HERREID, and SCHMIDT-NIELSEN, 1967 b) so that its respiratory losses of water are expected to be less. Although the echidna can maintain a constant body temperature at low environmental temperatures, its thermal evaporative cooling at high temperatures is poor, since it neither sweats nor pants under such conditions (SCHMIDT-NIELSEN, DAWSON, and CRAWFORD, 1966). The overall water balance of echidnas with a natural diet of termites has been calculated (BENTLEY and SCHMIDT-NIELSEN, 1967); it appears that, in their natural habitat, they probably require little or no water to drink. The ability of the kidney to form a hypertonic urine is critical to the maintenance of a positive water balance under such conditions. Dehydration results in a decreased volume of urine, that may have a concentration of 2300 m-osmole/l. This is achieved by tubular reabsorption of water and a decreased rate of glomerular filtration. The manner of controlling changes in urine volume and concentration is unknown but the pituitary contains the placental antidiuretic hormone, arginine-vasopressin, as well as oxytocin (SAWYER, MUNSICK, and VAN DYKE, 1960). The estimated (in a 3 kg echidna) concentration of vasopressin in the pituitary is about 10^{-9} M/kg body weight, which is less than that usually found in mammals (see FOLLETT, 1963; Table 3.1). The molecular ratio of vasopressin/oxytocin is 3, a value, as we shall see, that is similar to that of marsupials, but greater than in most placentals. The injection of oxytocin into a lactating echidna was shown by GRIFFITHS (1965) to produce a striking increase in

97

the secretion of milk. Oxytocin may thus function as a hormone controlling 'milk let down' as in other mammals, but the action of vasopressin on urine flow has not been tested. Large doses of oxytocin and vasopressin respectively decrease and increase the blood pressure of the platypus (FEAKES *et al.*, 1952), but this is generally considered to be a pharmacological response which is, incidentally, more similar to that of reptiles than other mammals.

The adrenal glands in monotremes do not show the distinct delineation into a cortex of interrenal tissue and a medulla of chromaffin cells. Indeed the echidna adrenal is somewhat akin in its morphological arrangement to reptiles, while the platypus has the potentialities of form seen in marsupials and placentals (WRIGHT, CHESTER JONES, and PHILLIPS, 1957).

WEISS and MCDONALD (1965) have identified corticosterone in the adrenal venous and peripheral blood of the echidna and this is present at about 5 times the concentration of cortisol. This type of ratio, in the rates of secretion of these two corticosteroids, is more similar to reptiles than to that which is usual in mammals. There are, nevertheless, exceptions; a placental mammal like the rat secretes corticosterone and no cortisol. The rates of secretion of corticosteroids in the echidna are much lower than in other mammals.

After adrenalectomy echidnas can survive for extended periods of time with no persistent changes in the plasma or urinary electrolyte levels (MCDONALD and AUGEE, 1968). However, when the environmental temperature is reduced (to 5°) the onset of torpor (hibernation?) is more rapid in the absence of the adrenals (AUGEE and MCDONALD, 1969). This torpor, which is a characteristic response in the echidna, is accompanied by reduced oxygen consumption and lowered plasma levels of glucose and fatty acids. The injection of cortisol (or glucose) into the adrenalectomized animals prevents it. Thus in the echidna, adrenocorticosteroids appear to exhibit a 'glucocorticoid' role but their relationship to electrolyte metabolism is not yet clear.

4. The Marsupials (Pouched Mammals)

Two families of marsupials live in the Western Hemisphere: the Didelphidae (opossums) and the Caenolestidae (shrew-like animals living in South America). Most marsupials, however, live in the Australian region, where the contemporary families occupy the variety of ecological niches on the continent, in a fashion that parallels the manner in which placentals occupy similar places elsewhere. Various species have adopted insectivorous, herbivorous and carnivorous diets. They may be arboreal like the possums, fossorial like marsupial moles or they may live in a more conventional way on the surface. Large areas of Australia are desert with little and irregular rainfall, combined with a high environmental temperature and a low humidity. Other regions on the continent are covered with tropical rain forest, while there are also many temperate areas with more equitable climates. The marsupial fauna has occupied all these regions with success, often to the chagrin of the sheep farmers whose flocks may fare poorly compared to the indigenous species. In many of the more barren regions sheep and cattle, in contrast to marsupials, require supplements of trace elements, protein and water in order to survive and breed.

Physiological knowledge about most of the 75 genera of Australian marsupials is lacking, but a modest amount of information is available for some species, especially members of the family Macropodidae. The macropods consist of 17 genera of kangaroos and wallabies, nearly all of which are herbivorous (the diet of musky rat kangaroos includes worms and insects) and have a grazing manner of life. They range in size from 500 g in the musky rat kangaroo to 70 kg in the larger members of the genus *Macropus*.

MARTIN in 1902 reported that the metabolic rate of several marsupials was only about one-third as great as would be expected in placentals of comparable size. That marsupials do indeed have a generally lower basal rate of oxygen consumption than placentals has recently been shown in a variety of species (DAWSON and HULBERT, 1969; ARNOLD and SHIELD, 1970). The basal rate of oxygen consumption is about one-third less than predicted in a placental of the same size. This is accompanied by a lower body temperature. It thus seems likely that evaporative water losses from marsupials may be less than in placentals.

Some marsupials have labile body temperatures that change in a characteristic cycle over the course of each day. This is well shown in the chuditch or western native cat, *Dasyurus geoffroii*. The chuditch is a carnivorous marsupial weighing 1 to 2 kg which has a wide geographic range throughout Australia where it even inhabits the hot desert regions. It is nocturnal in its behaviour coming out to hunt for food in the evenings. The body temperature of these marsupials changes in a parallel manner, declining to a minimum in late afternoon and rising as much as 4° C in the evening (ARNOLD and SHIELD, 1970). During the period of inactivity the rate of oxygen consumption was found to be only two-thirds as great as predicted in a placental of the same size. Such changes in the metabolic rate must result in a conservation of calories and probably also a reduction in the rate of evaporative water loss. These physiological adjustments could well assist the animals' survival, especially in its desert habitats (ARNOLD and SHIELD, 1970).

Information relevant to marsupial osmoregulation is available from three population groups, one living in a temperate and two in desert areas.

Professor HARRY WARING initiated such studies on a small (2 to 5 kg) wallaby, the quokka, *Setonix brachyurus*, that lives in a temperate, but seasonally dry, habitat on Rottnest island, 18 km from the coast of south west Australia (see Fig. 3.2). Today *Setonix* is principally confined to this island, though a few isolated pockets of the population remain on the mainland. The annual rainfall on Rottnest is 75 cm of which 68 cm falls between March and October, so that in the warm summer months, when the temperature may rise to 38°, little rain falls. The vegetation dries out during this period but isolated soaks of fresh water persist and are frequented by the quokkas. These animals have been studied in their natural environment and also in the laboratory where they are quite amenable to physiological investigation.

MARTIN in 1902 reported that marsupials do not regulate their body temperature adequately in hot environments, an observation that is relevant to their water loss in such situations. However, it was found that *Setonix brachyurus* regulated its body temperature as efficiently as placental mammals at temperatures up to 40° (BENTLEY, 1955; BARTHOLOMEW, 1956). This regulation is accompanied by sweating, salivation and panting, and results in a substantial loss of water. When *Setonix* is dehydrated, such evaporation of water is reduced by 40% (BENTLEY, 1960) an

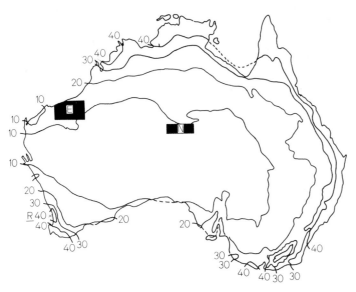

Fig. 3.2 Map of Australia showing the location and average annual rainfall (in inches) in the areas where the quokkas (*Setonix brachyurus*), euros (*Macropus robustus*) and red kangaroo (*Macropus rufus*) have been studied under field conditions. *E* 'Ealey' study area (euros), *N* 'Newsome' area (Red kangaroos), *R* Rottnest Island (quokkas).

adaptation reminiscent of that in placentals like the rat, camel and oryx. It seems likely that this decrease in water loss is mediated by a reduction in the metabolic rate, but neither this, nor its possible relationship to the secretions of the thyroid gland have been investigated. *Setonix* is not exceptional among marsupials in its ability to regulate its body temperature; members of this group are as efficient in this respect as placental mammals.

During the dry summer period, the concentrations of urine of the quokkas increase to levels 2 to 3 times as great as those observed in the wet winter period. Dehydrated quokkas can form a hyperosmotic urine with a concentration of 2 000 m-osmole/l (BENTLEY, 1955), which is comparable to the ability of placental mammals. Initiation of the formation of such a concentrated urine seems to reside with an antidiuretic hormone secreted by the neurohypophysis. The neurohypophysis of *Setonix*, like that in most placentals, has been shown, from pharmacological and chromatographic evidence, to contain arginine-vasopressin and oxytocin (FERGUSON and HELLER, 1965). The concentration of vasopressin is similar to that in placentals, about 3×10^{-9}M kg body weight. The ratio of vasopressin to oxytocin is high, 3.8, a character shared with other marsupials and the echidna. Exposure of *Setonix* to heat, with accompanying dehydration, results in the appearance of an antidiuretic substance, presumably vasopressin, in the plasma (ROBINSON and MacFARLANE, 1957). The injection of vasopressin into hydrated quokkas reduces the urine flow (PATRICIA WOOLLEY, quoted by WARING, MOIR, and TYNDALE-BISCOE, 1966), and increases its concentration (BENTLEY and SHIELD, 1962). There is thus strong presumptive evidence to suggest that urine flow in this marsupial

100

is regulated, as in placentals, by the release of antidiuretic hormone that acts on the kidney. The injection of vasopressin also increases the total sodium and chloride excretion in the urine of *Setonix* (P. J. Bentley; P. Woolley, quoted by Waring *et al.*, 1966), but the physiological significance of these observations is not clear.

These actions of neurohypophysial peptides in *Setonix* are similar to those observed in the American marsupial, *Didelphis virginiana*. The posterior pituitary gland of this opossum contains arginine-vasopressin and oxytocin (Sawyer *et al.*, 1960). Posterior pituitary extracts have an antidiuretic action and increase excretion of urinary chloride in this species (Silvette and Britton, 1938).

The young *Setonix* at birth weighs about 0.25 g, about one-fifteen thousandth of the maternal weight, and is deficient in many morphological and physiological characters when compared with the parent, or with new-born placental mammals (see Shield, 1961). Such deficiencies include an inability to form a hyperosmotic urine or respond to the injection of vasopressin (Bentley and Shield, 1962). It is not until an age of more than 120 days that they respond to dehydration, or injected vasopressin, and form a hypertonic urine in the adult manner. This inability primarily appears to be due to an insufficiently developed kidney, probably inadequate differentiation of the loops of Henle. There is also a deficiency of neurohypophysial peptides and these do not appear in reasonable levels in the gland until the young are about 135 days old. In the protected environment of the mother's pouch, the young *Setonix* is not normally subjected to stresses like dehydration, so that such mechanisms are not essential at that time.

Setonix lives in coastal areas where dietary supplies of sodium are adequate, and it has even been suggested that it may drink sea-water during the summer drought period and so possibly acquire a considerable excess of salt. These animals can concentrate urinary electrolytes to about 800 m-equiv/l, which is higher than the level of 560 m-equiv/l in sea-water. Some quokkas, in extreme circumstances when no other fluid is available, drink small amounts of sea-water and make a net gain of osmotically-free water (Bentley, 1955). However, it seems unlikely that this species normally drinks sea-water though a similar small wallaby, the Tammar, *Macropus eugenii*, apparently can do so (Kinnear *et al.*, 1968). Quokkas sweat in hot environments, especially from areas on their fore and hind paws. The sodium loss by this route is unknown, but Lorna Green (1961) has suggested that the reason that they lick their paws at such times is to retrieve salt secreted by the sweat glands. The first studies to be carried out on the endocrine regulation of sodium and potassium metabolism were those of Buttle *et al.* (1952). Earlier work on the American opossum, *Didelphis virginiana*, suggested that the adrenal cortex may not be as important to marsupials as to placental mammals, since this species could, in certain circumstances, survive adrenalectomy for prolonged periods, with little or no change in electrolyte level. Jenny Buttle and her collaborators found that adrenalectomy in the quokka was usually fatal within 2 days. Death was accompanied by elevated plasma potassium and decreased plasma-sodium concentrations. The mean survival time could be increased two or three times by either giving the quokkas 1% sodium-chloride solutions to drink, or by injecting them with a corticosteroid, deoxycorticosterone acetate. Cortisol and corticosterone have recently been identified in the blood of *Setonix* (Ilett, 1969). Australian brush-tailed possums, *Trichosurus vulpecula*, also die within a few days following adrenalectomy, the changes in plasma sodium and potassium levels being similar to those observed in the quokka (REID and McDONALD, 1968). These possums could be kept alive by giving them injections of cortisol, aldosterone being ineffective, though the dose of cortisol could be reduced if both of these steroids were given simultaneously. The adrenal cortex of Australian marsupials thus seems to play a vital role in their ability to regulate sodium and potassium levels. Injections of aldosterone decrease renal sodium excretion in the American opossum but have no consistent effect on urinary potassium (BECK, BROWNELL, and BESCH, 1969). As we shall see, other studies on marsupials have identified and implicated corticosteroids in their regulatory mechanisms.

The water metabolism of two large species of kangaroo has been studied in far hotter and drier areas than those occupied by *Setonix*. TIM EALEY led a series of ecological and physiological investigations on a wild population of the hill kangaroo, or euro, *Macropus robustus*, living in a hot dry desert region of north west Australia (Fig. 3.2). ALAN NEWSOME carried out a similar investigation on the red kangaroo, *Megaleia* (or *Macropus) rufus*, living in an arid central region of the continent near Alice Springs. The average annual rainfall in both the areas is about 25 cm, most of it falling during the summer months. The rain, however, is sporadic and unreliable, and both areas may undergo periods of drought when little rain falls for periods of three or four years. The summer temperature is high, the average daily maximum often being greater than 40°C for many weeks and may rise to 49° C. The relative humidity usually is low so that the potential evaporation is high. Despite these osmotically adverse conditions the kangaroos in these areas continue to survive and multiply. A man lost in these regions in the height of summer can expect to survive for less than one day (ADOLPH, 1947) so the adaptability of the kangaroos is impressive. Both the euro and the red kangaroo may form a highly concentrated urine, which can be 2700 m-osmole/l in the red kangaroo (A. NEWSOME, quoted by SCHMIDT-NIELSEN, 1964a) and 2200 in the euro (EALEY, BENTLEY, and MAIN, 1965). Such an ability must aid water conservation by these animals but is not a predominant factor in their survival. EALEY and his collaborators observed that euros can withstand dehydration equivalent to 30% of their body weight, and that they reduce evaporation by sheltering in rocky caves during the heat of the day. Euros drink periodically and even during hot dry periods only come in to obtain water on an average of once every three days. During the intervening period they incur a water deficit equivalent to about 12% of their body weight.

NEWSOME (1965 a and b) found that the red kangaroos also only drank sporadically, apparently gaining most of their water from their diet, which consists preferentially of succulent green shoots of the herbage. The animals limit evaporation during the day by seeking the shelter of bushy woodlands and avoiding the open plains. The red kangaroo, like the quokka, stores arginine-vasopressin and oxytocin in its neurohypophysis, and the ratio of these conforms to that usually found in marsupials of 4 or 5 to 1 (FERGUSON and HELLER, 1965). The estimated concentration of vasopressin in the neurohypophysis is 10^{-9}M/kg body weight, which is low compared to other species, but this could result from the fact that the animals had been hunted and shot before the pituitary was obtained. It can be reasonably surmised that vasopressin has a similar role, in regulating the urine volume and concentration in such large kangaroos, as it does in placental mammals. Considering the environmental osmotic problems this role could be substantial, and could necessitate prolonged periods of secretion by the neurohypophysis.

The few marsupials and the single monotreme that have been examined exhibit an elevated ratio of vasopressin/oxytocin (V/O) concentration in their neurohypophyses; it is 6.2 in the Australian possum, 3.8 in the quokka, 4.8 in the red kangaroo (FERGUSON and HELLER, 1965), 2.8 in the American opossum and 3 in the echidna (SAWYER et al., 1960). The V/O ratio in most placentals is 1 to 2 with the exception of the camel and llama (Tylopoda) in which it is 3 to 4. It is unknown whether such variations reflect differences in the function of these peptides, but they nevertheless are of phyletic interest. The relatively higher storage levels of vasopressin, compared to oxytocin, could reflect genetic differences in the rates of synthesis or release of these peptides, or they could result from the formation of an unidentified, structurally different, hormone. The oxytocin activity in the neurohypophyses of the species with a high V/O ratio has not been exhaustively characterized, and it is possible that this peptide is not identical in molecular structure with oxytocin. Thus a peptide with a lower specific oxytocin activity could result in an elevated V/O ratio. Neurohypophysial peptide hormones similar, but not identical, to oxytocin have been identified in a number of vertebrates including the reptiles, amphibians and fish, and these were initially also thought to be oxytocin.

The grasses of the central Australian regions are deficient in sodium compared to those in coastal areas (DENTON, 1965) so that conservation of this solute could be of particular importance to many kangaroos. However, there is no definitive information about this for the red kangaroos living in the NEWSOME study area. The urine (NEWSOME, quoted by SCHMIDT-NIELSEN, 1964a) of these animals has a Na/K ratio of 0.13, while the euros in the EALEY study area exhibit a ratio of 0.33. This suggests that relatively more sodium may be available to the euro than the red kangaroo. Two populations of the grey kangaroo, Macropus giganteus, living on grasses with contrasting contents of sodium, have been studied by COGHLAN and SCOGGINS (1967). One group of these kangaroos lives in the coastal areas of south east Australia, where the grasses contain more than 150 m-equiv sodium/kg dry weight, and the other occupies the Snowy Mountain plains, where the herbage contains less than 10 m-equiv sodium/kg dry weight. Aldosterone, cortisol and corticosterone have been identified in the blood of the red kangaroo (WEISS and McDONALD, 1967) as well as the grey kangaroo. The red kangaroos were studied in the laboratory, and while aldosterone was secreted at rates comparable with those seen in placentals, the cortisol levels were somewhat lower. Rates of corticosteroid secretion were measured in grey kangaroos caught in the field and it was found that animals living in the sodium-deficient Snowy Mountains area have blood aldosterone levels 8 times as great as those from the sodium-replete coastal regions. There was little difference in the blood levels of cortisol or corticosterone. The zona glomerulosa in the adrenal cortices of the animals from the Snowy Mountains plains was larger than that in the other group.

Corticosteroids have been detected, and measured, in the adrenal venous and peripheral blood of several Australian marsupials. In addition to the kangaroos, aldosterone has also been detected in the blood of the wombat, Vombatus hirsutus

(WEISS and McDONALD, 1966a; COGHLAN and SCOGGINS, 1967) but has not been found in the Australian possum, *Trichosurus vulpecula* (CHESTER JONES et al., 1964a; WEISS and McDONALD, 1966b), though it is present in the American o-possum, *Didelphis virginiana* (DAVIS et al., 1967). Cortisol is the predominant corticosteroid in all of the Australian marsupials so far examined, the ratio of cortisol/corticosterone (compounds F/B) usually ranges from 3 to 12. This contrasts with the echidna (Monotremata) in which the F/B ratio is less than 0.5 (WEISS and McDONALD, 1965). Placental mammals, however exhibit widely varying F/B ratios; it is, for instance, less than 0.05 in the laboratory rat and 20 in a monkey (BUSH, 1953). The possible phyletic and functional significance of these differences is not known.

The control of secretion of corticosteroids in marsupials appears to conform to the placental pattern, but the information is incomplete. DAVIS et al. (1967) found that the injection of placental corticotrophin into the American opossum, *Didelphis virginiana*, produced a 'striking' increase in secretion of aldosterone, cortisol and corticosterone. Renin extracted from opossum kidneys had similar effects when injected into these animals, and if hypophysectomized opossums were used, the action of the renin was more selective with respect to aldosterone. Sodium depletion of the opossum resulted in increased granulation of the renal juxtaglomerular cells, suggesting increased production of renin. The kidneys and blood of the Australian brushtailed possum, *Trichosurus vulpecula*, have also been shown to contain renin (REID and McDONALD, 1969). The circulating levels of this hormone increased in possums which were sodium-deficient and there was also a granulation of the renal juxtaglomerular cells in such animals. Thus in both the American opossum and the Australian possum the control of corticosteroid secretion appears to conform to the placental pattern involving the corticotrophin and renin-angiotensin axes. WEISS and McDONALD injected ovine and porcine corticotrophin into Australian wombats, possums and kangaroos, and found increases in the rate of secretion of cortisol and corticosterone. The activity of marsupial corticotrophin has not yet been described.

5. The Placental Mammals

Generically and numerically the placentals are the predominant group of mammals. All of the marine and most aquatic mammals belong to this group. They are present in all of the world's deserts. The placentals have thus adapted to a wide range of differing osmotic circumstances.

The Rodentia make up about one-third of all placental genera. This order has a very cosmopolitan distribution, occupying diverse habitats including the deserts of Africa, Asia, America and Australia, where, with the possible exception of the latter area, they are the principal mammals present. Osmoregulatory problems in such dry hot areas result in the exclusion of most mammals, except certain rodents, equipped with behavioural and physiological adaptability. KNUT and BODIL SCHMIDT-NIELSEN investigated the water metabolism of the banner-tailed kangaroo rat, *Dipodomys spectabilis*, and revealed the manner in which this little rodent lives in a desert without drinking water. The kangaroo rat belongs to the family Heteromyidae which is confined to the new world; it lives in the deserts of the western United States and parts of Mexico. Its ability to survive and thrive on a diet of air dried seeds and vegetation, without requiring water to drink, is the result of several adaptations that have subsequently been shown to exist in other rodents occupying comparable ecological situations elsewhere. The principal adaptation allowing the kangaroo rat to live under such conditions is its small evaporative water loss. Rodents do not sweat or pant, and thus have a limited ability to undertake thermal cooling; they may allow their body temperature to rise somewhat in hot conditions (SCHMIDT-NIELSEN, 1964b). Obligatory losses of water that take place by evaporation from the respiratory tract, are reduced in the kangaroo rat and white rat, by utilizing a countercurrent heat exchange mechanism in the nasal passages to cool the expired air (JACKSON and SCHMIDT-NIELSEN, 1964). Such cooling reduces its content of water vapour. In hot situations, which can result in a critical rise of body temperature, many rodents may spread saliva over their fur in order to facilitate thermal cooling. However, this is not usually necessary as species that live in deserts tend to avoid extremes of temperature by sheltering in protected situations such as burrows. Rodents, especially those living in arid areas, conserve additional water by forming a highly concentrated urine and relatively dry faeces. Not all desert rodents can, however, achieve a positive water balance while eating a dry diet, and some must obtain food containing a substantial amount of free water. The pack rat, *Neotoma*, like the kangaroo rat, lives in the deserts of the western United States but eats cactus plants, while the North African sand rat, *Psammomys obesus*, subsists on succulent halophytic vegetation (SCHMIDT-NIELSEN, 1964a, b).

Neurohypophysial peptides, arginine-vasopressin and oxytocin, have been identified in the four species of rodents that have been examined, including the white rat, *Rattus norvegicus*, the kangaroo rat, *Dipodomys merriami*, and the guinea pig, *Cavia porcellus* (see SAWYER, 1968). Arginine-vasopressin has also been identified in five strains of mice (*Mus musculus*), while in the peru strain, lysine-vasopressin has been found (STEWART, 1968). Neurohypophysial peptides containing lysine instead of arginine at position 8 in the molecule (see Table 2.4) have hitherto been found only in the Suiformes (pigs, peccaries and hippopotami).

The neural lobe of rats can be destroyed by cutting the supraoptico-hypophysial tract and allowing the peripheral tissue to degenerate. Such animals form large volumes of dilute urine which they are unable to concentrate (diabetes insipidus). VALTIN and his colleagues (1962) have identified a strain (Brattleboro) of rats with hereditary diabetes insipidus due to the specific absence of vasopressin, although oxytocin is still present (SAWYER, VALTIN, and SOKOL, 1964). Vasopressin has been identified in the blood of the white rat, as well as in that of the kangaroo rat (AMES and VAN DYKE, 1952), the mouse and the guinea pig (J. HELLER, 1961). Rats on a normal diet, with drinking water available, have low circulating levels of vasopressin (5 μU/ml peripheral plasma), but this increases 10-fold after 24 hours without drinking water. The circulating levels of vasopressin in *normal* kangaroo rats, which do not require drinking water, are far higher than in dehydrated white rats. Injection of vasopressin into hydrated white rats and mice reduces the urine volume and increases its concentration, and such an antidiuresis has also been shown in the kangaroo rat (COLE, CHESTER JONES, and BELLAMY, 1963). The *prima facie* evidence thus suggests that neurohypophysial peptides have a physiological role in controlling urine volume and concentration in rodents, just as they are thought to do in other orders of placental mammals.

The ability to limit urinary water loss may vary among species in relation to the dryness of the normal habitat. Extrarenal water loss, however, predominates, and in white rats kept in the laboratory without water it is 5 times greater than the urinary loss, while in the kangaroo rat it is 2.5 times as great (see DICKER and NUNN, 1957). Under such conditions these rodents form hyperosmotic urine, which in the white rat may be 9 times more concentrated than the plasma, and in the kangaroo rat 14 times that level (SCHMIDT-NIELSEN, 1964a). In the absence of neurohypophysial antidiuretic hormones, urine that is isoosmotic to plasma is formed, so that the total water loss under such conditions would be expected to increase considerably. Urinary loss would then be more than twice as great as the extrarenal loss. The role of such hormones in forming a concentrated urine should be kept in perspective as the primary mechanism for such water conservation exists in the kidney itself, while it is the hormones' role to activate this.

The diversity in the habitats occupied by different rodents, and the variation in their osmoregulatory problems, suggests that there may be associated differences in neurohypophysial function. The known differences in ability to concentrate the urine are related to the physiology of the kidney, *not* to quantitative differences in circulating vasopressin levels, though the requirements for such peptides may conceivably differ in contrasting osmotic circumstances. As we have seen, either of two vasopressins may exist in mammals and it is pertinent to see if such a difference affects the water metabolism. Differences in the antidiuretic potency of neurohypophysial peptides, if extreme, could result in a condition of diabetes insipidus, or if more moderate, necessitate differing circulating levels to produce comparable effects. The antidiuretic potencies of lysine – and arginine – vasopressin differ in various species. In the dog, which normally possesses the peptide containing arginine, the lysine analogue is only one-sixth as potent as the natural hormone, while in man it is about one-half as active. The pig possesses lysine-vasopressin, but its kidney, nevertheless, is only two-thirds as sensitive to this as it is to the exogenous arginine analogue (MUNSICK, SAWYER, and VAN DYKE, 1958). The antidiuretic potencies of arginine – and lysine – vasopressin have been examined in laboratory rats and the latter here also has only about two-thirds the activity of the other peptide (SAWYER, 1958). Such differences in the relative potencies of these natural peptides are not great enough to affect the water metabolism of the species possessing them. It is interesting that in some of the Suiformes, either or both such peptide hormones may be present (FERGUSON and HELLER, 1965) in animals living together in the same area. DOUGLAS FERGUSON (1969) has measured the distribution of these peptides in a group of 18 warthogs, *Phacochoerus aethiopicus*, from Rhodesia. Eight of these beasts contained arginine-vasopressin only, two had lysine-vasopressin only, while eight were heterozygous and contained both peptides. This distribution appears to be random, and suggests that the possession of a particular vasopressin confers no selective advantage to any one warthog over another with a different peptide. It is possible that in other populations of mammals, living in different circumstances, this could be of advantage, but it appears to be unlikely.

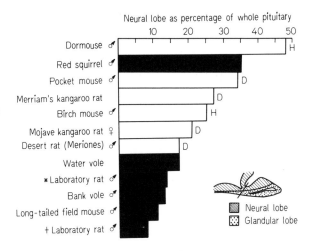

Fig. 3.3 Comparison of the relative size of the neural lobe of the pituitary gland in several species of rodents. *H* hibernating rodents; *D* desert-living rodents; □ non-hibernating rodents of temperate zone habitat. (From HOWE and JEWELL, 1959; Redrawn From ENEMAR and HANSTROM, 1956)

The size of the neurohypophysis and the amounts of peptide hormones stored in the gland vary in different species. ENEMAR and HANSTROM (1956) compared the size of the neural lobe, relative to the whole pituitary, in several rodents occupying different habitats (Fig. 3.3). The desert species and those that hibernate, tend to have relatively larger neural lobes than animals normally occupying temperate regions. In the desert dwelling spiny mice, *Acomys cahirinus* and *A. russatus*, and laboratory mice and rats, water deprivation results in a hypertrophy of the neural lobe as well as an increased activity in the supra-optic nuclei (CASTEL and ABRAHAM, 1969). This appears to reflect an increased ability to synthesize vasopressin . Differences also occur in the amounts of vasopressin stored in the neural lobe of different species; the kangaroo rat (on a body weight basis) stores 6 times as much vasopressin as the laboratory rat and 20 times that of the aquatic mountain 'beaver', *Aplodontia rufa* (Table 3.1).

When water is withheld from laboratory rats they become progressively dehydrated, and increased amounts of vasopressin appear in the blood. Some of this peptide (about 10%) appears in the urine, and after an initial stimulation, there is a progressive and large reduction of the hormone level in the neurohypophysis (DICKER and NUNN, 1957). The neurosecretory material in the neural lobe may be completely depleted under such conditions (HOWE and JEWELL, 1959). The kangaroo rat does not suffer progressive dehydration when deprived of drinking water, but large amounts of antidiuretic hormone are, nevertheless, normally excreted in the urine, indicating a high rate of endogenous secretion. The stores of peptides in the neurohypophysis of the kangaroo rat under such conditions are large, five times more than in a normally hydrated rat (AMES and VAN DYKE, 1950). These observations suggest that the kangaroo rat is better able to maintain its glandular levels of vasopressin than the laboratory rat, even though it is secreting such hormones at a higher rate. It thus seems that these desert rodents have greater ability to synthesize such peptides. The histological appearance of neurosecretory material in the neurohypophysis reflects the presence of hormones and the secretory activity

107

Table 3.1 *Quantities of vasopressin in the neurohypophysis of rodents*

	wt g	m-μ mole/kgBW
Mus musculus[1] (mouse)	20	7
Rattus norvegicus[1] (rat)	250	3.7
Aplodontia rufa[2] (Mountain beaver)	1000	0.8
Cavia porcellus[3] (Guinea pig)	400	3.2
Dipodomys merriami[4] (Kangaroo rat	40	18

[1]from FOLLETT (1963)
[2]DICKER and EGGLETON (1964)
[3]DICKER and TYLER (1953)
[4]AMES and VAN DYKE (1950)

of the gland; depletion of vasopressin in the rat is accompanied by disappearance of neurosecretory material. The African desert rat, *Meriones meriones*, kept on a dry diet without water, does not suffer depletion of neurosecretory material (HOWE and JEWELL, 1959). Spiny mice, *Acomys*, from Israel suffer an initial depletion of neurosecretory material when first deprived of water, but this starts to return to normal after about 2 weeks (CASTEL and ABRAHAM, 1969). Two other desert rodents that live in North Africa, the jerboa, *Jaculus jaculus*, and the gerbil, *Gerbillus gerbillus*, exhibit increased neurosecretory activity in the hot summer period as compared to the cooler winter, and this probably reflects differences in their requirements for vasopressin at these times (KHALIL and TAUFIC, 1964).

Neurohypophysial peptides appear to be involved in the control of renal water losses in rodents. The limited amount of information available suggests that such hormones are of more importance to those species living in arid as compared to more temperate areas. Such species probably have an increased ability to synthesize and store such peptides, while normally maintaining a high rate of secretion.

Removal of the adrenal gland from laboratory rats results in death, usually within a few days. This results largely from a depletion of the sodium and an excessive accumulation of potassium, due to an inadequate integration of renal electrolyte excretion in the absence of the adrenal corticosteroids. Adrenalectomy has also been carried out in the kangaroo rat, *Dipodomys spectabilis*, with the same results, that are characteristic of the mammals (COLE *et al.*, 1963). The desert gerbil, *Gerbillus gerbillus*, dies within seven days of adrenalectomy, but this period can be extended by injecting the animals with cortisone and feeding them a diet with a high content of sodium chloride (Burns, 1956). The adrenal cortex of laboratory rats secretes aldosterone and corticosterone, but little, if any, cortisol. Bush (1953) found that the ratio in the secretion of cortisol/corticosterone (compounds F/B) was less than 0.05. The banner-tailed kangaroo rat, on the other hand, also forms cortisol and has an F/B ratio of about 1.4. Bush, after comparing F/B ratios in a number of species of placentals which normally have a differing diet and way of life, concluded that such differences are probably genetically determined variations of the biosynthetic pathways in the adrenal cortex, and cannot be related to differences in the role of the corticosteroids.

The rates of corticosteroid secretion in desert rodents have not been directly determined. Indirect evidence, however, suggests that the kangaroo rat may have low circulating levels of cortisol and corticosterone but a high concentration of aldosterone (Cole et al., 1963). These conclusions are tentative and largely based on the renal deficiencies in excreting water and sodium, a facility in secreting potassium and a ready response to anti-aldosterone drugs. Aldosterone could contribute to the kangaroo rat's ability to concentrate urine by facilitating gradients of sodium concentration associated with the counter current-multiplier system of the renal tubule. Crabbe (1962) has shown that aldosterone increases the ability of man to concentrate urine and has suggested such a mechanism. Adrenalectomized rats have a decreased ability to concentrate urine (Sigler, Forrest, and Elkington, 1965) an observation that would be consistent with the above hypothesis.

Corticotrophin initiates release of corticosterone and aldosterone in rats, but the role of the renin-angiotensin system has been in doubt as injection of angiotensin has little effect on release of these hormones. Kinson and Singer (1968) have confirmed these observations but find that they are confined to sodium-replete rats, for if sodium-deficient animals are used, angiotensin stimulates release of aldosterone (but not corticosterone). The role of corticotrophin and renin in controlling corticosteroid secretion in the Rodentia thus appears to conform to the usual placental pattern. There may, nevertheless, be some differences in the physiological and biochemical details, that result in variations in sensitivity to these trophic hormones.

While there is probably more information about osmoregulation in diverse species of the Rodentia than in any other mammalian order, this is unfortunately still very incomplete. Research has been mainly directed towards laboratory species like the rat and a small number of more exotic animals that live principally in desert areas. The comparative information about osmoregulation in the Rodentia is thus somewhat lopsided, as, with the exception of *Aplodontia*, we have no such knowledge about species living at the aquatic extremes of the range. Such water dwellers are numerous among the rodents. These species are not only inherently interesting, but may provide information that can be used to assess better the significance of physiological and endocrine variations thought to be associated with the osmoregulatory problems of the animals. Thus measurements of evaporative water loss, urine concentrations, storage and secretion of neurohypophysial peptides and adrenocorticosteroids in aquatic rodents would yield interesting data for comparison with similar measurements in desert rodents.

The sea provides an ecological niche for many mammals and although fragmentary information about the water and salt metabolism of such species is available, the relationship of this to the endocrine glands is almost totally lacking. Arginine-vasopressin and oxytocin have been identified in the fin-back whale (CHAUVET, CHAUVET, and ACHER, 1963) and corticosteroids, including aldosterone, have been identified in the adrenal cortex of the sperm whale (RACE and WU, 1961). Marine mammals have rather special osmotic problems so that an extension of these observations would be most interesting.

Chapter 4

The Birds

Birds are the most cosmopolitan of the vertebrates. The ability to fly confers on them a mobility that is matched in other vertebrates only by the bats. Some avian species have, however, lost the ability to fly, or do so to little effect. Flightless birds include the ostriches, kiwis, cassowaries, emus and rheas, while many other species like the domestic fowl are aviatorially inept. Some birds, like the penguins and auks, use their wings for propulsion under water. Birds thus may live a predominantly terrestrial or marine existence or they may also utilize the skies above such areas. Geographically, birds are found on the most remote oceanic islands, in dry continental desert regions, and in tropical and temperate areas where water is abundant. They can indeed live, and thrive, in a variety of osmotic environments.

The osmotic problems of birds are basically like those of mammals; their osmotic anatomy is similar, they occupy the same geographical regions and habitats and they are also homoiothermic. Their needs for water and salts are thus potentially comparable, but differences in their physiological and morphological characteristics may modify their respective requirements. Most birds are smaller than mammals, so that they more often experience the difficulties inherent in a large surface area to body weight ratio. The body temperature in birds is usually in the region of 40 to 42°, which is 3° or 4° higher than that of most mammals. This reflects a metabolic rate that is higher than that of mammals, and so modifies heat exchanges with the environment. Flight results in metabolic burdens which are completely unfamiliar to mammals (except possibly the bats), and this is reflected in increased respiratory gas exchange, with an accompanying additional water loss. The mobility conferred by flying may, however, be a considerable osmotic advantage, as it allows birds to travel rapidly for long distances to suitable feeding and watering places. This may be a relatively local commuting, or it may take place between major geographical regions, in which seasonal changes in available food and water may occur.

The birds exhibit a number of other physiological features, relevant to their osmoregulation, that they share with either the mammals or their phyletic progenitors the reptiles. Both the birds and mammals, in contrast to the reptiles, can form a hypertonic urine. However, in birds the maximal concentrations are not as great as usually seen in mammals, but they conserve additional renal water by converting most of their catabolic nitrogen to uric acid. Most reptiles are similarly uricotelic. While in the mammals the kidney is the principal route for excretion of salts, some birds and reptiles can also secrete such solutes from cephalic 'salt' glands. Birds, like reptiles, but in contrast to mammals (except the monotremes), are oviparous, and it can be conjectured that this could influence the pattern of their osmoregulation. The rapid production of a clutch of eggs containing all of the water and salt necessary for an extended period of embryonic growth would seem to result in a more acute need for water and salt than embryonic development *in utero*. Tending and incubating such eggs temporarily restricts the movements of birds so that breeding can only occur in a place and time of adequate proximal supplies of food and water.

111

1. Water loss

Water loss in birds takes place by evaporation, and in the urine and faeces. *Evaporative water loss* has been measured in a number of birds (Fig. 4.1), and in equitable laboratory conditions of about 25°, varies from 5% to 35% of the body weight in a day. Water consumption *ad lib.* provides an indication of the total amount of water normally lost in a day, though it does not take into account water derived from the food. BARTHOLOMEW and CADE (1963) have summarized information on drinking in 21 species of birds, varying in weight from 10 to 140 g, and find that they drink water equivalent to 5 to 100% of their body weight each day. The differences between drinking and evaporation give an approximation (which will be somewhat low, as water in the food is not allowed for), of the normal rate of water loss in urine and faeces. Such estimates vary, in different species, from 2 to 20% of the body weight a day, but is more usually about 5 to 6%. The minimal total water losses compatible with life, and which are seen when water is restricted, are probably lower than these values.

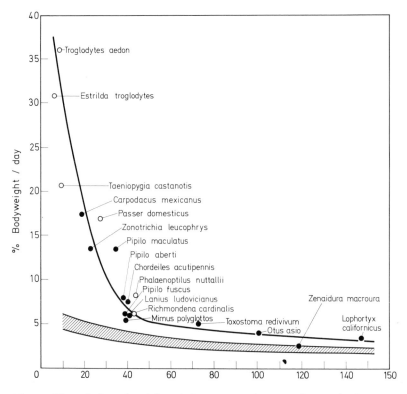

Fig. 4.1 The relation of weight-relative evaporative water loss to body weight of birds at ambient temperatures near 25°. The cross-hatched curve indicates the theoretical relation between body weight and the production of metabolic water by birds in a basal condition utilizing exclusively carbohydrates (upper boundary), or exclusively proteins (lower boundary). The values for fats are intermediate. (From BARTHOLOMEW and CADE, 1963).

112

Few birds can survive indefinitely on a diet of dry food with no water to drink. Most birds progressively lose weight on such a regimen, and this substantially represents a loss of body water. Death ultimately occurs, but the loss that is fatal varies in different species. The California quail dies after losing 50% of its body weight, the mourning dove after a loss of 37%, while the house finch cannot survive a loss of 15% in weight (BARTHOLOMEW and CADE, 1963). The reasons for these differences are not clear but they may be related to the rates of dehydration, for the house finch loses water five times as fast as the California quail and the mourning dove at an intermediate rate. Measurements of the concentrations of the body fluids at death would indicate whether or not these species have different tolerances to elevated osmotic pressures in their tissues.

a) Evaporative water loss

Relative evaporative water loss in birds is inversely related to their body weight (Fig. 4.1), an observation that substantially reflects differences in the rates of their metabolism, and surface area to body weight ratios. Sweat glands are absent in birds, and it is generally assumed that evaporative water losses take place predominantly from the respiratory tract. The exact magnitude of the cutaneous water loss is unknown, and the absence of sweat glands *per se* does not indicate that water loss through the skin does not occur. The laboratory rat has no sweat glands, but still loses substantial amounts of water through the skin.

Water loss occurs from the respiratory tract, as an unavoidable consequence of the exchange of oxygen and carbon dioxide. The high body temperature of birds potentiates such losses, as the expired air contains more water vapour than it would do if saturated at lower temperatures. The temperature of the expired air has, however, not been measured and it is possible that it may be reduced by a counter-current heat exchange in the nasal passages as seen in rodents by JACKSON and SCHMIDT-NIELSEN (1964). Water loss in desert rodents is only about 0.55 mg per ml of oxygen consumed while in man the loss from the respiratory tract is 0.9 mg per ml of oxygen used (SCHMIDT-NIELSEN, 1964a). Such water loss in birds is generally higher than that in mammals. The budgerygah, *Melopsittacus undulatus*, which lives in the desert areas of Australia, is a bird of similar size to the kangaroo rat but loses water at nearly twice the rate: 0.94 mg per ml oxygen used (GREENWALD, STONE, and CADE, 1967). DAWSON and SCHMIDT-NIELSEN (1964) conclude from the available information that such evaporative losses in birds range from 0.9 to 2.1 mg water per ml of oxygen used and may be even greater.

When the environmental temperature rises, the evaporative water loss is increased, and as the normal body temperature is approached, this may be physiologically accelerated in response to the need for temperature regulation. The need for such adjustments is related to the balance of metabolic heat production by the bird, and the rate at which this is dissipated to the external environment. Such exchanges are dependent on the difference between the temperature of the animal and the environment, which affects the rate and direction of dissipation of heat by radiation, conduction and convection. The high body temperature of birds will favour such heat loss, more than in animals with a lower body temperature that

113

approaches more closely that of the environment. Physiologically the rates of heat loss by radiation, conduction and convection can be altered by changes in cutaneous blood flow, modifying posture such as by spreading out the wings, extending the foot swim web, and fluffing or flattening the plumage. Such physical heat loss may play a predominant role in the ability of the bird to maintain its body temperature in hot situations. It can be supplemented, to various degrees, by facilitation of the evaporation from the respiratory tract. The temperature at which such processes are initiated varies in different species (KING and FARNER, 1964) depending on the normal body temperature, metabolic rate, the surface area and the birds insulation. A high body temperature could favour life in hot desert regions, as this would facilitate heat loss to the environment, but a comparison of the body temperatures of birds indicates there are no such differences of adaptive significance to a desert life.

At high ambient temperatures increased evaporative water loss inevitably occurs, and this is brought about physiologically by an increase in the ventilation of the lungs; initially with a change in tidal volume and then, at about 42°, by panting which results from increased respiratory movements, and in many birds by an accelerated movement of the hyoid apparatus called 'gular fluttering' (BARTHOLOMEW, LASIEWSKI, and CRAWFORD, 1968). At an environmental temperature of 40° such evaporative water loss may account for one to two-thirds of the heat dissipation. Flying results in a considerable increase in metabolic rate and accompanying increases in evaporative water loss. These processes have been carefully measured by VANCE TUCKER (1968) using budgerygahs flying in a wind tunnel. Evaporative water loss and oxygen consumption at 20° are 5 times as great in flight as at rest. When the ambient temperature was increased during flying, the oxygen consumption stayed the same, but evaporative water loss increased and was three times as great at 36° as at 20°.

Evaporative water loss in birds can be altered somewhat in different physiological circumstances. Birds, like mammals, withstand elevated body temperatures only to a very limited extent and cannot survive a body temperature of more than about 45° for any period of time. Toleration of hyperthermia, which results from storage of heat, with a reduction of evaporative water loss, can occur but is limited in its range. This can be extended by adopting a pattern of nocturnal hypothermia, followed by diurnal hyperthermia (see KING and FARNER, 1964), but the importance of such heat storage in water conservation is not clear. When birds are deprived of water, like mammals, they may reduce evaporative water loss and as a result undergo an increased body temperature. This has been observed in desert birds like the ostrich (CRAWFORD and SCHMIDT-NIELSEN, 1967), the budgerygah (GREENWALD et al., 1967) and the zebra finch (CALDER, 1964). The physiological causes of such adjustments in water loss are uncertain, but if the metabolic rate changes, as in nocturnal hypothermia, the thyroid gland could be involved. The thyroid function in such birds does not, however, appear to have been investigated from this aspect.

Birds living in arid areas adopt patterns of behaviour that reduce evaporative water loss. Such adaptations in the mammals include avoiding the heat of the day by seeking refuge in cooler places, such as burrows, and coming out to feed during the cool night hours. Few birds are, however, nocturnal and even fewer are fossorial

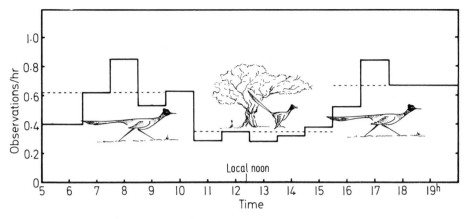

Fig. 4.2 Frequency of observation of wild active roadrunners in the Sonoran desert, an index of activity, as a function of standard zone time. The histogram shows the mean frequency of observation for the time periods indicated. The broken lines depict the mean values for morning (05:00-10:30), midday (10:30-15:30), and afternoon (15:30-20:00) periods. (From CALDER, 1968).

in their habits. Birds living in hot dry areas often reduce their activity during the hottest times of the day, when they seek refuge in the shade of trees, and bushes. CALDER (1968) has observed the daily activity of the roadrunner, *Geococcyx californianus,* in the dry arid areas of the southwest United States. This bird is a poor flyer and lives mainly on the ground where it feeds principally on insects. During the hottest times of the day it sits in the shade of trees and bushes and postpones its industrious search for food till the cooler periods of the morning and late afternoon (Fig. 4.2). Birds in flight incur even greater water losses at the hottest times of the day, and in arid areas they often curtail this activity accordingly.

b) Urinary and Faecal water losses

Water loss from these two channels is difficult to separate in intact birds, as urine and faeces are stored together in the posterior part of the gut and are voided simultaneously.

The excreta (urine plus faeces) of normally hydrated budgerygahs and zebra finches contains about 80 g/100 g wet weight, but when the fluid intake is restricted this decreases to nearly 50g/100g (CADE and DYBAS, 1962; CALDER, 1964). In these birds the total water loss in the excreta is equivalent to about 2% of the body weight in a day.

Excretion of water and solutes occurs through the kidney. Solute excretion results in water loss, the magnitude of which depends on the osmotic concentration that is attainable in the urine. Birds, like mammals, can form a urine that is hyper osmotic to the body fluids, and this is similarly due to the arrangement of the renal tubules to make a countercurrent-multiplier system (POULSEN, 1965). The maximum urinary concentrations that birds can achieve are, nevertheless, generally lower than those in mammals. Most species so far examined can concentrate their urine to give maximum urine to plasma osmotic concentration ratio of about 2. Even a desert bird like the roadrunner, *Geococcyx californianus*, or a sea bird like the pelican, *Pelecanus erythrorhynchos*, when deprived of water, only produces a urine with a concentration of about 700 m-osmole/l (CALDER and BENTLEY, 1967). A notable exception is the savannah sparrow, *Passerculus sandwichensis beldingi*, which inhabits saltmarshes in California, and which, when given saline solutions to drink, can produce a urine with a concentration of 2000 m-osmole/l, (POULSEN and BARTHOLOMEW, 1962). Many birds are not completely dependent on the kidneys for electrolyte excretion, as the nasal salt glands can secrete large amounts of highly concentrated fluids. The need for metabolic nitrogen excretion accounts for a substantial part of the renal osmotic water, and in birds such excretion occurs largely in the form of uric acid. Ducks excrete 54% of their catabolic nitrogen as uric acid, 29% as ammonia and less than 2% as urea (STEWART, HOLMES, and FLETCHER, 1969). It will be recalled that mammals form principally urea. Uric acid, in contrast to urea, is poorly soluble in water and is precipitated in the urine to form a milky suspension in which state it exerts negligible osmotic pressure. One gram of nitrogen, when excreted by a bird, requires only about 10 ml of urine for its excretion while the same amount of nitrogen in the form of urea requires 50 ml for its excretion in a man (H. SMITH, 1951).

α) Renal Water Conservation and the Neurohypophysis. After water is administered to the domestic fowl there is an increase in urine flow, accompanied by a decrease in its concentration from about 500 m-osmole/l to 80 m-osmole/l (DICKER and HASLAM, 1966). The glomerular filtration rate (GFR), as reflected by the endogenous creatinine clearance, markedly increases and this is closely related to the elevated urine volume. KORR (1939) also observed an increased GFR in hydrated chickens. SKADHAUGE and B. SCHMIDT-NIELSEN (1967) found that the urine increased in volume, and decreased in concentration following hydration of roosters, but they observed only a small (23%) increase in the GFR. The reasons for these differences in glomerular behaviour are not clear. The latter authors used inulin clearance to determine GFR and, while this is a more accurate method of measuring the GFR, the magnitude of the differences is still difficult to understand. The GFR of hydrated budgerygahs is about one-third greater than when they are dehydrated, but as the urine flow is nearly four times larger, tubular water reabsorption is predominant in this bird also (KRAG and SKADHAUGE pers. comm.). The exact role of changes in the GFR in alterations of the urine volume in the birds is uncertain. All the results, nevertheless, indicate that changes in the rates of water reabsorption from the renal tubules are of major importance. The GFR in mammals is more labile in some species, like the dog, than in others such as man, and experiments on more avian species may elucidate the role of the glomerulus in such processes.

Dehydration (SKADHAUGE and B. SCHMIDT-NIELSEN, 1967) or infusion of hypertonic sodium chloride solutions (DANTZLER, 1966) results in a decreased urine flow in the domestic fowl and such treatments are accompanied by decreases of the GFR as also seen in reptiles. DANTZLER infused 6% sodium chloride solutions into chickens and observed a 40% reduction in GFR (inulin clearance). This was found to result from a decreased number of functioning glomeruli (glomerular inter-

116

mittency), rather than a change in ultrafiltration at individual sites. It seemed likely that these effects were being mediated by a release of the neurohypophysial peptide vasotocin, but infusion of this hormone at modest rates did not alter the GFR. Urine flow in birds thus responds to changes in body hydration, and such processes involve tubular water reabsorption and some changes in the GFR, though the physiological importance of the latter is not clear.

Neurohypophysectomy in the domestic fowl results in the formation of large volumes of urine (SHIRLEY and NALBANDOV, 1956). A strain of chickens has recently been identified that exhibits a genetic predisposition to form excessive quantities of urine (DUNSON and BUSS, 1968). Such birds drink three times the normal volume of water (polydipsia), and secrete urine with an osmolarity of only about 100 m-osmole/l, as compared to the more usual concentrations of 500 m-osmole/l. These chickens are reminiscent of the Brattleboro strain of laboratory rats, which suffer from diabetes insipidus due to a lack of neurohypophysial antidiuretic hormone. At the time of writing such a deficiency has not been reported in this strain of domestic fowl, though injection of neurohypophysial peptides brings about an increase in the urine concentration.

The injection of neurohypophysial peptides into the normal domestic fowl results in decreased secretion of urine. BURGESS, HARVEY, and MARSHALL (1933) showed that 'Pitressin', derived from mammalian neurohypophyses, has such an effect and that this is due to *both* an increased reabsorption of water from the renal tubules and a decreased GFR. Pitressin also has an antidiuretic effect when injected into the domestic duck (HOLMES and ADAMS, 1963). Birds, including the domestic fowl and duck, store 8-arginine vasotocin and oxytocin (but not vasopressin) in their neurohypophyses (see Table 4.1). When vasotocin is injected into hydrated chickens it also produces an antidiuresis, while oxytocin has little effect in this respect (MUNSICK, SAWYER, and VAN DYKE, 1964). ERIK SKADHAUGE has carefully analyzed the renal sites of action of vasotocin. Small amounts of this peptide were injected into the renal portal vein of hydrated chickens and so made initial contact with the renal tubules, instead of the glomeruli. The decreased urine flow, which was observed, resulted from increased tubular reabsorption of water, with no measurable change in GFR or renal blood flow. Higher doses of vasotocin decreased the GFR and renal blood flow but this effect is considered to be of pharmacological, rather than physiological, significance. Neurohypophysial peptides in large doses decrease the blood pressure of the domestic fowl and a variety of other birds, including pigeons (WARING, MORRIS, and STEPHENS, 1956), penguins, emus and cormorants (WOOLLEY, 1959). Such a vascular action could mediate the decreased GFR that is observed when large doses of such peptides are injected. The dose of vasotocin that must be injected to produce a drop in the blood pressure of the domestic fowl is 10 to 20 times greater than that which has an antidiuretic effect (SKADHAUGE, 1969a). It seems likely that, under physiological conditions, vasotocin changes the urine volume in the fowl by altering renal tubular water reabsorption. The glomerulus may, however, respond in certain circumstances, especially as it shows lability in its activity.

The preceding information about the presence of vasotocin, and its effects when injected is only indirect evidence as to its role in the birds. Indeed neurohypophysial peptides have not been demonstrated in the circulation follow-

Table 4.1 *Neurohypophysial hormones in birds*

	Vasotocin	Oxytocin*	V/O ratio	Vasotocin mμ moles/kg BW
Galliformes				
Domestic fowl	+[1]	+[1]	6.7	1.2[2]
Domestic turkey[3]	+	+	1.2	0.2
Coturnix japonica[4] (Japanese quail)	+	+	4.2	approx 2
Columbiformes				
Columba livia[2] (Pigeon)	+	+	7	0.4
Anseriformes				
Domestic duck[5]	+	+		
Psittaciformes				
Melopsittacus undulatus[6] (Budgerygah)	+			
Charadriiformes				
Larus canus[7] (Gull)	Pressor and oxytocic activity		8.6	9.5
Passeriformes				
Zonotrichia leucophrys gambelii (White crowned sparrow)	+	+	3.5	

References: [1]MUNSICK *et al.* (1960); CHAUVET, LENCI, and ACHER (1960).
[2]HELLER and PICKERING (1961)
[3]MUNSICK (1964)
[4]FOLLETT and FARNER (1966)
[5]HIRANO (1966)
[6]HIRANO (1964)
[7]C. TYLER quoted by FOLLETT (1963)
[8]D. S. FARNER and W. H. SAWYER; quoted by SAWYER (1968)
* ACHER *et al.* have identified mesotocin in several birds (Eur. J. Biochem. 17: 509, 1970)

ing appropriate osmotic stimuli, though STURKIE and LIN (1966) have shown them to be present during oviposition in the domestic fowl. Water consumption in domestic fowl is very low prior to oviposition (0 to 2 ml/h) but increases to 50 ml/h immediately following expulsion of the egg (P. MONGIN, pers. comm.). The elevated vasotocin levels could thus be reflecting a release due to dehydration as well as an enhancement in the contractility of the oviduct. The levels of neurohypophysial peptides stored in the pituitary vary during different osmotic conditions, suggesting that they may have a role to play in such circumstances. When pigeons, *Columba livia*, are given 0.5 M sodium chloride solutions to drink for 5 days, the storage of neurohypophysial peptides in the gland is reduced to about 10% of the level seen in birds with fresh water to drink (ISHII, HIRANO, and KOBAYASHI, 1962). FOLLETT and FARNER (1966) found that when Japanese quail,

118

Coturnix coturnix japonica, are deprived of water, or given 0.2 M saline solutions to drink, the storage of vasotocin in the neural lobe is reduced by 80% and that of oxytocin is decreased by 60%. These changes were correlated with parallel decreases in the stainable neurosecretory materials in the neural lobe. Such histological changes, in response to dehydration or drinking saline solutions, have also been shown in the domestic fowl (LEGAIT, 1959), the white-crowned sparrow and the zebra finch (FARNER and OKSCHE, 1962). It is interesting that the budgerygah, which in some circumstances can survive for prolonged periods without drinking water, does not exhibit depletion of neurosecretory material after 7 days without water (UEMURA, 1964), a condition reminiscent of certain desert rodents.

Vasotocin is far more active than oxytocin in decreasing urine volume, and its action in contracting the avian oviduct is similarly greater. The half life of oxytocin in the circulation of chickens is about 9 min and that of vasotocin 13 min (HASAN and HELLER, 1968). Vasotocin is usually present at a higher concentration than oxytocin in the avian neurohypophysis (Table 4.1). Estimates of the total storage of vasotocin are sparse, but are similar to those in mammals. (c.f. Table 3.1). The single estimate of peptide storage in the gull, *Larus canus*, is especially high, and it may reflect the magnitude of its osmotic problems, which are associated with a marine environment. Comparison of such peptide storage in other species may provide information relevant to the importance of these hormones in their natural way of life.

2. Role of the Cloaca and Intestine in Water Conservation

Birds lack a urinary bladder; urine passes from the ureters into a posterior part of the gut called the cloaca. Faeces and products of the urinogenital system all ultimately pass this way. Anatomically the posterior part of the gut consists of several regions; the proctodaeum, urodaeum and coprodaeum that make up the cloaca. The large intestine is in communication with the coprodaeum. SKADHAUGE (1968) found traces of uric acid far up in the large intestine, and even in the ceca, of the domestic fowl and duck. When dyes were placed in the proctodaeum they were subsequently also found in these regions. Radiographic contrast media are excreted by the chicken kidney and x-ray examination has shown that within minutes of passing out of the ureters they pass far up into the intestine (NECHAY, BOYARSKY, and CATACUTAN-LABAY, 1968; AKESTER *et al.*, 1967). Thus the cloaca and large intestine of birds are in free communication, so that physiologically one cannot separate osmotic events that occur in these two regions.

That changes in the water and electrolyte content of the ureteral urine occur after it has passed into the cloaca has been considered likely for a long time, but until recently has been more the subject of speculation than conclusive experiments. The openings of the ureters can be transferred surgically to the exterior of the bird so that the urine bypasses the cloaca. HART and ESSEX (1942) found that domestic chickens with such a surgical rearrangement needed extra salt in their diet. More recently DICKER and HASLAM (1966) have shown that such birds drink twice the amount of water after such an operation. Both these observations could be due to added urinary losses, such as could result from a failure to reabsorb water and salt in the cloaca or large intestine. NECHAY and LUTHERER (1968) compared the composition of urine collected directly from the ureter of one kidney with that from the other kidney after it has passed through the cloaca in the normal way. Reflux of urine into the intestine was prevented with a ligature. Under such conditions no movement of electrolytes, and only a minor passive transfer of water from hypotonic urine were observed. However, if the intestine is left in communication with the cloaca (SKADHAUGE, 1968) it was found that dehydrated chickens reabsorbed about 50% of the water and sodium which passed out of the ureters. *In vivo* perfusion of the cloaca and large intestine was also performed by SKADHAUGE (1967), and he was able to demonstrate the reabsorption of sodium and water from such preparations. These processes were not altered by vasotocin.

The observations of SKADHAUGE on water and sodium reabsorption from the fluid in the cloacal and intestinal regions of chickens, are consistent with the increased drinking observed by DICKER and HASLAM in birds with ureters transplanted out of the cloaca. They also can account for the excessive losses of sodium observed by HART and ESSEX in such birds. The composition of the urine of birds, or more specifically the domestic fowl, would thus appear to be modified after it passes into the cloaca, but the actual site of the reabsorption is probably the coprodaeum and large intestine. The rate of absorption of sodium and chloride from this region of the gut can be regulated, as it has been observed to decrease by 50% in domestic fowl to which an excess of sodium chloride has been administerd (SKADHAUGE, 1967). Whether such processes are influenced by hormones is unknown. Corticosteroids from the adrenal cortex are logical candidates for such an action, but their possible action at such a site has not yet been examined in birds.

KNUT SCHMIDT-NIELSEN has suggested (SCHMIDT-NIELSEN *et al.*, 1963) that the functions of the cloaca and the nasal salt glands (see next section) may be integrated so as to conserve water. This would result from the ability of many birds to secrete salts at far higher concentrations in the nasal gland fluids than in the urine. Urinary salts and accompanying osmotic water can be reabsorbed from the cloaca and large intestine, and then, as seems likely in birds with functioning nasal glands, the salt may be excreted through this channel, at concentrations that would spare more water than would be possible in urine.

3. Conservation and Excretion of Salts

The principal avenues for salt loss in birds are the kidneys, gut and nasal salt glands. Birds lack sweat glands, and so are not subjected to the relatively large salt losses that may thus occur through the skin of many mammals.

a) Kidneys

Unavoidable salt losses occur in the urine of birds, just as in mammals, but these are probably also small.

The process of renal salt excretion has been examined in the domestic duck by HOLMES, FLETCHER, and STEWART (1968). When these birds were given fresh water to drink, the mean sodium concentration in the urine after 24 h fasting was 8.5 m-equiv/l and the standard error was 3.7, indicating that concentrations as low as 1 m-equiv/l can be attained. The total sodium loss in these ducks (weighing 2.2 kg) was only 1.25 m-equiv/day. Potassium excretion was about 4 times as great, probably reflecting a release resulting from tissue metabolism due to fasting. The low rate of sodium excretion is achieved by tubular reabsorption of 99.8 $\pm 0.1\%$ of the sodium that is filtered at the glomerulus; changes in sodium excretion do not normally appear to be mediated by changes in the GFR.

HOLMES and his collaborators collected the urine as it was voided from the cloaca, so that any changes that may have occurred after incubation of the urine in the posterior regions of the intestine are unknown. As discussed earlier, sodium can be reabsorbed from the urine after it is regurgitated up into the large intestine. Absorption of sodium from the prospective faeces presumably occurs simultaneously in the large intestine.

Many birds obtain an excess of salts in their diet. Marine birds, especially those feeding on invertebrates, must take in substantial quantities of salts with their food. On Rottnest island near the southwest coast of Australia I have observed ducks feeding in salt lakes that attain saturation levels of solutes in the summertime. These lakes contain little life, apart from large numbers of brine shrimps, *Artemia salina*, which presumably constitute a very salty part of the ducks' diet.

b) Nasal salt glands

When excess salts are ingested by birds they may be excreted by the kidneys, and in many species also by the nasal salt glands. The kidneys of birds usually can only form a urine with a moderate salt concentration, so that they have a relatively limited capacity to excrete large quantities of such solutes and still conserve water. SCHMIDT-NIELSEN and FÄNGE (1958) found that when pelicans were injected with solutions of sodium chloride, only 20% of the total salt excreted appeared in the urine, the rest being in the secretions of the nasal glands. Similarly, HOLMES and PHILLIPS (1964) found that while potassium could be excreted in relatively substantial amounts by the kidneys, only about 1% of an ingested load of sodium chloride appeared in the urine. It should be recalled, however, that most terrestrial birds lack a functional nasal salt gland, and so are dependent on their kidneys for excretion of any excess they may obtain in their diet. The concentrations of sodium and chloride which may be attained in the secretions of the kidneys and nasal glands are given in Table 4.2.

The size of the nasal salt glands in birds is related to their feeding habits. If young gulls, *Larus glaucescens,* or ducklings are regularly given salt solutions to drink, the nasal glands hypertrophy compared to those birds provided with fresh water to drink (HOLMES, PHILLIPS, and BUTLER, 1961; SCHMIDT-NIELSEN and KIM, 1964). Birds living in marine environments have larger nasal glands than those species which habitually obtain fresh water to drink (TECHNAU, 1936). STAALAND (1967) compared the sizes of the nasal glands in a number of charadriiforme birds. He found that in species that rarely live in marine situations, like the snipe, *Capella*

Table 4.2 Concentrations of Na^+ and Cl^- in the urine and nasal gland secretions of birds

(i) Urine — Chloride*

			m-equiv/l
Columba livia	(Pigeon)	Melopsittacus undulatus	(Budgerygah) 445
Zenaidura macroura	(Mourning dove)	Amphispiza bilineata	(Black-throated sparrow) 527
Carpodacus mexicanus	(House finch)	Passerculus sandwichensis	(Savannah sparrow) 960
Salpinctes obsoletus	(Rock wren)		

Left values:
		m-equiv/l
Columba livia	(Pigeon)	298
Zenaidura macroura	(Mourning dove)	327
Carpodacus mexicanus	(House finch)	370
Salpinctes obsoletus	(Rock wren)	372

(ii) Nasal gland secretion

		m-equiv/l Na^+ **			m-equiv/l Cl^- ***
Phalacrocorax sp.	(Cormorant)	600	Tringa ochropus	(Green sandpiper)	328
Anas platyrhynchos	(Mallard)	550	Gallinago gallinago	(Snipe)	378
Rhynchops nigra	(Black skimmer)	700	Tringa glareola	(Wood sandpiper)	477
Pelecanus occidentalis	(Brown Pelican)	750	Charadrius hiaticula	(Ringed plover)	712
Larus argentatus	(Herring gull)	800	Cepphus grylle	(Black guillemot)	716
Spheniscus humboldti	(Humboldt's penguin)	850	Larus ridibundus	(Black-headed gull)	792
Cepphus grylle	(Black guillemot)	850	Plautus alle	(Little auk)	828
	(Albatross)	900	Larus tridactylus	(Kittiwake)	857
Oceanodroma leucorhoa	(Leach's petrel)	1100	Fratercula arctica	(Common puffin)	887

* from SMYTH and BARTHOLOMEW (1966), except Budgerygah (GREENWALD et al., 1967)
** from SCHMIDT-NIELSEN (1960)
*** from STAALAND (1967)

122

gallinago, and the green sandpiper, *Tringa glareola*, that nasal glands weigh 0.1 to 0.3 mg per g body weight while in marine species, such as the herring gull, *Larus argentatus*, and the little auk, *Plautus alle*, they weigh about 1 mg per g BW. These differences probably reflect the level of extrarenal salt excretion which the birds normally experience. Such marine species also have an enhanced ability to secrete salt solutions from the nasal glands. These differences in nasal gland morphology and function depend on both the genetic and dietary habits of the birds. Marine birds possess larger nasal glands than terrestrial-freshwater species, irrespective of the salt content of the diet, but the development of these glands is enhanced by a high salt diet (SCHMIDT-NIELSEN and KIM, 1964).

The *adrenocortical hormones* can influence sodium excretion by the kidneys and nasal glands, though their exact physiological role is not clear. When birds, such as the domestic fowl or pigeon, are adrenalectomized they only survive for two or three days, but this period can be extended by providing them with salt solutions to drink, or injections of corticosteroids (PARKINS, 1931; MILLER and RIDDLE, 1942). Ducks adrenalectomized by PHILLIPS, HOLMES, and BUTLER (1961) suffered a decline in plasma sodium concentration but, in contrast to mammals, no change in potassium levels. When these ducks were given saline solutions *per os,* the sodium excretion and urine volumes were greater than in normal birds, or in adrenalectomized ones which had been maintained with cortisol injections. This suggests that corticosteroids enhance reabsorption of sodium from the avian renal tubules, just as in mammals. Injection of corticosterone or cortisol into intact ducks, decreased the renal excretion of sodium and decreased potassium levels in the urine, the latter effect contrasting with its action in mammals.

α) Mechanism(s) Controlling Secretion. The mechanism(s) controlling secretion from the nasal salt gland appear to be primarily nervous in origin, but are facilitated by the action of adrenocortical steroids. KNUT SCHMIDT-NIELSEN and his collaborators (SCHMIDT-NIELSEN, JORGENSEN, and OSAKI, 1958; FÄNGE, SCHMIDT-NIELSEN, and ROBINSON, 1958; SCHMIDT-NIELSEN and FÄNGE 1958) investigated the process of nasal salt secretion in cormorants, herring gulls and brown pelicans, and found that injection or ingestion of hypertonic solutions of sodium chloride evoked secretion of fluid by the gland. Hyperosmotic solutions of sucrose (which lowered the plasma sodium concentration!) also increased secretion, indicating that the primary stimulus is osmotic, rather than an elevated sodium concentration. This has recently also been found to be so in ducks (ASH, 1969). When ducks are given a saline load *per os,* there is an initial diuresis, lasting about one hour, during which the nasal gland starts to secrete (HOLMES, PHILLIPS, and BUTLER, 1961); in ASH's experience the initiation varies from 7 to 25 min.

The specific physiological events that link the osmotic stimulus with the initiation of secretion involve a nervous reflex. SCHMIDT-NIELSEN's group found that when hyperosmotic solutions were administered to the herring gull, the commencement of secretion from the nasal gland could be prevented if the birds were anaesthetized, suggesting that the osmoreceptors are located in the central nervous system. It was also found that stimulation of the parasympathetic nerve supply to the nasal glands, resulted in secretion of fluid within a minute and this reached maximal levels in 5 to 10 min. The volume and composition of these fluids were similar to those following administration of hyperosmotic solutions. The effects of either nerve stimulation or hyperosmotic solutions, could be blocked by injecting atropine which is a peripheral parasympathetic blocking agent. Secretion could also be induced with parasympathomimetic drugs such as acetylcholine. Ash, Pearce, and Silver (1969) have recently extended these observations in ducks. Stimulation of nasal gland secretion by infusion of hyperosmotic saline solutions was abolished when the nerve supply to the gland was excised. This clearly indicates that adrenocorticosteroids are not directly involved in the initiation of secretion. Furthermore the injection of carbachol can initiate secretion in such ducks. If the nerve is cut on the central side of its ganglion, or if transmission through this tissue is blocked by hexamethonium the response could still be initiated. This suggests that the receptors are not located in the central nervous system but possibly within the ganglion itself. All of these observations clearly indicate that the response is directly initiated by the parasympathetic nerves, probably through a reflex initiated at an osmoreceptor site somewhere in the nerve tract proximal to the gland.

The overall control of nasal gland secretion is probably more complex than this, and appears to involve also the adrenocorticosteroid hormones. Phillips *et al.* (1961) found that when ducks were adrenalectomized, initiation of nasal salt gland secretion in response to administration of saline solutions was abolished. However, if the birds were maintained with daily injections of cortisol they responded in the usual way. Adrenalectomy is a traumatic procedure that adversely affects a multitude of body functions, but it was also found that injections of cortisol, aldosterone and corticotrophin into normal ducks *potentiated* salt gland secretion (Holmes, Phillips, and Butler, 1961). Cortisol is not normally present in ducks and the actions of aldosterone are probably an indirect result of elevated plasma sodium levels due to a prominent renal action (Phillips and Bellamy, 1962). Corticosterone, which occurs naturally in high concentrations in duck plasma, also stimulates nasal salt gland secretion in saline-loaded ducks (Holmes, Phillips, and Chester Jones, 1963), but the levels of this steroid in the blood do not change when the birds are given such saline loads (Donaldson and Holmes, 1956; Macchi et al., 1965). This latter observation is difficult to reconcile with a primary excitatory action of these steroids on nasal gland secretion. When saline solution is given to ducks the extracellular (inulin) space of the nasal glands doubles, and it has been suggested that this may afford the tissue an added opportunity to react with the circulating corticosteroids (Bellamy and Phillips, 1966).

The blood flow through the nasal glands of ducks and gulls increases during secretion; this probably accounts for the increased inulin space referred to above (Fänge, Krog, and Reite, 1963). This effect is mediated partly by the parasympathetic stimulation, and possibly also by local vasodilators like carbon dioxide and kinin-like substances. Fänge *et al.* have shown that this vasodilatation is *secondary* to the onset of secretion. However, it may play an important part in the process, as Schmidt-Nielsen's group found that adrenaline, which has a vasoconstrictor action, reduced the response of the nasal gland. It is also interesting that when the rates of nasal gland secretion are reduced in herring gulls following deprivation of water, secretion can be increased by the injection of adrenergic blocking agents that promote vasodilatation (Douglas and Neely, 1969). It is possible that increased access of corticosteroids, during such vasodilatation, may play a role in maintaining the integrity of the processes of sodium transport. Prolactin may also influence the secretion from the avian salt gland since injections of ovine prolactin have been found to facilitate the flow of this secretion in ducks (PEAKER, PHILLIPS, and WRIGHT, 1970). It has been suggested that the release of this hormone may be associated with the migration of these birds from fresh water to marine estuaries where it may facilitate the salt gland activity when the birds commence to drink sea-water.

In summary, the process of salt secretion by the avian nasal glands would appear to be initiated by stimulation of osmoreceptors associated with the nervous system, and nerve impulses pass from these to the parasympathetic nerves supplying the gland. Secretion follows, and this leads to a secondary vasodilatation in the gland, which results in the admission of additional corticosterone to the secretory tissue. This steroid may be necessary for the optimal function of the sodium transport mechanisms involved, especially if they are to work at a high rate.

4. Adrenocortical Hormones and Salt Metabolism

A variety of corticosteroids have been isolated from the blood of birds, and from the media incubating slices of their adrenocortical tissues (*in vitro*). The levels of some of these steroids are very low so that there is doubt as to their exact identity. The principal steroids that have been isolated, and which probably play a physiological role, are corticosterone and aldosterone (Table 4.3). Aldosterone can be produced during corticosterone synthesis, so that there are even some doubts as to its hormonal role, especially as it has only been identified *in vivo* in chicken blood and at low concentrations. Corticosterone is the principle corticosteroid secreted *in vivo*.

Corticosterone is produced at increased rates in the presence of mammalian corticotrophin, both *in vitro* and *in vivo*. The avian hormone does not appear to have been tested for such an action. Nevertheless, the bird adrenal has a substantial degree of autonomy from the adenohypophysis. Indeed, at one time it appeared that it may be almost completely independent of the pituitary, as injections of corticotrophin, or hypophysectomy, often failed to produce changes in the weight of the adrenals, or the levels of cholesterol and ascorbic acid present (see for instance ZARROW and BALDINI, 1952; NEWCOMER, 1959). This contrasts with observations in mammals. These criteria of avian adrenocortical function are, however, somewhat misleading; the weight of the duck adrenal is unchanged after adenohypophysectomy but histological examination shows an atrophy of the inner parts of the gland. (WRIGHT, PHILLIPS, and HUANG, 1966). When chickens are hypophysectomized, the circulating levels of corticosterone decrease by 40%, while injection of corticotrophin (mammalian) elevates the levels of the steroid in the plasma (NAGRA *et al.*, 1963). Corticotrophin also has this effect in pheasants, and indirect evidence indicates that it has the same role in ducks (HOLMES, PHILLIPS, and BUTLER, 1961), in which it also increases the weight of the adrenal. When adrenocortical tissue slices from chickens, ducks, pigeons and western gulls are exposed to corticotrophin *in vitro*, there is an increased rate of corticosterone production (DE-ROOS, 1961) but aldosterone formation is little affected. In duck adrenal tissues,

Table 4.3 *Principal adrenocorticosteroids in birds*

	Corticosterone (i) adrenal venous blood	Aldosterone	Other components
Galliformes			
Domestic fowl[1]	+	+	cortisol (?)
Phasianus colchicus[2] (Ring-necked pheasant)	+		
Domestic turkey[2]	+		
	(ii) *in vitro* tissue incubation		
Columbiformes			
Columba livia[3] (Pigeon)	+	+	present
Anseriformes			
Anas platyrhynchos[3,4] (Mallard duck)	+*	+	18-hydroxy-corticosterone
Charadriiformes			
Larus occidentalis[3] (Western gull)	+	+	11-dehydroxy-corticosterone
Passeriformes			
Pipilo fuscus[3] (Brown towhee)	+	+	

* also *in vivo*[4]

References: [1]PHILLIPS and CHESTER JONES (1957)
[2]NAGRA, BAUM, and MEYER (1960)
[3]DEROOS (1961)
[4]DONALDSON and HOLMES (1965)

corticotrophin increases corticosterone production 164 times, but aldosterone causes only a 6-fold increase (DONALDSON, HOLMES, and STACHENKO, 1965). In mammals the secretion of aldosterone is largely related to activation of the renin-angiotensin system, but mammalian angiotensin II fails to alter aldosterone secretion from chicken adrenals *in vitro* (DEROOS and DEROOS, 1963). However, avian angiotensin behaves in a different pharmacological way to the mammalian peptide so that the insensitivity probably reflects structural differences between the peptides. The existence of a renin-angiotensin system has been shown in chickens and this is stimulated by sodium depletion (TAYLOR *et al.*, 1970). However, neither this treatment nor the infusion of kidney extracts (renin) from such birds altered the aldosterone or corticosterone levels in the blood. Thus regulation of secretion of corticosteroids and their role in mineral metabolism of birds is still not understood.

The site of the adrenal glands in different avian species has been related to their habitat; birds that normally frequent marine situations have larger adrenals than those from areas where they have fresh water to drink (HOLMES, BUTLER, and PHIL-LIPS, 1961). These differences are thought to reflect the sodium content of the diet. Young gulls given sodium chloride solutions to drink develop larger adrenal glands than gulls provided with fresh water. HOLMES and his collaborators have suggested that this may reflect a greater need for corticosteroids, in order to facilitate the excretion of salt from the nasal glands of such birds. The young gulls provided with saline solutions to drink grew less rapidly than the birds with fresh water. The same has also been observed in ducklings on such a regimen (SCHMIDT-NIELSEN and KIM, 1964). Administration of corticosteroids to young rats inhibits their growth (see HARTMAN and BROWNELL, 1949), so that the observations on birds suggest an enhanced rate of secretion of corticosteroids. The circulating steroid levels in such young birds, however, have not been determined. As we have seen, adult ducks provided with saline solutions to drink have corticosterone levels in their plasma similar to those of birds given fresh water (DONALDSON and HOLMES, 1965). The adrenal tissues from these two groups of ducks also produce similar amounts of corticosterone and aldosterone *in vitro*, but as the birds given saline have additional tissue, they potentially can produce more of the hormones.

The physiological significance of the relationships of the weight of the adrenal glands to the salt content of the diet, and the role of corticosteroids in the electrolyte metabolism in birds are not clear. In mammals, corticosteroids mediate sodium retention and conservation, but in birds they have two postulated roles in sodium metabolism that are opposite in effect, retention by the kidneys and excretion by the nasal glands. Numerous species of birds lack functional nasal glands and so presumably only utilize corticosteroids to mediate renal sodium conservation. Others, such as marine species, could use the hormones mainly to facilitate sodium excretion through the nasal glands. Yet others (like ducks) conceivably could require either process, depending on the dietary circumstances.

The role of the adrenocortical hormones in sodium metabolism of birds thus appears to be in a somewhat schizophrenic state as they may produce retention or excretion of sodium. One rationalization of this paradox would be that these hormones have a different role in marine sodium-replete species with nasal salt glands, from that in terrestrial species which potentially suffer sodium depletion. Intuitively I find this difficult to believe, principally because it also requires a complete realignment in the stimuli that inititiate discharge of the steroids. The usual stimulus for corticosteroid release in vertebrates is a decreased sodium level resulting in a reduction of plasma volume. A large sodium intake will have an inhibitory effect on this. However, we must conclude either that the available information has been misinterpreted or a novel and unique mechanism controlling release of corticosteroids exists in at least some birds that utilize it to promote sodium excretion. I would hesitantly make two suggestions. First, that adrenal hypertrophy seen in young sodium-replete birds is the result of excessive secretion of corticotrophin resulting from a depression of the inhibitory action of circulating corticosteroids on the adenohypophysis. Second, that normal basal levels of corticosteroids in the plasma are adequate to maintain the integrity of the secretory mechanisms in the nasal gland, provided there is assistance by a vasodilatation in this tissue.

5. Sources of Water and Salts

a) Food

Salts and water may be obtained from the food, the adequacy of which depends on the diet. A succulent herbivorous diet may supply a bird with all of the water and salts that it requires, and carnivorous or insectivous fare may also contribute substantially to its overall requirements. Many birds subsist on seeds, which when air dried, contain only about 10% preformed water, so that such species may be particularly dependent on drinking water. Water is formed as a result of metabolism of fats, carbohydrates and proteins in the diet so that these may be important components of the water intake. CADE and DYBAS (1962) have calculated that commercial bird seed yields metabolic water equivalent to about 45% of its weight. Birds gain a little more water from metabolism of proteins than mammals, reflecting the ultimate formation of uric acid instead of urea; they produce 0.499 g of water for each gram of protein metabolized compared to 0.396 g in mammals. There is little or no information about the salt balance in wild birds, but presumably herbivorous species are subject to the same deficiencies as mammals that live in regions where the sodium content of the soil is low. Seeds have a very low sodium content compared to the roots, leaves and stems of plants, so that granivorous birds may be expected normally to subsist on much less sodium than other species. The implications of this in the renal and adrenocortical physiology of such birds does not appear to have been explored.

b) Drinking

Most birds depend on drinking water in order to maintain a positive water balance. The amount of fresh water consumed for this purpose is inversely related to the body weight, and in some instances may be equivalent to 100% of the body weight in a day (BARTHOLOMEW and CADE, 1963). Some birds can consume large volumes of water very rapidly; the mourning dove can drink water equivalent to 14% of its body weight in 30 minutes (SMYTH and BARTHOLOMEW, 1966), a procedure that may reduce the risk of being prey at water holes, and also reflects the birds' ability to travel, and survive, for extended periods between drinks.

Birds in the laboratory can often be persuaded to drink brackish-saline solutions or even sea-water. The concentrations of such solutions that are acceptable, and the volumes imbibed, vary a great deal from species to species and from individual to individual. The stimulus to drink such solutions is simply withdrawal of fresh water and a diet of relatively dry food which leave the birds no other choice. Such an alternative may be reasonable or unreasonable, depending on the ability to excrete the excess salt. The frequency with which birds actually drink brackish or sea-water in their natural surroundings is uncertain. Marine birds, especially those that eat invertebrates, must involuntarily take in some sea-water during feeding. Terrestrial birds in some areas, such as salt marshes, have ample supplies of brackish water available, but the extent to which these are utilized is unknown. Even the savannah sparrow, *Passerculus sandwichensis beldingi*, which lives around salt marshes in California, and can be persuaded to drink salt solutions in the laboratory, has 'never' been observed 'drinking under natural conditions ... (and) ... depend(s) to a large extent on the water which condenses on the marsh vegetation from the frequent coastal fogs' (POULSEN and BARTHOLOMEW, 1962). In special circumstances, it is conceivable that birds may utilize their physiological ability to drink saline solutions, and this may be useful in 'osmotic' emergencies, like those that occur in laboratories.

The salt that is taken in with the drinking water may be excreted by either the kidneys or, especially in marine birds, by the nasal salt glands. The ability to gain free water from an ingested saline solution depends on the ability to produce urine or a nasal gland fluid with a higher concentration than that taken by drinking. The maximum concentrations that can be achieved by the kidney and nasal glands vary in different species (Table 4.2). It should be remembered that such concentrations are often attained only in special circumstances, and cannot necessarily be maintained during the secretion of large volumes of fluid. Sea-water usually has a concentration equivalent to a 560 m-equiv/l (3.3%) sodium chloride solution, so that a gain of free water from drinking such solutions is feasible, especially in many species with functional nasal salt glands. Such birds are nearly all marine species. Nasal salt glands have also been found in the ostrich, *Struthio camelus*, and the desert partridge, *Ammoperdix heyi*, though their physiological importance in these birds is not yet clear (SCHMIDT-NIELSEN *et al.*, 1963).

6. Reproduction, Migration and Osmoregulation

Reproduction results in an increased need for water and salts for the formation of eggs, and in many instances for the maintenance of the young. As described later, certain birds from dry areas, such as the budgerygah and zebra finch, do not breed regularly but await a stimulus, or signal, which is associated with rainfall (MARSHALL, 1961). Endocrines play an important role in the initiation and regulation of reproduction, but the manner in which rain can influence this is not clear. The effects may conceivably be indirect and result from an improved nutritional state, due to an increase in the supply of food, or, as shown by MARSHALL and DISNEY (1957) in the diochs, *Quelea quelea*, from Tanzania, by the sight of green grass. Other stimuli could include humidity or an optimal state of body hydration (CADE, TOBIN, and GOLD 1965).

Many birds are relatively self-sufficient when they hatch, and fend for food and water by themselves. Others may first spend a period in the nest where they are tended by adult birds. Food may be collected by the adults and provided directly to the young, but in other instances it is first swallowed and regurgitated in a moistened and partially digested condition. Columbiforme birds, such as pigeons and doves, secrete a special fluid into their crop which they feed to the young. This secretion, called 'pigeons milk', appears to be under the control of the adenohypophysial hormone prolactin, as injections of this stimulate growth and secretion of the special cells concerned in this process (RIDDLE and BRAUCHER, 1931). This hormone, or a related natural analogue, is concerned with the control of secretion of milk in mammals and the osmoregulation in certain fishes. Birds that feed their young with a secretion from their crop, or with regurgitated and partly digested food, are subjected to an additional drain on their supplies of body water and salts.

Birds often migrate long distances to breed in areas where the climate is more equitable, and supplies of food and water are more abundant. The process of migration would appear to place a considerable burden on the water metabolism of birds. YAPP (1956) indeed, considers water loss as the single greatest limiting factor in bird migration. TUCKER (1968) has conservatively calculated that a small bird, the budgerygah, can fly at least 500 km before becoming seriously dehydrated, and it seems likely that larger birds could travel considerably further than this. The ability of some birds to make long transoceanic flights is, nevertheless, quite a remarkable osmoregulatory feat.

The role of physiological factors in initiating the migration of birds has been subject to considerable speculation. The possible roles of the endocrine system have received their share of this attention which is reflected in the following quotation: 'There is little doubt that predisposition to migration is a hormone induced state probably a complex of several hormones' (HÖHN, 1961). Physiological changes that occur prior to migration may include, an activation of the reproductive system, the laying down of reserves of fat and the onset of restless nocturnal activity called *Zugunruhe*. Such changes all could involve hormonal processes but not necessarily as an initiating influence. FARNER (1955) does not consider that gonadal sex hormones play a direct role in avian migration, especially as sexually immature and castrate birds also migrate and injections of sex hormones do not mimic the associated preparatory metabolic changes. MARSHALL (1961), however, felt that, with the available evidence, the gonads could not be excluded unequivocally. Alterations in thyroid function have also been considered (see FARNER, 1955); the Zugunruhe is reminiscent of a hyperthyroid condition but this would not be consistent with the laying down of fat reserves. Migration, and its physiological preparation, is undoubtedly a complex process involving the interaction of many factors. Changes in endocrine function must inevitably occur, but just how directly these are involved is not at present clear.

7. Osmoregulation in Two Desert Birds; the Budgerygah and the Zebra Finch

The budgerygah, *Melopsittacus undulatus*, is a small parrot weighing about 30 g and the zebra finch, *Taeniopygia costanotis*, is a passerine species of about 12 g. Both these species eat seeds and live in the hot arid interior regions of Australia. These areas are shared with marsupials, some of which were discussed in the last chapter. The climate is characterized by a low and unpredictable rainfall combined with high temperatures and low humidities. There are seasonal droughts when little or no rain falls and these sometimes may last for three or four years.

DONALD FARNER and DOMINIC SERVENTY and their collaborators (OKSCHE *et al.*, 1963) reported that zebra finches kept outdoors in captivity in Perth, Western Australia, could survive for extended periods with little or no water to drink. CADE, *et al.*, (1965) found that 8 out of 16 domesticated zebra finches placed on a diet of seeds with no water to drink, survived for more than 90 days when kept in the laboratory at 21°. Budgerygahs also have a remarkable ability to survive without drinking water; a group of five domestic birds kept by CADE and DYBAS (1962) lived at least 38 days at 30° and substantially maintained their body weight. Small birds of this size usually die within 2 or 3 days if deprived of water, so that these two Australian species are particularly interesting in this respect.

As previously described, evaporation is the largest channel for water loss in birds, especially those weighing less than about 40 g. At an equitable 25° this can be equivalent to as much as 35% of the body weight in a day (BARTHOLOMEW and CADE, 1963). How then do these two small birds attain a positive water balance in such circumstances? The laboratory diets of seeds contain about 10% preformed water while that which can be produced by metabolizing the seeds is equivalent to about 45% of their weight. The diet in the field probably yields a similar amount of water, though I have found that budgerygahs in captivity will also eat (with re-lish!) more succulent food, such as lettuce and fruit. These birds in their natural habitat may do the same, especially when drinking water is not available. When provided with a granivorous diet, zebra finches normally drink water equivalent to 25% of their body weight in a day, while in the same time budgerygahs imbibe only 5 or 6% of their weight (CALDER, 1964; CADE et al., 1965; CADE and DYBAS, 1962).

When deprived of such drinking water, budgerygahs and zebra finches can, as we have seen, survive in the laboratory for extended periods of time, apparently deriving sufficient water for their basal sustenance from the seeds they eat. The available knowledge about the water balance in these birds is summarized in Table 4.4. The conclusion from these results is that such birds, even at 20°, experience a net water loss, but it must be emphasized that the values for pulmocutaneous water loss were collected when the animals were breathing *dry* air. When the birds are breathing more humid air, as they do in laboratory cages, the evaporative water loss would be substantially less. It is apparent that these birds when deprived of water, make certain adjustments, as the combined urinary and faecal water losses are reduced. KRAG and SKADHAUGE (1970) have recently made direct measurements of the ureteral urine in budgerygahs under both normal conditions and during de-hydration. The flow and concentration of this urine was similar in both circum-stances, so that the birds are normally in a 'fairly antidiuretic state'. The decreased total excretory water losses previously observed thus are presumably the result of a decreased faecal loss and/or an increased rate of water reabsorption from the cloaca of the dehydrated budgies. The evaporative water loss declines, especially at high ambient temperatures. The reasons for the latter are not clear, but they are not the result of a reduction in the resting level of oxygen consumption although they could reflect differences in activity. The losses of water from the respiratory tract are reduced in relation to the amount of oxygen consumed; this could be the result of a more efficient extraction of oxygen from the inspired air, or a reduction in the temperature of the expired air with a resulting decrease in its content of water vapour. In the budgerygah such savings of evaporative water are greater at ambient temperatures over about 36°, when they result in an impaired ability to regulate the body temperature. These birds, when deprived of water reduce their activity, and so conserve water that would be lost during the added respiratory activity. Both the budgerygah and the zebra finch do not start to utilize evaporation of water

131

for temperature regulation until the ambient temperature is greater than about 35°, while prominent increases are not seen until 40°. These birds are poorly insulated so that heat loss readily occurs by radiation, conduction and convection. TUCKER (1968) has calculated that in flying budgerygahs only 15% of the metabolic heat production is dissipated by evaporation at 20°, while even at 36° just 50% of the heat is lost in this manner.

In their native habitats budgerygahs and zebra finches come at dawn and dusk to drink at water holes and windmill tanks. When carrying out their normal ac-

Table 4.4 *Water metabolism of the Budgerygah and Zebra finch (values are g/24 h)*

	Budgerygah (30 g)	Zebra finch (12 g)
Water consumption; (drinking *ad lib.*)	1.5 to 3	3
Food intake; (seeds)		
a) Water *ad lib.*	4	2.9
b) Water deprivation	4	3
Water derived from food; (metabolic and preformed)	1.84	1.35
Water in excreta; (Faeces and urine)		
a) Water *ad lib.*	0.87	1.28
b) Water deprivation	0.65	0.54
Difference	0.22	0.74
Evaporation; (at rest breathing dry air)		
Water *ad lib.* a) 20°	3.3	2.5
b) 35°	4.2	2.6
Water deprivation; a) 20°	3	1.4
b) 35°	3.8	1.4
Water loss mg/ml O_2 used		
Water deprivation; a) 20°	0.94	0.87
b) 35°	1.59	1.47
Water balance in water deprived birds at 20° breathing *dry* air;		
Total water losses (excreta & evaporation)	3.65	1.94
Gain of water in food	1.84	1.35
Negative water balance	1.81 g/day	0.49 g/day

Values have been obtained or calculated as follows:
Budgerygah — CADE and DYBAS (1962); GREENWALD *et al.* (1967) and TUCKER (1968).
Zebra finch — CALDER (1964); CADE *et al.* (1965).

tivities they appear to drink regularly, the prolonged survival without water in the laboratory being substantially the result of their inactivity, and the lack of necessity to travel far in search of food. Their relatively low rate of water loss must allow them to cover considerable distances and survive the difficult 12 to 24 h period between drinks. This may be of considerable importance, especially when it is remembered that during the hottest time of the year in such arid areas even a man is only expected to survive for about a day without water (ADOLPH, 1947). Their tolerance to dessication has not been reported. The distance which these small birds normally cover in search of food and water is unknown, but TUCKER (1968) has measured the rates of water loss of budgerygahs flying at different ambient temperatures. When flying at 35 km an hour in a wind tunnel at 20° the net water loss (water loss minus gain of metabolic water) of a budgerygah is equivalent to 1.1% of its body weight in an hour, which, if it can withstand a total water loss amounting to 15% of its weight, would allow the bird to fly for 14 h and travel 490 km. At 30° the rate of water loss was about 25% greater. At 36° to 37° the water loss was 3 times as great as at 20°, and the birds could only fly for about 20 min as they became overheated. These very elegant experiments suggest that the budgerygah in its native habitat can travel considerable distances to feeding and watering places each day, but preferably not during the hottest times of the day. In the middle of the day budgerygahs are known to rest in the shady branches of trees, moving about mainly in the morning and late afternoon.

In the laboratory both budgerygahs and zebra finches can be persuaded to drink saline solutions. The latter species can maintain their body weight when drinking 0.2 M sodium chloride solutions while the budgies can tolerate 0.4 M solutions. In natural conditions it is unlikely that these birds resort to saline drinking, but they could do so in peculiar circumstances.

While physiological adjustments play an important role in the day-to-day life of budgerygahs and zebra finches, the most vital adaptation ensuring the survival of the species in arid regions is their breeding behaviour. Day length plays a predominant role in the regulation of reproduction of birds of many species, but other factors including rainfall also influence this (see FARNER and FOLLETT, 1966). Certain African and Australian birds, including the budgerygah and zebra finch, breed irregularly, usually only doing so in response to rain or factors such as food that are associated with rain (see MARSHALL, 1961). The exact nature of the environmental stimulus is unknown but the hypothalamic-neuroendocrine systems is activated (OKSCHE et al., 1963) and this sets in motion the rest of the endocrine and metabolic machinery associated with reproduction. Reproduction is associated with an increased need of food and water for egg production and feeding the young. For birds living in areas where rain and adequate food supplies are unpredictable, such a breeding pattern would be an advantage. Indeed the survival of the species in such areas is difficult to imagine if regular breeding, irrespective of the often adverse conditions, were to occur.

While osmotic conditions and the endocrine system interact to ensure successful reproduction in the budgerygah and zebra finch, the evidence for a role of hormones in the osmoregulation of these birds is largely circumstantial. The anatomy of the hypothalamo-hypophysial neurosecretory system in these two species has been investigated (OKSCHE et al., 1963; KOBAYASHI et al., 1961). Vasotocin has

133

been identified in the budgerygah neurohypophysis (see Table 4.1), and water deprivation for 7 days has no effect on the associated stainable neurosecretory material present (UEMURA, 1964). More severe dehydrating procedures deplete the neurosecretory products in the zebra finch (OKSCHE et al., 1963). It seems likely that the low water content of the excreta which is observed when these birds are deprived of drinking water reflects, at least partly, a renal action of vasotocin in limiting the formation of urine. The failure of water deprivation to deplete neurosecretory material in the neural lobe of the budgerygah is reminiscent of observations in desert rats, which were discussed earlier, and may reflect a high rate of hormone synthesis. Information about neurohypophysial function that would be interesting and relatively easy to obtain includes the effects of vasotocin on urine flow in these birds as well as determinations of the storage levels of this peptide in the neural lobe.

The regulation of salt metabolism in the budgerygah and zebra finch has not been investigated. These birds live in areas where the sodium content of the plants is often low, while seeds, which they normally feed on, also have a low sodium level. The role of the adrenocortical hormones in such processes is potentially important but no information is available about this either.

Chapter 5

The Reptiles

Reptiles are considered phyletically to represent the first truly terrestrial vertebrates. They originated in the early Mesozoic period from an amphibian-like ancestor. In those times they became the predominant tetrapod vertebrates, living
not only on the dry land, but also in the fresh water of lakes and rivers, and in
the sea. Four principal groups of reptiles have persisted to the present day. The
Chelonia (turtles and tortoises) have changed little since their origin in early Triassic times and today are represented by about fifty genera. These reptiles have a
world wide distribution; they are usually aquatic (or more strictly amphibious)
in their habits. Five species live in the sea although they must return to dry land
in order to lay their eggs. A number of Chelonians have adopted a life in arid desert
regions and these include the North American desert tortoise, *Gopherus agassizii*,
and the Mediterranean tortoise, *Testudo graeca*. The Crocodilia (9 genera) have
existed in a relatively unchanged form since they first appeared in the late Triassic
period. They are mostly aquatic, living in the vicinity of fresh water, but at least
one species, *Crocodilus porosus*, ventures into the sea for periods of uncertain duration. The Squamata, numerically the principal contemporary reptiles, consist of
two main groups; the Lacertilia (lizards) and Ophidia (snakes), which originated
in the Jurassic and Cretaceous periods respectively. Today they are each represented by about 300 genera. The lizards have the widest geographic distribution
of the reptiles and are even found on many oceanic islands. The marine iguana of
the Galapagos islands, *Amblyrhynchus cristatus*, spends much of its time feeding
on the algae in the sea. The snakes also have a wide distribution and one family,
the Hydrophiidae (15 genera) lives principally in the sea. Terrestrial snakes and
lizards live in habitats ranging from tropical rain forests to dry desert regions. The
remaining major group is the Rhynchocephalia, a relict branch of the reptiles with
one surviving species, *Sphenodon* (the tuatara), which is now confined to a wet
temperate environment on a few small islands situated off the coast of New Zealand.

   Compared with their amphibian ancestors, reptiles are considered to be relatively independent of the proximity of fresh water. Many species, however, continue to live in the vicinity of lakes and rivers, though, unlike almost all amphibians
they produce an egg that, by virtue of its shell and amnionic membrane, can be
laid on land. Some of the sea snakes, *Pelamis*, however, do not even need to return
to land for this purpose, as they give birth to live young in the sea. As compared
to the amphibians the reptiles have a relatively impermeable skin which limits exchanges of their water and salts with the environment. Many reptiles, especially
those living in arid areas, can form uric acid (instead of urea or ammonia) as the
principal end-product of nitrogen metabolism, so that they require little water for
the renal excretion of catabolic nitrogen. The three aforementioned factors, amniotic egg, relatively impermeable integument and uricotelism, distinguish the reptiles osmotically from their ancestral amphibian forms and must contribute substantially to the cosmopolitan radiation of this group into regions of potential
osmotic hostility.

The geographic dispersion of many terrestrial species across large expanses of the oceans is often difficult to envisage, but this has undoubtedly occurred among the reptiles (DARLINGTON, 1957), a feat that must largely reflect their relative osmotic independence from their environment. Reptiles are thought to have originated in the old world and migrated to the new world, with a subsequent pilgrimage from South America to Australia. DARLINGTON states that these movements can be explained 'without resorting to special land bridges or continental drift' presumably by making transoceanic voyages. Fossil representatives of an extinct group of land tortoises (Meiolaniidae) from South America have been found on Lord Howe Island and Walpole Island in the eastern Pacific Ocean. It seems likely that these reptiles floated across the many hundreds of kilometers of ocean to occupy these remote islands. Turtles have a remarkable propensity to survive such conditions. When at Duke University I placed three turtles, *Pseudemys scripta*, in a tank of sea-water, two of these animals died within 14 days but the third was still alive after 31 days. I replaced it in fresh water after this time, and it proceeded to carry on with its normal existence. Among the snakes, the genus *Natrix* has a remarkably wide geographic distribution and is found in Europe, Asia, Africa, Australia, America and islands such as Cuba and Madagascar. These snakes are primarily aquatic, but have probably attained their cosmopolitan status by crossing the seas. PETTUS (1958) has compared the fluid metabolism of two races of *Natrix sipedon* living in the southern United States. One of these groups normally lives in fresh water and the other in brackish water; but the former cannot usually survive transfer to a saline medium. The success of such adaptation seems to reside solely in a predisposition to refrain from drinking salt solutions. Providing that reptiles maintain the relative impermeability of their integument, dispersal of terrestrial species through the seas is physiologically conceivable.

Reptiles have a water content equivalent to about 70% of their body weight, an amount similar to that of birds and mammals, though less than that of the Amphibia. The concentration of the body fluids conforms to the usual tetrapod pattern, being about 250 to 300 m-osmole/l. Variations in the concentration and electrolyte content of the body fluids do, however, occur; reptiles in a marine environment have a slightly higher plasma concentration than those living on land (Table 5.1) though they remain hypoosmotic to sea-water. Reptiles can tolerate quite large changes in the concentrations of their body fluids; the sodium concentration in the plasma of the lizard, *Trachysaurus rugosus*, may rise from 150 to nearly 200 m-equiv/l during the dry Australian summer (Table 5.1) and comparable increases have also been observed in the lizard, *Amphibolurus ornatus*, and the desert tortoise (BRADSHAW and SHOEMAKER, 1967; DANTZLER and SCHMIDT-NIELSEN, 1966). Decreased plasma electrolyte levels can also be readily tolerated; the softshell turtle, *Trionyx spinifer*, normally has a plasma sodium concentration of about 150 m-equiv/l, but in healthy hibernating individuals this may decrease to 80 m-equiv/l (Table 5.1). Birds and mammals do not seem to be able to withstand such variations in the concentrations of their body fluids, but in some reptiles such tolerance may be an important factor in their ability to survive adverse conditions when the possibility of osmotic regulation is limited.

Reptiles are poikilotherms and so, unlike birds and mammals, do not utilize evaporative water loss for thermal cooling. Nevertheless, they do attempt to

Table 5.1 *Concentrations of sodium and potassium in the plasma of reptiles from various habitats*

| | Habitat | Plasma concentration (m-equiv/l) | |
		Sodium	Potassium
Chelonia			
Chelonia mydas[1] (Green turtle)	Marine		
a) Kept in sea-water		158	1.5
b) Kept in fresh water		130	2.6
Trionyx spinifer[2] (Softshell turtle)	Fresh water		
a) Normal		144	—
b) Hibernating		90	—
Lacertilia			
Trachysaurus rugosus[3] (Bobtail lizard)	Semi-desert		
a) Winter		152	3.5
b) Summer		196	4.6
Uromastyx acanthinurus[4] ('Dob')	Desert (Sahara)	150	4.4
Amblyrhynchus cristatus[5] (Marine iguana)	Marine	178	11
Ophidia			
Natrix cyclopion[6] (Water snake)	Fresh water	134	5.9
Pelamis platurus[7] (Yellow-bellied sea snake)	Marine	264	11

References: [1]HOLMES and McBEAN (1964); [2]DUNSON and WEYMOUTH (1965); [3]BENTLEY (1959b); [4]GRENOT (1968); [5]DUNSON (1969); [6]ELIZONDO and LeBRIE (1969); [7]DUNSON (1968).

maintain their body temperature at levels that are optimal for their activity and discreetly utilize external heat sources for this purpose. This is performed principally by conforming to a certain pattern of behaviour; warming by periodic basking in the sun and opposing the ventral body surface to warm surfaces, and cooling by seeking shade and refuges, such as rock crannies and burrows, where the temperature is more moderate. Modification of the body temperature may be assisted physiologically by changing the conductivity of the body by means of circulatory adjustments. Some species can change the colour of their skin so as to alter the rate of absorption of solar radiation (see SCHMIDT-NIELSEN, 1964a). Such adjustments are probably mediated principally by the nervous system, though in some reptiles melanocyte stimulating hormone mediates colour change (WARING, 1963). Evaporation of water from the respiratory tract and skin increases when the temperature of the body and the environment rises, but this is physically obligatory and cannot be considered as part of a regulatory response to facilitate cooling.

The principal avenues for the exchange of water and solutes between the reptile and its environment are basically the same as in mammals and birds. The physiological significance and magnitude of the exchanges through the different channels however, differ. The main osmoregulatory organs are the kidneys, as in all vertebrates, but in certain reptiles the action of these organs may be augmented by cephalic 'salt' glands, and the cloaca and urinary bladder.

1. Water Exchanges

a) Skin and Respiratory Tract

In a terrestrial environment reptiles lose water by evaporation from the external integument, and the lungs and pulmonary passages. This loss increases when body temperature and environmental temperature increase; thus the bobtail goanna, *Trachysaurus rugosus*, when kept in dry air at 25°, loses water by evaporation at the rate of 0.7 g/100 g day while at 37.5° it is 2.9 g/100 g day (WARBURG, 1965a). This water loss is small, relative to that in birds, mammals and the Amphibia, and makes little contribution to thermal cooling; the temperature of the goannas when equilibrated to such conditions only being one or two degrees less than that of the environment.

The increased evaporation of water at high environmental temperatures is due to the decrease in the saturation deficit for water in the surrounding air, and an increase in the body temperature of the animal. The metabolic rate of reptiles increases two to three times for every 10° rise in body temperature (BENEDICT, 1932), so that there is an increased rate of gas exchange in the lungs that further facilitates water loss.

Different species of reptiles have been observed to lose water by evaporation at various rates (BOGERT and COWLES, 1947). The reasons for such differences were not initially clear as they could involve several factors such as: the rate of oxygen consumption, the ratio of surface area to body weight or the particular properties of the skin of the different species. This was not further investigated until SCHMIDT-NIELSEN in 1964 (a) suggested, after examining the fragmentary information that was then available, that cutaneous evaporation in reptiles may be greater and more variable than was hitherto considered likely. When cutaneous and pulmonary water losses were measured separately in a variety of species of reptiles (at 23°) the movement through the integument was indeed found to contribute from 66 to 87% of the total evaporative loss (see Table 5.2; BENTLEY and SCHMIDT-NIELSEN, 1966; SCHMIDT-NIELSEN and BENTLEY, 1966). At higher temperatures (35° and 40°) pulmonary water loss was increased relative to that from the skin but was still about 50% of the total. Similar results have been described in a variety of lizards (see CLAUSSEN, 1967). The rates of water loss across the skin vary considerably in different reptile species, in a manner that suggests that animals normally occupying

Table 5.2   *Evaporative water loss from the respiratory tract and skin of various reptiles in dry air at 23°*

	Respiratory H_2O Loss total mg/g day	mg/ml O_2	Cutaneous H_2O Loss mg/cm² day	as % total	O_2 Consumption ml/g day	Habitat
Crocodilia						
Caiman sclerops (Caiman)	9.6	4.9	33	87	1.8	Aquatic
Chelonia						
Pseudemys scripta (Slider turtle)	4.3	4.7	12	78	0.9	Aquatic
Terrapene carolina (Box turtle)	2.6	4.2	5.3	76	0.6	Temperate forest
Gopherus agassizii (Desert tortoise)	0.4	1.5	1.5	76	0.26	Desert
Lacertilia						
Iguana iguana	3.4	0.9	4.8	72	2.6	Tropical forest
Sauromalus obesus (Chuckawalla)	1.1	0.5	1.3	66	1.2	Desert
Ophidia[1]						
Natrix taxispilota (Brown water snake)	3.6	1.6	16.7	88	2.1	Aquatic
Pituophis catenifer (Sonora gopher snake)	3.1	1.3	3.7	64	2.3	Desert

References: [1]from PRANGE and SCHMIDT-NIELSEN (1969) others from BENTLEY and SCHMIDT-NIELSEN (1966) and SCHMIDT-NIELSEN and BENTLEY (1966)

Table 5.3 *Cutaneous water exchange in reptiles kept in fresh water or 3.3% sodium chloride solution (= sea-water)*

	mg/cm² day	
	fresh water (gain)	3.3 % NaCl (loss)
Crocodilia		
Caiman sclerops[1] (Caiman)	26	14
Chelonia		
Trionyx spinifer[2] (Softshell turtle)	26	38
Pseudemys scripta[2] (Slider turtle)	6	10
Lacertilia		
Sauromalus obesus[2] (Chuckawalla)	Not detectable	

[1]BENTLEY and SCHMIDT-NIELSEN (1965)
[2]BENTLEY and SCHMIDT-NIELSEN (1970)

dry habitats suffer a smaller loss in this way than those from wetter situations. Thus, the aquatic crocodilian, *Caiman sclerops*, loses water by evaporation through the skin 25 times as rapidly as the desert lizard, *Sauromalus obesus*.

The skins of various species of reptiles also exhibit differences in their osmotic permeability to water. In fresh water, which is hypoosmotic to its body fluids, the crocodilian, *Caiman sclerops*, takes up water across its integument at the rate of 26 mg/cm² day (Table 5.3) which is 20 to 50% the rate of osmotic water uptake across the skin of amphibians. The softshell turtle, *Trionyx spinifer*, takes up water osmotically at a similar rate, but the slider turtle, *Pseudemys scripta*, takes up water only about 20% as rapidly. The softshell turtle is covered with a soft pliable skin while about two-thirds of the surface of *Pseudemys* consists of hard keratin plates which probably have a more restricted permeability to water. When these reptiles were kept in hyperosmotic sodium chloride solution, similar in concentration to sea-water, they lost water (Table 5.3). The desert lizard, *Sauromalus*, has a very restricted permeability with respect to evaporative loss and we could not detect any movement across the skin of these animals when they were bathed for a day in fresh water or 3.3% sodium chloride solution. TERCAFS (1963) has, however, shown that the skin of the lizard, *Uromastyx*, is osmotically permeable to water *in vitro*, an observation that is not comparable with our *in vivo* experiments. The skin of many Amphibia exhibits an increased osmotic permeability to water in the presence of neurohypophysial hormones but there is no evidence that suggests that this occurs in reptiles.

140

Water loss from the respiratory tract depends on the rate of oxygen consumption and the ability to extract this gas from the inspired air. The oxygen consumption of reptiles depends on their activity and body temperature. In addition, the basal metabolic rate seems to vary considerably in different reptiles. An inspection of the rates of oxygen consumption of the different reptiles in Table 5.2 indicates a range from 0.26 ml/g day in the desert tortoise to 2.6 ml/g day in the iguana. Such differences are partly reflected in the respective rates of water loss from the respiratory tract of such animals. The desert tortoise thus loses water through this channel at the rate of 0.4 mg/g day while the iguana loses 3.4 mg/g day. The relationship between oxygen consumption and respiratory water loss is, however, not an exact one as there is considerable variation in the ability of the different animals to extract oxygen from the inspired air. Usually this corresponds to a reduction of oxygen from the atmospheric volume of 20.9% to only 17 to 20%. The chuckawalla, *Sauromalus obesus,* can on occasion breathe sporadically at the rate of only 2 or 3 breaths an hour and reduce the oxygen concentration in its lungs to 5% of the volume in the atmosphere (SCHMIDT-NIELSEN, CRAWFORD, and BENTLEY, 1966). Such behaviour would reduce respiratory water loss, but whether this occurs normally as a response to promote water conservation is uncertain. The thyroid hormones control the rate of oxygen consumption in birds and mammals but this does not normally seem to be so in reptiles, though in certain circumstances, as when the temperature is elevated, they may be involved (LYNN, McCORMICK, and GREGOREK, 1965).

Evaporative water losses in reptiles are probably temporally regulated, in conjunction with the body temperature, by changes in behaviour. Reptiles may be diurnal or nocturnal in their habits and avoid extreme temperatures by seeking the shelter of cool refuges. WARBURG (1965a) found goannas, *Trachysaurus rugosus,* in burrows 2 to 3 metres deep. The temperature in these holes never exceeded 28°, even though the external air temperature was as high as 40° and that of the surface soil 50°.

b) Urinary and Faecal Water Loss

Like in birds, it is difficult to distinguish clearly between urinary and faecal water losses in reptiles, as both products pass through a cloaca prior to their expulsion. The proportion of the animals' total water loss that is lost through such channels varies, and in terrestrial situations this is partly dependent on the temperature, which influences the rate of evaporation more than other water losses. At a temperature of 23° (in dry air) the crocodilian, *Caiman sclerops,* loses about 20% of its daily total water deficit through the cloaca (BENTLEY and SCHMIDT-NIELSEN, 1965 and 1966). The lizard, *Trachysaurus,* which lives in semi-arid areas, loses 40% of its overall water loss in this way at 23° but only about 25% at 35°. At 30° *Uta hesperis,* a lacertilian from southern California, only loses 5% of its total daily water decrement in the urine and faeces (CLAUSSEN, 1967). While in aquatic situations evaporation is expected to be small, a net water loss may still occur osmotically across the skin if the environmental fluid is hyperosmotic, like the sea. Additional water may also be excreted through the cephalic 'salt' glands in such species. When the estuarine turtle, *Malaclemys centrata,* is kept in sea-water, it loses about equal

141

amounts of water through cloacal and extrarenal channels (BENTLEY, BRETZ, and SCHMIDT-NIELSEN, 1967a).

The normal rates of urine flow differ considerably in different reptiles (Table 5.4) and can vary from amounts equivalent to about 0.5% of the body weight in *Malaclemys* kept in sea-water, to 6% in the gecko lizard and 9% in the caiman when kept in air. Water loss from the kidney and gut can constitute an avenue of water loss, which in circumstances when such supplies are restricted, may be physiologically critical. Reptiles, unlike mammals and birds, cannot form a urine that is hyperosmotic to their body fluids and so cannot conserve water by increasing urine concentration in this way. However, some do have the ability to convert increased amounts of their waste nitrogen into uric acid, which requires less water for its renal excretion than the alternative products, ammonia and urea. The desert tortoise,

Table 5.4 *Variations in the urine flow of reptiles in different states of hydration*

	Normal	Dehydrated	Water loaded	Saline loaded	Habitat
		ml/kg hr			
Crocodilia					
Caiman sclerops[1]					Aquatic
a) In fresh water	3.5				
b) In air	1.1				
Chelonia					
Pseudemys scripta[2]	1.3	Aprox. Anuric	3.6	0.4 to Anuric	Aquatic
(Slider turtle)					
Malaclemys centrata[3]					Estuarine
(Diamondback terrapin)					
a) In fresh water	1.0				
b) In sea-water	0.2				
Gopherus agassizii[2]	2.0	Aprox. Anuric	8.3	0.9 to Anuric	Desert
(Desert tortoise)					
Lacertilia					
Hemidactylus sp.[4]	2.6	1.3	12.1	2.4	Desert
(Gecko)					
Phrynosoma cornutum[4]	2.0	0.8	1.8	0.6	Wet tropical
(Horned toad)					
Trachysaurus rugosus[5]	0.24	Anuric	11.3	Anuric	Semi-dessert
(Bobtail lizard)					
Ophidia					
Natrix sipedon[6]	Varies with vasotocin injections				Aquatic
(Water snake)					

[1]BENTLEY and SCHMIDT-NIELSEN (1965); [2]DANTZLER and SCHMIDT-NIELSEN (1966); [3]BENTLEY *et al.* (1967); [4]ROBERTS and SCHMIDT-NIELSEN (1966); [5]BENTLEY (1959 b); [6]DANTZLER (1967 a).

Table 5.5 *Glomerular filtration rate of various reptiles in different states of hydration*

	Normal	GFR ml/kg h Saline loaded	Water loaded	Dehydrated	Habitat
Crocodilia					
Crocodilus acutus[1]	9.6	6.1	13.0	7.3	Fresh water
Crocodilus porosus[2] (Salt water crocodile)	1.5	—	18.8	2.8	Sometimes marine
Chelonia					
Pseudemys scripta[3] (Slider turtle)	4.7	—	10.3	2.8	Fresh water
Chelonia mydas[2] (Green turtle)	14.3	—	—	—	Marine
Gopherus agassizii[3] (Desert tortoise)	4.7	—	15.1	2.9	Desert
Lacertilia					
Hemidactylus sp.[4] (Gecko)	10.4	3.3	24.3	11.0	Wet tropical
Phrynosoma cornutum (Horned toad)	3.5	2.1	5.5	1.7	Desert
Trachysaurus rugosus[5] (Bobtail lizard)	1	1	3	Zero	Semi-Desert

[1]B. SCHMIDT-NIELSEN and SKADHAUGE (1967); [2]B. SCHMIDT-NIELSEN and DAVIS (1968); [3]DANTZLER and SCHMIDT-NIELSEN (1966); [4]ROBERTS and SCHMIDT-NIELSEN (1966); [5]Calculated from SHOEMAKER *et al.* (1966) and BENTLEY (1959 b).

Gopherus agassizii, normally excretes relatively more waste nitrogen as uric acid than the aquatic turtle, *Pseudemys scripta* (DANTZLER and SCHMIDT-NIELSEN, 1966). Some tortoises also may alter the relative quantities of these nitrogenous end-products in response to the state of the reptiles' hydration (KHALIL and HAGGAG, 1955), uric acid predominating when water is restricted.

The urine flow of reptiles exhibits considerable lability (Table 5.4). When water is administered to the lizard, *Trachysaurus rugosus,* its urine flow may increase nearly 50 times as compared to that normally observed, while if the water intake is restricted, the urine volume may be too small to measure. In reptiles, variations in urine flow are related to changes in both the GFR (Table 5.5) and reabsorption of water from the renal tubules. This contrasts with mammals, in which the latter process normally predominates. The possible amplitude of the changes in GFR differs in various species and may be only 2 or 3-fold in a lizard such as the horned toad, and more than 10-fold in the crocodile (Table 5.5). As indicated by differences in the inulin or creatine urine/plasma concentration ratios, renal tubular water reabsorption may also vary. In mammals this normally amounts to more than 99% of the water that is filtered across the glomerulus, but in the horned toad this is only 45% in normally hydrated animals though it may increase to 60% during dehydration (ROBERTS and SCHMIDT-NIELSEN, 1966). In the Australian lizard, *Trachysaurus,* tubular water reabsorption is only 40% in hydrated individuals, but it may increase to more than 95% when the urine flow is low (SHOEMAKER, LICHT, and DAWSON, 1966). Thus, the relative importance of the changes in GFR and renal tubular water reabsorption varies in different species of reptiles and is also dependent on the physiological circumstances. At high rates of urine flow, increases in the GFR may play a predominant role in controlling urine volume, but when the urine volume is small, tubular water reabsorption may be relatively more important. Ideally the two processes probably work in conjunction with each other. Changes in GFR may be due to an increased rate in filtration across individual glomeruli, or can also result from changes in the numbers of active units (glomerular intermittency). The latter process has been shown to occur in the snake, *Natrix sipedon* (LeBRIE and SUTHERLAND, 1962; DANTZLER, 1967a) and the turtle, *Pseudemys scripta* (DANTZLER and SCHMIDT-NIELSEN, 1966). The site of the tubular water reabsorption has been examined by 'stop-flow' procedures in *Natrix sipedon* (DANTZLER, 1967b) and the results suggest that during antidiuresis, water is reabsorbed from the distal parts of the nephron.

While it is clear that both changes in the GFR and reabsorption of water by the renal tubule mediate alteration in the urine flow of reptiles, the immediate physiological reasons for such changes are, at the best, speculative. Changes in the GFR can be envisaged as resulting to some extent from dilution or concentration of plasma protein, which could, respectively, increase or decrease the forces for ultrafiltration across the glomerulus. Haemodynamic factors, resulting from changes in the blood pressure, can also conceivably alter glomerular activity. Tubular water reabsorption can be influenced by the rate at which the filtrate is delivered to the absorption sites. Water transfer across the renal tubule may occur as an osmotic accompaniment of solute movement or, as in mammals, result from a change in the properties of the tubular epithelium, allowing osmotic equilibration between the two sides to occur more rapidly. We do not know which of these factors occurs

144

physiologically in reptiles, but injections of peptides from the neurohypophysis can initiate decreases in glomerular activity and increase fluid absorption from the renal tubules.

α) Role of Neurohypophysis. Three peptides with such actions on the kidney have been identified in the neurohypophyses of different reptile species. HELLER (1942) showed that the pituitary of the European grass snake, *Tropidonotus natrix,* contained a biological activity similar to that in mammals in so far as it had an antidiu-

Table 5.6 *Distribution of neurohypophysial peptides in reptiles*

	Vasotocin	Mesotocin	Oxytocin
Crocodilia			
Caiman sclerops[1] (Caiman)	+	—	—
Chelonia			
Testudo graeca[2] (Tortoise)	+	Oxytocin — like	
Chelonia mydas[1] [3] (Green turtle)	+	+ (?)	+ (?)
Pseudemys scripta[1] (Slider turtle)	+	Oxytocin — like	
Lepidochelys kempi[3] (Ridley turtle)	+	+ (?)	+ (?)
Caretta sp.[3] (Loggerhead turtle)	+	+ (?)	+ (?)
Lacertilia			
Iguana iguana[4] (Iguana)	+	Oxytocin — like	
Ophidia			
Tropidonotus natrix[2] [3] (Grass snake)	+	+ (?)	+ (?)
Crotalus atrox[5] (Rattler)	+	+	—
Naja naja[6] [8] (Cobra)	+	+	±
Vipera aspis[7] [8] (Viper)	+	+	—
Elaphe quadrivergata[8] (Elaph)	+	+	—

[1]SAWYER *et al.* (1961); [2]HELLER and PICKERING (1961); [3]FOLLETT (1967); [4]SAWYER (1968); [5]MUNSICK (1966); [6]PICKERING (1967); [7]ACHER, CHAUVET, and CHAUVET (1968); [8]ACHER, CHAUVET, and CHAUVET (1969 a).

retic effect when injected into rats. However, its properties were not identical with those of mammalian neurohypophysial peptides, as it exerted a far more potent 'water balance effect' (water retention) when injected into frogs. The activity in the neurohypophysis of the grass snake was later (HELLER and PICKERING, 1961) shown to be pharmacologically identical with 8-arginine-oxytocin (vasotocin). A second peptide, similar in its properties to mammalian oxytocin, was also found in the grass snake. The caiman neurohypophysis was found to contain vasotocin, but a second activity could not be identified in this species, though both were found in the green turtle, *Chelonia mydas* (SAWYER, MUNSICK, and VAN DYKE, 1961). Vasotocins have since been identified in other groups (Table 5.6), including the snakes, in which their structure has been confirmed by amino acid analysis (PICKERING, 1967; ACHER, CHAUVET, and CHAUVET, 1968). The second peptide activity that was found initially in most of the reptiles examined has been pharmacologically identified as 8-ileu-oxytocin (mesotocin) in the rattlesnake (MUNSICK, 1966) and found chemically to be such in the viper, cobra and elaph (ACHER *et al.*, 1968; ACHER, CHAUVET, and CHAUVET, 1969a). PICKERING (1967) found that amino acid analysis, indicated both the presence of mesotocin and oxytocin in the neurohypophysis of the cobra and FOLLETT (1967) has presented tentative pharmacological evidence to suggest that this may also be so in other reptiles. ACHER *et al.* (1969 a) emphasize that they are unable chemically to detect oxytocin in the viper, cobra or elaph. PICKERING also found an additional peptide, similar to vasotocin, in the cobra, but this has not yet been exhaustively characterized. Few results indicating concentrations of such peptides in reptiles are available, but in the grass snake the level of vasotocin is about $4 - 9 \times 10^{-9}$ M kg body weight (calc. from HELLER and PICKERING, 1961; FOLLETT, 1967), a value comparable with those in other tetrapods. There is usually 2 – 4 times as much vasotocin as the other active peptides.

Mammalian neurohypophysial peptide preparations have been shown to exert an antidiuretic effect when injected into the alligator and into *Trachysaurus rugosus* (Table 5.7). Extracts of reptilian neurohypophyses also exert such an action when injected back into the contributing species. DANTZLER (1967 a) has made a careful analysis of the actions of vasotocin, mesotocin and oxytocin on the renal function of the water snake *Natrix sipedon.* All of these neurohypophysial peptides can produce an antidiuresis, but mesotocin and oxytocin are only effective in large doses that are always associated with a reduced GFR and a decrease in blood pressure. Vasotocin reduces urine flow in small doses that have no action on blood pressure and acts both by reducing the GFR and by increasing tubular water reabsorption. The renal tubular action of vasotocin has a lower dose-response threshold than its glomerular effects. The results suggested that the decreased GFR was due to a reduction in the number of active glomeruli. It is uncertain whether vasotocin has this pattern of action throughout the Reptilia for, as we have seen, there is a considerable diversity in the renal processes of this group. Injection of large amounts of Pitressin into the alligator thus was found to reduce the GFR without affecting tubular absorption but such experiments should be repeated using small doses of vasotocin.

The evidence for a physiological role of neurohypophysial peptides in controlling urine flow in reptiles remains circumstantial, and less complete than in any other tetrapod group. Unfortunately the effects of neurohypophysectomy on urine

146

Table 5.7   *The effects of neurohypophysial peptides on urine flow, the GFR and renal tubular water reabsorption in various reptiles*

		Mechanism		
	Antidiuresis	GFR decr.	Tubular H$_2$O Reabsorption	Peptides Used:
Crocodilia				
Alligator mississippiensis[1,2] (Alligator)	+	+	—	Pitressin, Pitocin
Chelonia				
Pseudemys scripta[3] (Slider turtle)				Neural lobe
a) Small dose	+	—	+	extract from
b) Large dose	+	+	—	*Pseudemys*
Lacertilia				
Trachysaurus rugosus[4] (Bobtail goanna)	+	?	?	Pitocin, Pitressin, extract from lizard
Ophidia				
Natrix sipedon[5, 6] (Water snake)	+	—	+	Vasotocin
	+	+	—	High dose mesotocin or oxytocin

+ = effect
— = no effect
[1]Burgess *et al.* (1933)
[2]Sawyer and Sawyer (1952)
3Dantzler and Schmidt-Nielsen (1966)
[4]Bentley (1959 b)
[5]LeBrie and Sutherland (1962)
[6]Dantzler (1967 a)

flow have not been reported, nor has the presence of such peptides been demonstrated in the circulation.

It is noteworthy that neurohypophysial peptides exert other actions in reptiles. They may reduce the blood pressure as in *Trachysaurus rugosus* (Woolley, 1959) and *Natrix sipedon* (Dantzler, 1967 a) but high doses are required, so that these effects are probably only of pharmacological significance. Vasotocin *(in vitro)* contracts the oviduct of the turtle, *Pseudemys scripta*, when it is in an ovulating condition (Munsick *et al.*, 1960). LaPointe (1969) has shown that vasotocin also contracts the oviduct of the viviparous island night lizard, *Klauberina riversiana*, *in vitro*. Oxytocin and mesotocin were very much less effective than vasotocin. It is possible that this action of vasotocin reflects a physiological role for this peptide in reptilian oviposition and parturition just as oxytocin assists the latter process in mammals.

β) *Role of Cloaca and Urinary Bladder.* Urine does not pass from the reptiles kidney directly to the outside of the animal, but is first accumulated in a cloaca and in some species may pass from here into a urinary bladder. A urinary bladder is absent in Crocodilia, Ophidia and some Lacertilia, but is present in many of the latter, as well as in the Chelonia. As the urinary and faecal pellets in many reptiles contain little water, its reabsorption presumably occurs across the cloaca and large intestine, as in birds.

The Brazilian snake, *Xenodon sp.*, suffers an increased rate of water loss when the openings of its ureters are cannulated so as to allow water to by-pass the cloaca (JUNQUEIRA *et al.*, 1966). The increased fluid loss was equivalent to 2.6% of the body weight in a day, and this may represent the amount which is normally reabsorbed in the region of the cloaca and large intestine of these snakes.

Direct measurements of fluid (water and sodium) absorption from the cloaca of a large Australian lizard, *Varanus gouldii*, have been made (BRAYSHER and GREEN, 1970). The coprodaeum of this lizard was isolated *in vivo* from the rest of the cloaca with the aid of rubber bulbs. When isotonic saline was introduced into this segment it was absorbed at the rate of 8 ml/kg of the lizards body weight each hour. This was increased to 21 ml/kg h following the injection of vasotocin. Sodium absorption was facilitated, but the relative roles of this and differences in colloid osmotic pressure (MURRISH and SCHMIDT-NIELSEN, 1970) in promoting the water movement are not clear. The large intestine of the tortoise, *Testudo graeca*, is osmotically permeable to water *in vitro*, but high concentrations of vasotocin have no effect on water movement (BENTLEY, 1962 b). The urinary bladder of chelonians is also permeable to water and in *Pseudemys scripta* water movement may take place both along an osmotic gradient and as an accompaniment to active sodium transport from the mucosal to serosal surface (BRODSKY and SCHILB, 1965). The permeability of the bladder to water is, in contrast to frogs and toads, unaffected by vasotocin (CANDIA, GONZALES, GENTILE, and BENTLEY, unpublished observations). The urinary bladder of the terrestrial tortoise, *Testudo graeca*, is also osmotically permeable *in vitro*, and is also unaffected by vasotocin (BENTLEY, 1962 b). DANTZLER and SCHMIDT-NIELSEN (1966) examined the permeability of the bladder in the desert tortoise, *Gopherus agassizii*, to water (*in vivo*), and found that when water was introduced into the bladder, it was reabsorbed at the rate of about 20 ml/h in either hydrated or non-hydrated animals. The urinary bladder of turtles may hold large amounts of fluid; I have observed volumes equivalent to 20% of the body weight in the bladder of the diamondback terrapin, *Malaclemys centrata*. Such stores of water could facilitate survival in circumstances in which water supplies are restricted. As reptiles only lose water at a slow rate. rapid increases in the osmotic permeability of the bladder, like those mediated by vasotocin in frogs and toads, are probably unnecessary, so that the absence of such an action of this hormone is not surprising. Many lizards also store fluid in a urinary bladder and an investigation of the functioning of this organ, particularly in species from desert areas, would be interesting.

2. Salt Exchange

a) Skin

The relative impermeability of the integument considerably restricts the movements of salts across the body surface of reptiles. However, precise measurements of the movements of ions like sodium and potassium are lacking. When the caiman is placed in distilled water it loses sodium across its skin, the net loss amounting to about 1 μ-equiv/cm^{2}h (BENTLEY and SCHMIDT-NIELSEN, 1965). This is about half the rate of loss in frogs and salamanders kept under similar conditions. Unlike the skin of the amphibians, no electrical p.d. could be recorded across the caiman integument *in vitro* when it was bathed on both sides with physiological saline, though a small p.d. (4 mV, outside negative) was seen when the outside media was diluted. The latter can be accounted for by different rates of diffusion of sodium and chloride, an observation that at least confirms the permeability of the skin to ions. When the caiman is placed in 3.3% sodium chloride solution (= to sea-water) it gains large amounts of sodium chloride but this occurs as a result of drinking, little measurable uptake occurring across the skin. I have also measured cutaneous sodium loss in the softshell turtle when it is bathed in distilled water and it loses this ion at about the same rate as the caiman. When the diamondback terrapin, *Malaclemys centrata*, is kept in sea-water, circumstantial evidence obtained from sodium levels in the blood and urine indicates that a small accumulation of sodium is taking place through the skin (BENTLEY, *et al.*, 1967a). In normal physiological circumstances cutaneous exchanges of solutes are probably only of minor significance, but as apparent from the above experiments, the information is fragmentary. Species like the caiman and softshell turtle have a more permeable integument than most reptiles so they must be considered atypical, while the conditions (bathed in distilled water) for such measurements are hardly physiological. By utilizing labelled isotopes of sodium and potassium it should be possible to measure movements of electrolytes across the skin of a wider range of reptiles, under conditions that can be more properly considered physiological.

b) Kidney

Excretion and conservation of electrolytes occur during the process of urine formation. If excess water is accumulated, as a result of feeding, drinking or cutaneous uptake, it will be excreted by the kidney with a simultaneous, and unavoidable, loss of some salt. The ability of reptiles to form a hypoosmotic urine by reabsorption of such salts from the glomerular filtrate minimizes such losses. When the diamondback terrapin is fasted and maintained in fresh water, the urine concentration is about 60 m-osmole/l, while that of the sodium present is less than 1 m-equiv/l (BENTLEY *et al.*, 1967a). I have found the concentration of the urine of the softshell turtle, kept under the same conditions, to be similar, while *Pseudemys scripta* forms a urine with a higher sodium (about 5 m-equiv/l) level. Terrestrial reptiles do not appear to perform as well as these aquatic species. I have observed sodium concentrations as low as 15 m-equiv/l in the lizard, *Trachysaurus rugosus*, but ROBERTS and SCHMIDT-NIELSEN (1965) found that in three species of lizards (gecko, horned toad and Galapagos lizard), the urine sodium level was always about 100 m-equiv/l, even when the animals had been hydrated. Solute reabsorption from the renal tubule varied from less than 50% in the Galapagos lizards to 85% in the geckos. The ability of many reptiles to restrict urinary sodium loss is not impressive, and may be subject to some adaptation (genetic or physiological?) commensurate with the environmental conditions to which the animals are normally subjected.

By analogy with other vertebrates, it appears likely that in the reptiles adrenocortical steroids may increase the reabsorption of sodium across the renal tubule and facilitate tubular secretion of potassium. However, there is little evidence available to indicate that this is so. Some experiments to elucidate the role of such steroids on urinary excretion of these ions in reptiles are clearly needed. Injections of vasotocin not only decrease urinary water loss in reptiles, but in the snake, *Natrix sipedon*, they also increase renal tubular absorption of sodium while at the same time facilitating potassium secretion (DANTZLER, 1967a). This peptide thus mimics the actions of corticosteroids on the mammalian renal tubule. The action of vasotocin seems to be too rapid for it to be acting by releasing such steroids in the snake, so it would appear to have a direct effect. The physiological significance, if any, of this interesting response is not known.

Many reptiles, especially species that live in the sea, may gain excessive quantities of solutes. The reptile kidney has a limited ability to regulate the concentrations of electrolytes, especially sodium, in the body fluids, as the concentration of the urine never exceeds that of the plasma. Indeed, when hyperosmotic solutions are administered to reptiles like the lizard, *Trachysaurus rugosus*, or the desert tortoise, *Gopherus agassizii*, they may become anuric and completely fail to excrete any of the solute (BENTLEY, 1959b; DANTZLER and SCHMIDT-NIELSEN, 1966). Even hypoosmotic solutions of sodium chloride are poorly excreted; when 100 mM NaCl solutions are given to *Trachysaurus rugosus* they excrete less than 20% of this through the kidney in 24 hours. The renal responses to potassium chloride are slightly more rapid and this may be due to the ability to secrete potassium across the renal tubule (SHOEMAKER *et al.*, 1966). The latter response may be important in herbivorous reptiles that gain excess potassium from their diet.

The relative inability of the reptilian kidney to excrete salts, may be compensated for in two ways. As will be described shortly, many reptiles have the ability to excrete sodium and potassium extrarenally through cephalic 'salt' glands. In other reptiles, which lack such accessory excretory glands, an ability to withstand considerable increases in the solute concentrations of the body fluids may exist. As we have seen, this has been shown in natural conditions in the Australian lizards, *Trachysaurus rugosus* and *Amphibolurus ornatus*. Such animals retain electrolytes and await the arrival of more adequate supplies of water which they then utilize for the renal excretion of such solutes. BRADSHAW and SHOEMAKER (1967) found that the plasma sodium levels of *Amphibolurus* could rise to as much as 300 m-equiv/l during periods of summer drought. During a summer rainstorm these lizards were observed to dart about and drink the water rapidly as it fell; ten hours later plasma samples were collected and the electrolyte concentrations were found to have returned to normal. Diamondback terrapins kept in sea-water accumulate sodium, but when they are placed in fresh water they drink and rapidly excrete the excess solute through extrarenal channels, presumably the orbital 'salt' gland (BENTLEY et al., 1967a). Some reptiles lack extrarenal channels for salt excretion and have a poor tolerance to the presence of such solutes in their body fluids; we have found that young caiman die within 24 h after being placed in 3.3% sodium chloride solutions (BENTLEY and SCHMIDT-NIELSEN, 1965). The reptile kidney, in the absence of adequate osmotically-free water has a very poor ability to excrete sodium.

c) Cloaca and Urinary Bladder

The urine of reptiles, like that of birds, passes into a cloaca where it may be accumulated prior to periodic expulsion to the outside. In certain species it may pass from the cloaca to be stored in a urinary bladder. The composition of the urine, including its electrolyte content, may be altered by both of these organs. Urine is stored for about 4 h in the cloaca of the caiman, during which time the sodium concentration decreases by 50% (BENTLEY and SCHMIDT-NIELSEN, 1965). This probably affords these crocodilians a substantial reduction in renal sodium loss. Such an economy has also been observed in the crocodile, Crocodilus acutus (B. SCHMIDT-NIELSEN and SKADHAUGE, 1967). The cloaca of the caiman, in vitro, has an electrical p.d. of about 10 mV (mucosa negative) across its wall which is consistent with the process of active sodium transport. The cloaca and large intestine of the Ophidia and the Chelonia cannot be readily separated for in vitro observations, but active sodium transport also appears to occur in this region of the gut of various snakes and in Testudo graeca (JUNQUEIRA et al., 1966; BAILLIEN and SCHOEFFENIELS, 1961; BENTLEY, 1962b). In Testudo this process is not affected in vitro by the presence of vasotocin or aldosterone. As we shall see in the next chapter sodium transport across the large intestine of toads can be facilitated by such hormones, suggesting that the actions of these substances on the cloaca and large intestine of reptiles may be worth further investigation.

The urinary bladder of the chelonians, Pseudemys scripta and Testudo graeca, also has the ability to transport sodium actively from the mucosal to the serosal surfaces (BRODSKY and SCHILB, 1960; BENTLEY, 1962b) thus further aiding conservation of sodium from the urine. In Testudo graeca this process is unaffected, in vitro, by the presence of vasotocin, an observation that contrasts with the effects of this peptide hormone in the Amphibia. However, like in frogs and toads, aldosterone increases the rate of sodium transport across the bladder of the Greek tortoise, while if these reptiles are pretreated with an antialdosterone drug (spirolactone), sodium transport is reduced. These observations need to be extended, especially to other species of reptiles, as they suggest that endogenous aldosterone may physiologically facilitate sodium transport across the urinary bladder of reptiles.

The urinary bladder and cloaca of turtles may also be involved in the accumulation of sodium from the external environment. Pseudemys scripta exchange sodium with bathing fluids at the rate of 0.04 to 10 μ-moles/100g h (DUNSON, 1967). When the cloaca is blocked, this rate of exchange is reduced by 70%, indicating that the processes of sodium transfer in the cloaca and urinary bladder may be involved. Turtles are known to irrigate their cloacal regions with the fluid that bathes them, so that an active transport of sodium in such regions could mediate a net accumulation of this ion by turtles. In the softshell turtle DUNSON considers that active sodium transport across the pharynx may be responsible for such solute collection.

d) 'Salt' Glands

"'So they went up to the Mock turtle who looked at them with large eyes full of tears'; 'for I am the crocodile' and he wept crocodile-tears to show it was quite true'". These two quotations from LEWIS CARROLL and RUDYARD KIPLING introduce a delightful account by SCHMIDT-NIELSEN and FÄNGE (1958) of their discovery of the salt secreting supraorbital ('tear') glands in reptiles. The diamondback terrapin, *Malaclemys centrata*, and the loggerhead turtle, *Caretta caretta*, secrete highly concentrated salt solutions from such tear (Harderian) glands, situated in the orbit. The marine iguana, *Amblyrhynchus cristatus*, of the Galapagos islands utilizes a nasal gland for this purpose, just as in the birds. Crocodiles, however, have not been shown to secrete such salty 'tears'. Other marine reptiles that utilize 'salt' glands for electrolyte excretion are: the yellow-banded sea snake, *Laticauda semifasciata*, and the yellow-bellied sea snake, *Pelamis platurus*, in which the particular tissue concerned appears to be an Harderian gland (DUNSON and TAUB, 1967; DUNSON, 1968).

SCHMIDT-NIELSEN and his collaborators (1963) also found functioning nasal salt glands in a number of terrestrial lizards; the tropical iguana, *Iguana iguana*, the desert iguana, *Dipsosauras dorsalis*, and the North African desert lizard *Uromastyx aegyptus*. Such secretory glands have also been demonstrated in the North American lizards: the chuckawalla, *Sauromalus obesus*, the false iguana, *Ctenosaura pectinata*, and the blue spiny lizard, *Sceloporus dorsalis*, (TEMPLETON, 1964 and 1966), as well as a lizard, *Uromastyx acanthinurus*, that lives in the Sahara (GRENOT, 1967). Such functioning secretory glands are not, however, present in all terrestrial reptiles, nor when present, do they all function with similar efficiency.

The fluid produced by such salt glands in reptiles is many times more concentrated than the blood plasma while the electrolyte levels in the secretions from marine species are even higher than those in sea-water (Table 5.8). Reptiles with such accessory salt secreting organs can thus excrete the salts from a varied marine diet, and probably even drink some sea-water without incurring a deficit in their body water. While in marine reptiles the sodium concentrations present in the salt gland secretions are much higher than those of potassium (Na/K as much as 20/1), the terrestrial lacertilians usually secrete a relatively potassium-rich fluid, Na/K< 0.1. Such an arrangement is consistent with the relative abundance of these ions in the diets of marine and terrestrial species.

The stimuli that inititiate secretion by the salt glands of reptiles appear to be similar to those seen in birds. Injections of hyperosmotic solutions of sodium chloride, potassium chloride or sucrose and the cholinergic drug methacholine can all bring about secretion. It is interesting that even in terrestrial lizards, in which potassium is the predominant ion present, injections of sodium chloride still have a kalitropic action (TEMPLETON, 1964 and 1966). However, after such administra-

Table 5.8 *Sodium and potassium secretion from the 'salt' glands of reptiles*

	Concentration m-equiv/l		Rate of secretion μ-equiv/kg h	
	Na	K	Na	K
Chelonia				
Malaclemys centrata[1] (Diamondback terrapin)	784	—	70	—
Caretta caretta[1] (Loggerhead turtle)	878	31	—	—
Chelonia mydas[2] (Green turtle)	685	645	1340	50
Lacertilia				
Ctenosaura pectinata[3] (False iguana)	21	21	0.6	19
Sauromalus obesus[3] (Chuckawalla)	44	430	3	27
Dipsosaurus dorsalis[4, 5] (Desert iguana)	494	1387	22	51
Uromastyx aegyptus[4] (Desert lizard)	639	1398	—	—
Amblyrhynchus cristatus[6] (Marine iguana)	969	123	2550	510
Ophidia				
Laticauda semifasciata[7] (Yellow-banded sea snake)	—	—	730	33
Pelamis platurus[8] (Yellow-bellied sea snake)	—	—	2180	92

[1]SCHMIDT-NIELSEN and FÄNGE (1958); [2]HOLMES and McBEAN (1964); [3]TEMPLETON (1964); [4]SCHMIDT-NIELSEN et al. (1963); [5]TEMPLETON (1966); [6]DUNSON (1969); [7]DUNSON and TAUB (1967); [8]DUNSON (1968).

tion of sodium chloride a slow adjustment may occur, through the course of several days and, as shown in the false iguana, the K/Na ratio can drop from 8/1 to 1/1 (TEMPLETON, 1967). Such an effect could be mediated by hormones.

In marine reptiles the quantities of salt excreted by the salt glands far exceeds those excreted in the urine; it makes up more than 90% of the total loss in the green turtle and sea snakes (HOLMES and McBEAN, 1964; DUNSON, 1968) and 75% in the marine iguana (DUNSON, 1969). Measurement of the rates of solute excretion in terrestrial lizards indicate that they are less efficient, only 20% of an administered dose of sodium chloride being excreted through the nasal glands of the desert iguana and less than 1% in the blue spiny lizard (TEMPLETON, 1966). Potassium is excreted relatively more efficiently in these lizards, the respective proportions being 40% and 10% of the given dose. Marine reptiles also appear to be able to secrete salts

153

at far higher rates than has so far been demonstrated in terrestrial species (Table 5.8). The propensity of terrestrial lizards to form nasal gland secretions seems to be rather variable, even within the same species. A lizard, *Iguana iguana,*('Fred') kept in SCHMIDT-NIELSEN's laboratory at Duke University regularly secreted large amounts of salt from his nasal gland, while others of this species, which were brought into the laboratory, often failed to perform in this way. I have observed similar differences in the behaviour of individual chuckawallas, *Sauromalus obesus.* It is not clear why such variation occurs but it may reflect the previous environmental conditions under which the animals lived. The state of development of such 'salt' glands may vary; green turtles kept in sea-water for several months have orbital glands 40% larger than those that are maintained in fresh water (HOLMES and McBEAN, 1964).

The precise mechanism that controls secretion from salt glands in reptiles is unknown but by analogy may be similar to that in birds. Osmotic stimuli and cholinergic drugs have a similar action in both groups of vertebrates, so that secretion is possibly activated (as in birds) by changes in plasma concentration stimulating a parasympathetic nerve supply to the glands. In birds the adrenocortical steroid hormones influence such secretions and this may also be so in reptiles. HOLMES and McBEAN (1964) found that injections of amphenone, which reduces the rate of formation of corticosteroids in mammals and reptiles (PHILLIPS, CHESTER JONES, and BELLAMY, 1962a), reduced extrarenal salt excretion in the green turtle. That the action of this drug is indeed mediated in this way was indicated by the ability of corticosterone to overcome its effect and enhance salt excretion. Amphenone also reduced excretion of sodium relative to that of potassium, implying that the corticosteroids enhance sodium more than potassium levels in the nasal secretions of this reptile. TEMPLETON and his collaborators (1968) have shown that adrenalectomy in the desert iguana, *Dipsosaurus dorsalis,* results in copious secretion of fluid from the nasal gland, but in contrast to that normally seen in this species, it has a high Na/K ratio. The injection of aldosterone into such lizards resulted in a decrease of this ratio towards normal levels. Thus in *Dipsosaurus* aldosterone affects the composition of the salt gland secretion in a manner which parallels its action on the mammalian kidney viz: sodium retention and potassium excretion. It is apparent that the effects of corticosteroids on the green turtle and the desert lizard are not consistent with each other. However, it will be recalled that the usual primary pattern of sodium relative to potassium excretion in the salt glands of the two species are opposite to each other, and this could influence the pattern of regulation.

e) Role of Adrenocortical Tissue

Reptiles, like all vertebrates, possess adrenocortical tissue (see CHESTER JONES, 1957) but its physiological role is inferred more often by analogy with other groups of vertebrates, than from direct experimental evidence. Adrenalectomy is a difficult procedure in reptiles and results in death within a few days in lizards like *Trachysaurus rugosus* and *Dipsosaurus dorsalis* (BENTLEY, 1959b; TEMPLETON *et al.,* 1968). In *Trachysaurus* this operation results in an elevated plasma potassium level, as in mammals, but the sodium concentration does not change. The failure of the latter to change may reflect the very low rate of urine formation in these reptiles. The painted turtle, *Chrysemys picta*, responds to adrenalectomy in a more classical manner since, after a period of two weeks, the plasma sodium concentration decreases while the potassium rises (BUTLER and KNOX, 1970). ELIZONDO and LEBRIE (1969) tied off the blood supplies to the adrenal glands of water snakes, *Natrix cyclopion*, and allowed the tissue to atrophy. Urine formation was maintained and both the sodium and potassium levels in the plasma declined. Change in sodium concentration appeared to reflect a 20% decrease in the rate of sodium reabsorption across the proximal segment of the renal tubule. Hypophysectomy causes atrophy of adrenocortical tissues in the lizard, *Agama agama*, and the grass snake, *Natrix natrix*, but this is not associated with any significant changes in the electrolyte levels in the body (WRIGHT and CHESTER JONES, 1957). In mammals hypophysectomy has since been shown to have little effect on the secretion of aldosterone, so that the latter observations on the possible role of adrenocorticosteroids in reptilian mineral metabolism are inconclusive. In addition, it should be remembered that physiological processes in poikilotherms are far slower than those in homoiotherms, so that any derangement in electrolyte metabolism that follows adrenalectomy may be slow to appear, especially if there is little turnover of sodium and potassium. While it appears likely that corticosteroid hormones influence renal sodium and potassium excretion as in other tetrapods, there is little information to show that this is so. It has recently been shown that the composition of the urine of water snakes, *Natrix cyclopion*, in water diuresis fails to change in response to injections of aldosterone, corticosterone or the aldosterone inhibitor 'Aldactone' (ELIZONDO and LE BRIE, 1969). However, when these snakes are in a saline diuresis due to the administration of sodium chloride solutions, aldosterone injections result in a 20% increase in the rate of sodium reabsorption across the proximal renal tubule (LE BRIE and ELIZONDO, 1969). As described previously, corticosteroids also increase the rate of sodium transport across the tortoise urinary bladder and interference with the adrenal tissue alters the secretion of sodium and potassium from the salt glands of the green turtle and the desert iguana.

Direct identification of corticosteroids in the blood of reptiles has not been systematically made, though corticosterone has been found in the blood of the grass snake, *Natrix natrix*, (PHILLIPS and CHESTER JONES, 1957) and the lizard, *Anolis carolinensis* (LICHT and BRADSHAW, 1969). Identification of such steroids in other reptiles is based on experiments utilizing *in vitro* incubation of adrenocortical tissues, sometimes in the presence of the substrate progesterone, and the subsequent isolation of the products formed. These results (Table 5.9) indicate that diverse reptiles have the ability to form corticosterone, as well as aldosterone, along with a few other steroids including 11-deoxycorticosterone. It is, of course, not clear from such experiments whether these steroids are secreted *in vivo*, but it seems likely that some of them are.

Information about the regulation of secretion from adrenocortical tissues in reptiles is sparse. The presence of a corticotrophic principle in the pituitary is implied from the experiments of WRIGHT and CHESTER JONES (1957). Adrenocortical tissue in *Agama agama* and *Natrix natrix* degenerated following hypophysectomy, but this could be prevented by injection of mammalian corticotrophin. The alliga-

Table 5.9   *Distribution of adrenocortical steroids in reptiles.*
(Determined by in vitro incubation of adrenal tissues)

	Corticosterone	Aldosterone	Other components
Crocodilia			
Alligator mississippiensis[1] (Alligator)	+	+	Two present
Chelonia The turtles:			
Pseudemys spp.[2]	+	+	Present
Emys orbicularis[2]	+	+	Two present
Chrysemys picta[3]	+	+	Present
Lacertilia			
Lacerta viridis[4] (Green lizard)	+	+	Present
Anolis carolinensis[5] (Chameleon)	+*		
Ophidia			
Natrix natrix[2, 4]* (Grass snake)	+	+	Three present

* also identified in blood PHILLIPS and CHESTER JONES (1957)
[1]GIST and DEROOS (1966).
[2]MACCHI and PHILLIPS (1966).
[3]SANDOR, LAMOUREUX, and LANTHIER (1964).
[4]PHILLIPS *et al.* (1962 a).
[5]LICHT and BRADSHAW (1969).

tor pituitary has been shown to contain a substance that increases corticosteroid production by chicken adrenocortical tissue, in a manner that is similar to the action of mammalian corticotrophin (GIST and DEROOS, 1966). This evidence suggests that the reptilian pituitary contains a corticotrophic hormone analogous to that in other vertebrates. However, the importance of its role is not clear, for while mammalian corticotrophin strongly stimulates corticosteroid production from the adrenal tissue of many vertebrates *in vitro*, it exerts little effect on reptilian tissue under such conditions (PHILLIPS *et al.*, 1962a; MACCHI and PHILLIPS, 1966; GIST and DEROOS, 1966). This is difficult to understand, and may indicate that such a control mechanism is poorly developed in reptiles. These observations have also been attributed alternatively to the possibility that reptilian corticotrophin differs markedly in its structure from other corticotrophins (PHILLIPS *et al.*, 1962a) or results from the absence of suitable precursors under *in vitro* conditions (GIST and DEROOS, 1966).

LICHT and BRADSHAW (1969) have studied the relationship between the adenohypophysis and circulating levels of corticosterone in the chameleon lizard, *An-*

olis carolinensis, and with more fruitful results than have been achieved *in vitro.* Removal of segments of the adenohypophysis resulted in the disappearance of corticosterone from the circulation. The administration of dexamethasone also had this effect, suggesting the presence of a negative feed-back system controlling the release of corticotrophin. When mammalian corticotrophin was injected into lizards so treated with dexamethasone, corticosterone reappeared in the plasma. Extracts from the adenohypophyses of *Anolis* as well as the desert lizard, *Dipsosaurus dorsalis,* the slider turtle, *Pseudemys scripta,* and the caiman have a similar action, indicating the presence of a corticotrophic principle in their pituitaries.

The possibility of the presence of an additional control mechanism involving renin and angiotensin does not yet appear to have been investigated in the reptiles. However, renin-like activity has been identified in the kidneys of the freshwater turtle, *Chrysemys picta,* and the desert tortoise, *Gopherus agassizii,* (CAPELLI *et al.,* 1970). Such a system could have an important role, for in other vertebrates angiotensin has a potent stimulating action on the production of corticosterone as well as aldosterone. As in the instance of the neurohypophysis, knowledge of the physiology of the hormones from the adrenocortical tissue of reptiles is less complete than in any other tetrapod group.

3. The Accumulation of Water and Salts

Water and salts are acquired by reptiles in the usual manner, from their food and drink.

Since amphibians do not drink, the reptiles represent phyletically the first tetrapods to utilize this osmotically useful habit. Certain aquatic reptiles, such as the caiman and softshell turtle, also acquire substantial amounts of water by osmosis through their skin, but the skin of most reptiles seems to be relatively impermeable, so that water uptake in this way is not usually important. The possibility that some terrestrial reptiles may accumulate water through their skin has been the subject of speculation which has not been confirmed *in vivo.* The mountain devil, *Moloch horridus,* an agamid lizard that lives in the desert regions of Australia, was once popularly thought to absorb water through its skin. When these lizards are placed in water it flows *onto* the integument, giving it the appearance of damp blotting paper, but absorption *through* the skin does not take place. Instead, water can be taken from the skin into the mouth, a process accompanied by 'chewing' movements of the jaws (BENTLEY and BLUMER, 1962). These lizards can accumulate water in this way but are never seen to drink in the more conventional manner. Such a procedure may be useful in desert reptiles, as it would allow them to gather water from sources too sparse to collect from otherwise. Drinking may occur only sporadically in desert reptiles; as described earlier the Australian lizards, *Amphibolurus ornatus,* have been observed to rush about drinking avidly when an infrequent shower of rain falls. Such rapid sporadic drinking probably allows many reptiles to make periodic osmotic adjustments. SCHMIDT-NIELSEN (1964 a) has described the behaviour of a desert tortoise which, in a single session, drank water equivalent to 40% of its body weight and DARWIN (1839) has commented on the prominent drinking habits of the tortoises of the Galapagos Islands. It is possible that certain marine reptiles can obtain water by drinking sea-water, but for many species in a marine situation the propensity to refrain from drinking may be even more vital to their survival. When an aquatic species, like the caiman, or a terrestrial one like the box turtle, *Terrapene carolina,* are placed in sea-water they drink, and as a result die in a short time (BENTLEY and SCHMIDT-NIELSEN, 1965; BENTLEY *et al.,* 1967 a). The diamondback terrapin and many *Pseudemys scripta* refrain from drinking such a bathing medium and can then survive for extended periods. We (BENTLEY *et al.,* 1967 a) only observed evidence for drinking (0.2 ml) in a single diamondback terrapin in sea-water, but when such turtles were returned to fresh water after 14 days in such a fluid, they rapidly imbibed it and excreted the accumulated salt. DUNSON (1969) has also noted that some marine reptiles, including the banded sea snake, drink fresh water when it becomes available, while other more fully marine species like the green turtle, the yellow-bellied sea snake and the Galapagos marine iguana do not appear to do so. Some accumulation of sea-water through the mouth is probably inevitable by marine reptiles during feeding, but this is probably only minimal. The ability of a reptile to distinguish between the potability of fresh water and sea-water, while appearing a somewhat mundane skill, may be vital to its survival in adverse osmotic conditions.

Water and salts can be accumulated from the food, some of the water being in the free form while some will arise as a result of metabolism in the body. The diets of different species of reptiles is varied and influences their osmoregulation. Many reptiles, especially many lacertilians, are herbivorous; such fare usually contains plenty of free water and salts, especially potassium. The marine iguana, however, grazes on seaweed which also contains large amounts of sodium and is isoosmotic with sea-water. Many lizards also eat small invertebrates and an occasional tetrapod. The Ophidia and Crocodilia are principally carnivorous, a diet from which they can also obtain sufficient water and salts. Sea snakes appear mainly to eat fish, a food that has the advantage of being generally hypoosmotic to sea-water (DUNSON, 1968). Many terrestrial Chelonia are herbivorous or omnivorous while the marine turtles subsist largely on invertebratess that are isoosmotic with sea-water. Such varied diets present different osmotic problems and in many instances may supply adequate or even excess water and salt.

4. Water Metabolism of the 'Dob'

GRENOT (1967; 1968) has given a very interesting account of the life, habits and osmoregulation of a lizard that lives under the extremely dry conditions of the Sahara desert. This agamid, *Uromastyx acanthinurus,* is locally called the 'Dob' or 'Fouette-Queue' (whip tail). The aridity of the Sahara is legendary, and in the area of Beni-Abbes, where these lizards were studied, the daily air temperature can rise to 47° while that on the earth's surface may be 70°. Rainfall is uncertain and is as little as 0.3 mm in a year. The 'Dob' lives among outcrops of rocks, on which, by either basking in the sunshine or seeking the cool refuge of crevices and cracks, it can regulate its body temperature and at the same time reduce evaporative water loss. The importance of such behaviour was shown by the high mortality rate of lizards tethered in the sun without shelter. When these lizards were kept without food or water for 10 days they lost as much as 24% of their body weight and so can presumably withstand substantial dehydration. In their native habitat, however, the serum sodium and potassium levels of the 'Dob' are steady, indicating that they can maintain a normal state of hydration. In order to do this they require water equivalent to about 3 to 5% of their body weight in a day which is obtained from the plants and occasional insects which they eat; they do not drink. Such a diet contains substantial quantities of salts, particularly potassium, and these are excreted from active nasal salt glands. This secretion is rich in potassium, (the K/Na ratio is as great at 50/1) and this is augmented during dehydration. The rate of secretion also depends on the diet of the lizard. Calculations from GRENOT's results indicate that these lizards normally secrete 500 to 1000 μ-equiv/kg body wt of potassium from their nasal glands in a day. The composition of the plants that they eat has not been reported, but if they contained 100 μ-equiv of potassium/g wet wt the amount secreted by the salt glands would only be equivalent to that in 5 to 10 g (kg body wt) of vegetation. Presumably they eat more than this so that urinary excretion of potassium must also be substantial. Information about the role of the neurohypophysis and the corticosteroid hormones in the regulation of water and salt excretion of these lizards would be most interesting, but so far has, alas, not been described.

Chapter 6

The Amphibia

The Amphibia bridge the phyletic gap between the fishes and terrestrial tetrapod vertebrates. This transition has involved considerable osmoregulatory, as well as respiratory, change so that water and salt metabolism may be expected to be particularly varied and interesting. Both for these reasons, and the relatively free liaison between physiologists and amphibians, the group has received considerable scientific attention.

The Amphibia arose from a crossopterygian fish in Devonian times. This transition is thought to have occurred in fresh water, where the Amphibia lived for a considerable period of time subsequently. There is only one surviving crossopterygian fish, the coelacanth, *Latimeria chalumnae,* which was not found until 1938. This coelacanth is a marine species, several of which have been caught off the southeast coast of Africa. The contemporary Amphibia are represented by three orders: the Apoda, the Urodela and the Anura. The Apoda, or caecilians, are small shy worm-like animals, of which there are about 17 genera, living exclusively in tropical equatorial areas of America, Africa and Asia. Unfortunately very little is known about their physiology and there is no information about their osmoregulation. The Urodela, or the Caudata, contain the newts and salamanders of which there are 43 genera, mostly confined to temperate regions in the northern hemisphere and completely absent from Australia. The Anura, are the most numerous of the three orders, containing more than 200 genera which are widely distributed around the world. 'Nothing – neither climates, nor deserts, nor mountains nor even narrow gaps of salt water – has prevented the spread of dominant frogs over the world in the course of time' (DARLINGTON, 1957). Anurans are present on all the major land masses (except Antarctica) where they occupy habitats ranging from freshwater ponds and streams, to tropical forests and arid deserts. A few species even venture into the cold polar regions north of the arctic circle. Many are primarily aquatic but they are usually more comfortable in fresh than brackish water. At least one species, *Rana cancrivora,* the crab-eating frog, lives in sea-water among mangroves along coastal areas of south east Asia. DARLINGTON (1957) concludes that frogs must have crossed narrow gaps of salt water, but notes that they usually have a low tolerance to such solutions and this presumably accounts, as commented upon by DARWIN (1859), for their absence from most oceanic islands. Amphibians readily lose water through their permeable skin so that desert regions may also be expected to restrict their movements. Nevertheless, while such areas may slow their migrations, they certainly do not stop them.

The amphibians exhibit a number of well defined physiological differences from their piscine ancestors and reptilian descendants. Such divergences may have a considerable influence on the osmoregulation of the respective groups so that I have attempted to summarize some of them (Table 6.1). Perhaps the major factor that affects amphibian osmoregulation is the adoption, by most of the adult forms, of atmospheric, instead of aquatic, respiration. The Amphibia are assisted in this process by gaseous exchanges across their skins (the plethodont salamanders have no lungs and depend solely on their skin for this purpose). As a result the skin is also more permeable to water. It has been suggested that the crossopterygian ancestors of the Amphibia underwent a reduction of their cutaneous scales in order to facilitate aeriform respiration (SZARSKI, 1962). A relatively impermeable skin was not phyletically restored until the emergence of the reptiles. The other major factor influencing the osmotic life of amphibians is the necessity to lay their eggs in water and for the larvae to undergo an aquatic period of development before they metamorphose into a form which can live on land. Reptiles have adopted an amniotic and cleidoic egg which is normally laid on land, and so have achieved a considerable further measure of osmotic independence. There are many other differences in the osmoregulatory pattern of fishes, amphibians and reptiles (Table 6.1) that may further hinder or facilitate their osmotic homeostasis. These will be discussed in subsequent sections.

Water makes up about 80% of the total body weight of most amphibians (see for instance SCHMID, 1965) and this is a higher proportion than the 70% seen in

Table 6.1 *Some physiological characters of Amphibia (compared with those of reptiles and bony fishes) that influence their osmoregulation*

	REPTILIA	AMPHIBIA	OSTEICHTHYES
Poikilothermic	+	+	+
Branchial respiration (gills)	—	— or +	+
Egg amniotic	+	—	—
Kidney:			
(i) Hyperosmotic urine	—	—	—
(ii) Antidiuretic to neurohypophysial peptides	+	— or +	— or diuresis
Urinary bladder derived from cloaca	+	+	—
Nitrogen excretion:			
Uric acid	+	—	—
Urea	+	+	+
Ammonia	+	+	+
Drinks water	+	—	+
Skin: restricted permeability	+	—	+
Extrarenal salt excretion	+	—	+
Extrarenal salt conservation	—	+	+
Neurohypophysis:			
neural lobe present	+	+	—
Secretes:			
(i) Vasotocin	+	+	+
(ii) Mesotocin	+	+	— or in a few
(iii) Isotocin	—	—	+
Adrenocortical tissue secretes:			
(i) Corticosterone	+	+	+
(ii) Aldosterone	+	+	—
(iii) Cortisol	—	some (?)	+

most other tetrapods. Nevertheless, this may vary; two Australian tree frogs (family Hylidae), *Hyla moorei* and *H. caerulea*, have water contents equivalent, respectively, to 71% and 79% of their body weights (MAIN and BENTLEY, 1964). The osmotic concentrations of the body fluids of amphibians are also usually less than those of other tetrapods, about 250 m-osmole/l; there is also less sodium present in the plasma, about 115 m-equiv/l. This may also vary a great deal depending on the particular species, its habitat and osmotic condition (Table 6.2). The North American spadefoot toad, *Scaphiopus couchi*, can survive in deserts during extended periods of drought. McCLANAHAN (1967) has shown that when this toad aestivates in burrows during such dry periods, the concentrations of solutes in its body fluids rise considerably, from 305 m-osmole/l to as high as 630 m-osmole/l. These frogs emerge after rain when they hydrate and restore the solutes to more normal levels. The crab-eating frog, *Rana cancrivora*, also increases the concentrations of its body fluids to very high levels when living in sea-water (GORDON *et al.*, 1961). Both of these anurans tolerate sodium concentrations up to about 250 m-equiv/l in their plasma, while large quantities of urea are also accumulated and may reach a level of 500 mM. Considerable decreases in plasma solute levels are also tolerated (Table 6.2). Amphibians, even more so than reptiles, thus exhibit a considerable ability to tolerate different solute levels in their body fluids.

Tolerance to changes in the osmotic concentrations of the body fluids may vary in different species and account for their distinctive abilities to withstand differing degrees of dehydration. In 1943 THORSON and SVIHLA compared the abilities of a number of North American frogs and toads to survive dessication. Their results indicated that amphibians normally living in more aquatic habitats withstand loss of their body fluids poorly, compared to those which live in drier areas. Thus, the aquatic frog, *Rana grylio*, dies after losing water, by evaporation, equivalent to

Table 6.2 *Plasma solute concentrations in the Amphibia*

| | Plasma Concentration | | | |
	Sodium m-equiv/l	Potassium m-equiv/l	Urea m-mole/l	Osmolarity m-osmole/l
Scaphiopus couchi[1] (Spadefoot toad)				
(i) In fresh water	159	4.9	39	305
(ii) During aestivation	228	5.4	286	630
Rana cancrivora[2] (Crab-eating marine frog)				
(i) In fresh water	125	9	40	290
(ii) In sea-water (80%)	252	14	350	830
Bufo marinus[3] (Giant toad)				
(i) In damp soil	108	2.7	—	243
(ii) Hydrated (vasotocin injection)	94	2.6	—	205

[1]McCLANAHAN (1967); [2]GORDON *et al.* (1961); [3]TOYOFUKU and BENTLEY (1970).

about 30% of its body weight, while the spadefoot toad, *Scaphiopus hammondi*, can survive a loss of 50% in its body weight. A similar relationship between the habitat and survival of dessication has been shown by SCHMID (1965) in North American anurans and we have also observed this to be so in some Australian tree frogs (MAIN and BENTLEY, 1964). WARBURG (1967) found that some frogs from dry regions in central Australia withstand dehydration better than those from wetter habitats but he has emphasized the importance of the speed of water loss in determining the animals' survival. Frogs seem to be able to live longer when the rate of dehydration is slow, than when it is fast, and although I have made no precise measurements of this, I would agree that this is probably an important factor. Differences in ability to withstand dehydration are not invariably seen among different groups of anurans. The Australian genera *Neobatrachus* and *Heleioporus* each have 4 or 5 species that have adopted habitats ranging from wet forests to dry deserts and they all survive a water loss equivalent to 40 to 45% of their body weight (BENTLEY, LEE, and MAIN, 1958). These Australian frogs (family Leptodactylidae) thus all tolerate water losses as great as those of the best adapted North American species or of Australian Hylidae. The Australian deserts are far older than those in North America and the leptodactylids are an ancient anuran family that probably gave rise to the now more widely distributed Bufonidae. The Australian leptodactylids may thus have evolved into a basically xerophilous group which in the course of long periods of time have adapted more uniformly to dry conditions. As we shall see, these leptodactylids also have a remarkable ability to rehydrate rapidly and can store large volumes of water in their urinary bladders.

Urodeles do not live in conditions as arid as those experienced by many anurans. It is interesting that 5 species of salamander from temperate areas in North Carolina died after losing water equivalent to only 18 to 30% of their body weight (SPIGHT, 1968). It would be interesting to know if any urodeles can withstand the extremes of dehydration tolerated by some anurans.

Amphibians may live in three contrasting types of osmotic environment; fresh water, dry land or, very occasionally, in salt water. Larval amphibians, certain urodeles, such as the mudpuppy and the congo eel, and anurans like the South African clawed toad are almost entirely aquatic. Many species habitually live in ponds and streams from which they make periodic excursions and are thus truly amphibious. The fresh water where such species live, or visit, is hypoosmotic to their body fluids so that there is a continual accumulation of water occurring across their skin. This may be equivalent to 30 to 40% of their body weight in a day and is excreted by the kidneys as a dilute (hypoosmotic) urine. This urine contains only small, but nevertheless, significant amounts of solutes and the loss of some of these, especially sodium, may be physiologically significant. In addition, as shown by the classical studies of KROGH (see KROGH, 1939), amphibian skin is permeable to sodium and chloride, so that on simple physico-chemical grounds one would expect further losses to occur across the integument. While in certain unphysiological circumstances, like bathing in distilled water, such losses can be demonstrated, they are usually prevented by an active transport, by the skin, of sodium (and sometimes chloride) from the bathing medium.

DARWIN (1859) noted that frogs have a poor tolerance to salt solutions and only two or three species have been discovered for which this is not true. When leopard frogs, *Rana pipiens*, are placed in sea-water they die in 30 to 60 min. SCHMIDT-NIELSEN and I have found that under these conditions they lose water through the skin, but the main reason for their demise is not water loss, but a massive accumulation of salt that takes place through their skin and as a result of drinking. Even when these frogs are placed in saline solutions similar in concentration to their body fluids (0.7% sodium chloride) they often do not survive for longer than 24 h. In more dilute solutions they live for longer periods, gaining weight due to the accumulation of salt and water (ADOLPH, 1933). Nevertheless, some anurans do habitually enter brackish water and at least one, the crab-eating frog, *Rana cancrivora*, lives in sea-water. The European green toad, *Bufo viridis*, has been found in ponds where the salt concentration, while less than that of sea-water, is a substantial 2% (GISLEN and KAURI, 1959). There are other reports of amphibians being found in brackish water, but it is often difficult to judge how typical such an event is in the life of the animal. The sight or sound of an approaching naturalist can conceivably strike sufficient terror into a frog to prompt it, unwittingly, to leap into a pool of salt water. GORDON (1962) collected *Bufo viridis* in different areas of Europe ranging from Naples to Belgrade and found that these toads, in common with Roumanian ones previously studied by STOICOVICI and PORA (1951), could live for extended periods of time in solutions with a sodium chloride content of 2% or, sometimes, even more. The most basically important osmotic adjustment *Bufo viridis* and *Rana cancrivora* make towards their life in saline solutions, is to initiate and maintain the concentrations of their body fluids at levels that are always, sometimes only slightly, hyperosmotic to the bathing fluids. This elevation in the osmotic pressure of the body's fluids results from an elevation of the sodium levels and, particularly in the instance of *Rana cancrivora*, an accumulation of urea (Table 6.2). The tadpoles of *Rana cancrivora*, however, like all amphibian larvae cannot form urea, but utilize a completely different mechanism for their life in sea-water. They maintain their body fluids, like those in bony fishes, at a concentration which is hypoosmotic to sea-water, possibly excreting excess salt through their gills (GORDON and TUCKER, 1965).

When amphibians utilize their inherent ability to live on dry land, they are potentially subject to an osmotic stress that is initiated in a manner different from that in sea-water, that is the concentration of their body fluids as a result of evaporation through the integument. Evaporation through the skin is characteristically rapid in the Amphibia as compared to other tetrapods. Nevertheless, amphibians, especially anurans, have successfully occupied arid areas in which high environmental temperatures and a low content of water vapour in the air, considerably facilitates evaporation. Such species utilize a number of mechanisms that somewhat mitigate these conditions. Like all vertebrates living in such places they have adopted patterns of behaviour that reduce the times of exposure to the extremes of the environment; they seek refuge in cracks in rocks, and burrow, often deeply, into the soil. In addition a variety of physiological processes affect the osmoregulation of such amphibians.

The amphibians, like other tetrapods, are aided in their osmoregulation by the action of some hormones, the most important being those secreted by the neurohypophysis and the adrenocortical tissues.

a) The Neurohypophysis

In contrast to most fishes, the neurohypophysis of amphibians is enlarged posteriorly to form a neural lobe. This persists throughout the tetrapods. Among the lungfishes, the size of the neurohypophysis may be somewhat increased in this manner, while a few amphibians, like the mudpuppy, may have a neurohypophysis that retains a more fish-like form (see WINGSTRAND, 1966). In 1921 BRUNN showed that when extracts from the neurohypophysis of mammals, which have an antidiuretic action in this group, are injected into frogs kept in water, there is an increase in weight due to an accumulation of water. This action has been called the BRUNN effect or 'water balance effect'. HELLER in 1941 found that while the pituitary of English frogs, *Rana temporaria*, contains a substance that causes an antidiuresis when injected into rats, this was not identical with the activity present in the mammalian neural lobe. The amphibian extract, when injected into frogs, was far more potent in its ability to produce a 'water balance effect' than similar mammalian extracts and HELLER called this substance the 'amphibian water balance principle'. Nearly 20 years later this was shown to be identical to arginine-vasotocin (PICKERING and HELLER, 1959), a peptide that is also present in birds, reptiles and most of the fishes, but is absent from mammals. A second substance, with properties similar to those of oxytocin, was also found to be present in amphibian neurohypophysial extracts. This was pharmacologically similar to 8-isoleucine oxytocin (FOLLETT and HELLER, 1964 a). It was chemically identified as this by ACHER and his collaborators (Table 6.3). This peptide differs from oxytocin by

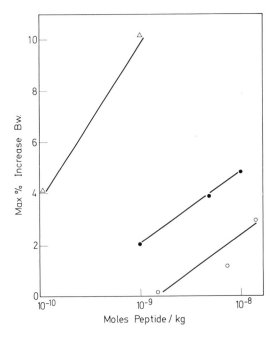

Fig. 6.1 Water balance effects of vasotocin (△), oxytocin (●) and mesotocin (○) in toads (*Bufo marinus*).
Note: Log – dose scale.
(From BENTLEY, 1969b).

Table 6.3 *Distribution of neurohypophysial peptides in the Amphibia*

	Vasotocin	Other	Ratio Vasotocin/Mesotocin	
Anura				
Rana temporaria (English frog)				
a)[1]		HELLERS 'water balance principle'		
b)[2]	+	Oxytocin-like		
Rana esculenta [3,4,7] (European edible frog)	+	Mesotocin	2.7[3]	2.9[4]
Rana catesbeiana [6] (American bullfrog)	+	Oxytocin-like		
Rana pipiens [7,8] (Leopard frog)	+	Mesotocin[7] ?oxytocin[8]	1.8[6]	
Bufo bufo [2,3,9] (European toad)	+	Mesotocin[3,9]	3.3[3]	
Bufo americanus [6]	+	Oxytocin-like		
Xenopus laevis [3,4] (South African clawed toad)	+	Mesotocin	1.7[3]	6.7[1*]
Urodela				
Necturus maculosus [3] (Mudpuppy)	+	Mesotocin	2.2	
Triturus alpestris [3] (Alpine newt)	+	Oxytocin-like		
Salamandra maculosa [10] (Fire-salamander)	+	Oxytocin-like		

The neurohypophyses of the tiger salamander (*Ambystoma tigrinum*), congo eel (*Amphiuma means*) and the mud eel (*Siren lacertina*) also exhibit biological activity (increased permeability of toad bladder) consistent with the presence of neurohypophysial hormone[11]. The neurohypophysis of the crab-eating frog (*Rana cancrivora*) contains amphibian water-balance activity and rat 'pressor' and oxytocic activity[12].
* very young toads (about 10 g)
[1]HELLER (1941); [2]ACHER et al. (1960); HELLER and PICKERING (1961); [3]FOLLET and HELLER (1964a); [4]ACHER et al. (1964 a, b); [6]SAWYER et al. (1961); [7]ACHER, CHAUVET, and CHAUVET (1969 b); [8]MUNSICK (1966); [9]ACHER et al. (1968); [10]BENTLEY and HELLER (1965); [11]BENTLEY (1969 b); [12]DICKER and ELLIOTT (1969, 1970 a).

the presence of a single amino acid, isoleucine being substituted for leucine, and was called mesotocin by ACHER. Oxytocin has been identified pharmacologically in the neurohypophysis of a single amphibian, *Rana pipiens* (MUNSICK, 1966), but ACHER *et al.*, (1968) have also found mesotocin to be present in this frog. Mesotocin is present among reptiles, but is absent from most fishes with the interesting exceptions of some dipnoans (lungfishes) (FOLLETT and HELLER, 1964 a; SAWYER, 1969). Vasotocin, mesotocin and oxytocin, when injected, can all promote water retention in various amphibians, but vasotocin is by far the most active of these peptides. In *Bufo marinus* it is more than 200 times as effective as mesotocin (Fig. 6.1).

b) Adrenocortical Tissue

Amphibian adrenocortical tissue appears to be able to secrete principally two steroids, aldosterone and corticosterone (Table 6.4) both of which are also found in birds and reptiles. Other steroids have been identified in incubates of amphibian adrenocortical tissues but, except for cortisol, they have not been identified in the circulation. The latter steroid has been found in adrenal effluent blood of the anuran, *Xenopus laevis*, and the urodele, *Amphiuma tridactyla* (CHESTER JONES, PHILLIPS, and HOLMES, 1959). Both aldosterone and corticosterone have been identified in the blood of toads, *Bufo marinus*, and bullfrogs, *Rana catesbeiana* (CRABBE, 1961 a; JOHNSTON *et al.*, 1967). The phyletic spread of such observations among the Amphibia is thus somewhat limited, especially in urodeles. In view of the presence of cortisol as well as corticosterone, but not aldosterone, among the fishes a closer examination of corticosteroids in the Amphibia would be most interesting.

Table 6.4 *Distribution of the adrenocorticosteroids in the Amphibia*

	Aldosterone	Corticosterone
Anura		
Rana catesbeiana[1, 2*] (American bullfrog)	+	+
Rana pipiens[3] (Leopard frog)	+	+
Bufo marinus[4*] (Toad)	+	+
Xenopus laevis[5*] (South African clawed toad)	?	? Cortisol
Urodela		
Amphiuma tridactyla[6*] (Congo eel)	?	+ Cortisol

* Identified in the blood; others from adrenal tissue incubation.
[1]CARSTENSEN *et al.* (1959; 1961); [2]JOHNSTON *et al.* (1967); [3]KRAULIS and BIRMINGHAM (1964); [4]CRABBE (1961 a); [5]PHILLIPS and CHESTER JONES (1957); [6]CHAN and EDWARDS (1970); [7]CHESTER JONES *et al.* (1959).

1. Water Exchange

a) Skin

Skin plays a major role in the osmoregulation of the Amphibia, due not only to its high permeability to water and salts, but also to the fact that such permeation may vary in different species and be subject at times to physiological regulation.

α) *In water.* Amphibians do not drink, except in certain non-physiological situations as when they are placed in hyperosmotic salt solutions. Instead, water moves across the skin, a process that occurs down a gradient of osmotic concentration. When the solutions on either side of the skin are osmotically equal, little water is transferred. If the external solution is that of an impermeant solute like sucrose, virtually no water transfer occurs. If a permeant solute, such as sodium, is present small movements of water, consistent with any solute transfer, may occur in order to maintain osmotic equality. Such transfer of water is normally minor. The movement of water across the skin of frogs and toads is directly proportional to the osmotic gradient between the two sides of the skin (SAWYER, 1951). The movement of water is thus considered to be a passive process, not requiring the direct intervention of any metabolic energy. Normally it is from the outside solution into the animal, but if amphibians inadvertently enter hyperosmotic solutions they lose water; in *Rana pipiens* placed in sea-water, water passes out across their skin at the rate of 55 $\mu l/cm^2 h$ (P.J. BENTLEY unpublished observations).

Water transfer across the skin is also influenced by the temperature. The Malayan toad, *Bufo melanostictus*, when kept in water at 12°, gains water through its skin at the rate of 6.4 $\mu l/cm^2 h$; at 29° this is more than doubled and is 17 $\mu l/cm^2 h$, while at 37° it is 34.5 $\mu l/cm^2 h$ (DICKER and ELLIOTT, 1967). The latter temperature cannot be considered a physiological one, but some toads regularly experience temperatures approaching 30° and the speed of their hydration, or rehydration, would be increased under these conditions.

The rate of water transfer across the skin differs by species. SCHMID (1965) has measured the rates of osmotic water transfer across the skins *(in vitro)* of six different species of North American frogs and toads. He found that the skins of aquatic species like the mink frog, *Rana septentrionalis*, had a lower permeability to water (6 $\mu l/cm^2 h$) than those from more terrestrial species like the Dakota toad, *Bufo hemiophrys* (20 $\mu l/cm^2 h$). Frogs from habitats judged to be intermediate in their 'dryness' showed correspondingly different permeabilities. The skin from anurans with low permeability to water was found to have a higher lipid content than those with a higher permeability (SCHMID and BARDEN, 1965). Other values, collected from the literature are given in Table 6.5 and generally support the conclusions of SCHMID. Such a relationship is also seen among the terrestrial urodeles. Such species as the firesalamander and the alpine newt have a greater osmotic permeability than aquatic species, such as the mudpuppy and the congo eel. It is also interesting that the skin of the aquatic tadpoles of bullfrogs is less permeable than that of the adults.

When dehydrated amphibians return to water they regain water by absorbing it across their skin. They can also take up water in this way from damp surfaces such as soil. The ability of the Amphibia to gain water from damp earth presumably depends on the moisture content of the soil and the various forces (osmotic, hygroscopic and capillary) that tend to hold water there. The water content of the soil may be quite variable, depending on such things as when rain last fell, the presence of surface vegetation and the proximity of ground water. SPIGHT (1967 a) found that the six species of salamanders studied by him could absorb water from a soil sample that contained about 10% water. When the moisture content was lower, 3 to 4%, the salamanders slowly lost water, showing that the forces controlling water retention by the soil may also have a dehydrating action on the animals. However, it was apparent that these salamanders, which were mostly native to

Table 6.5 *Permeability of amphibian skin to water (osmotic gradient 200 to 220 m-osmole/l) in the absence and presence of vasotocin*

	$\mu l/cm^2\ h$		
	Normal	Vasotocin present	Habitat
Anura			
Bufo marinus[1] (American toad)	28	85	Terrestrial
Bufo bufo[2] (European toad)	19	44*	Terrestrial
Rana catesbeiana[1,3] (Bullfrog)			
(i) Adult	13	61	Amphibious
(ii) Tadpole	3	5	Aquatic
Rana pipiens[4] (Leopard frog)	5	19**	Amphibious
Rana esculenta[2] (European frog)	9	27*	Amphibious
Xenopus laevis[1,2] (African clawed toad)	8	9	Aquatic
	7	7*	
Rana cancrivora[5] (Crab-eating frog)	8	8	Fresh water or marine
Urodela			
Ambystoma tigrinum[6] (Tiger salamander)	4	7	Terrestrial
larva	2	5	Aquatic
Salamandra maculosa[7] (Fire salamander)	11	7	Terrestrial
Amphiuma means[1] (Congo eel)	4	4	Aquatic
Siren lacertina[1] (Mud eel)	4	5	Aquatic
Necturus maculosus[5] (Mudpuppy)	3	4	
Triturus alpestris[5] (Alpine newt)	7	8	Terrestrial

* oxytocin
** Pituitrin
[1]BENTLEY (1969 b); [2]MAETZ (1963); [3]BENTLEY and GREENWALD (1970); [4]SAWYER (1951); [5]DICKER and ELLIOTT (1969, 1970); [6]BENTLEY and HELLER (1964); [7]BENTLEY and HELLER (1965).

North Carolina, should be able to absorb water from the soil in the agricultural areas of this region at any time of the year. If perchance they burrowed into soil with a low moisture content, the expected losses would be still far less than would usually take place by evaporation at the surface.

When amphibians are dehydrated they usually absorb water through their skin more rapidly than when they are normally hydrated. There is considerable variability in the magnitude of such differences; they may be small, as seen in aquatic species like *Xenopus laevis*, or large as in the terrestrial toad, *Bufo carens*. Such observations on these two species led EWER (1952 b) to suggest that such a response may be better developed in species that are normally terrestrial in their habits as compared to more aquatic species. Parallel with these observations it was found that when neurohypophysial peptides were injected, *Bufo* retained large amounts of water, while *Xenopus* completely failed to respond. These differences are reminiscent of the observations of STEGGERDA (1937) who found that while terrestrial species of the Amphibia accumulated large quantities of water after being injected with preparations of mammalian neurohypophysial peptides, aquatic ones retained little. An ability to rehydrate rapidly could be important to some amphibians, particularly to those living in areas where water is only available sporadically and quickly disappears either by evaporation or soaking into the soil. It is also possible that entry into open pools of water may increase the risk of attack. Thus, a facilitated rate of absorption would reduce exposure to predators. The rates of rehydration in a large number of North American and Australian amphibians have been measured, and compared in relation to the availability of water in the habitats where they normally live. It is evident that there is considerable divergence in the rates of rehydration, but just how uniformly this may be related to the habitat, thus representing a physiological adaptation, is not clear.

Rehydration in a large variety of North American frogs and toads from the families Ranidae, Bufonidae, Hylidae and Pelobatidae has been measured (THORSON, 1955; CLAUSSEN, 1969), but no relationship between the rate of water accumulation and the 'dryness' of the habitat was found. A number of Australian frogs, especially those from the family Leptodactylidae have also been examined for evidence of such a correlation. We (BENTLEY et al., 1958) confined our studies to species within the genera *Heleioporus* (5 species) and *Neobatrachus* (4 species). The habitats of species within these two genera range from areas in which rain falls regularly, on an average of 120 days a year, to desert, or semi-desert regions where the annual rainfall may be spread over only a few days. Frogs of the genus *Heleioporus* all rehydrated at similar rates (about 50 μl/cm^{2}h) after being dehydrated to 75% of their normal weight. However, species of *Neobatrachus* showed considerable differences in their abilities to absorb water after such dehydration; *N. pelobatoides* which lives in areas where rain usually falls on 60–120 days a year gain water at the rate of 33 μl/cm^{2}h, while *N. wilsmorei* which lives in dry areas towards the interior of the continent regain water at the rate of nearly 100 μl/cm^{2}h. It is notable that such differences in ability to rehydrate were paralleled by the rates at which these frogs accumulated water after being injected with a neurohypophysial peptide, oxytocin. The failure of *Heleioporus* to show any evidence of a correlation between water uptake and the aridity of the habitat could reflect the ability of this group to avoid the excesses of the climate by very efficient burrowing, thus making rehydration ability somewhat redundant. WARBURG (1965 b) has also found that frogs that live in the arid and semi-arid central regions of Australia rehydrate more rapidly than those species that live in more temperate areas. Few urodeles have been examined for their ability to rehydrate following dehydration but SPIGHT (1967 b) has measured this in four species of salamanders. He found only relatively small increases in the rates of water absorption by dehydrated animals, the maximum rate of water uptake being observed in *Ambystoma opacum*, and this was only 10 μl/cm^{2}h. Urodeles also only accumulate water slowly after being injected with neurohypophysial peptides (see HELLER and BENTLEY 1965).

Whether or not the rate of water absorption through the skin following dehydration can be related to the habitat of the species is not clear. Such a correlation certainly does not exist among all species. Nevertheless, considerable diversity does exist, and this, reflects differences in the permeability of the skin that range from about 4 μl/cm^{2}h in dehydrated *Xenopus* to as much as 420 μl/cm^{2}h in the ventral pelvic skin in *Bufo punctatus* (McCLANAHAN and BALDWIN, 1969). In addition, the permeability of the skin may be augmented during dehydration, in such a manner as to suggest the presence of a regulatory mechanism. In view of these observations I feel that it is reasonable to suppose that such differences arose and persisted in response to ecological stresses on the Amphibia during their evolutionary history, even though they may not today be precisely related to the life of all contemporary species.

The rate at which amphibians may gain water is directly related to the rate that water passes across their skin. This may depend on several factors. As shown by SCHMID's (1965) studies, the permeability to the skin of anurans may differ even in the absence of dehydration, and this will, of course, contribute to the differences observed in the rates of rehydration. The permeability of the skin to water may, as we have seen, be augmented in dehydrated animals and this could result from several effects. Due to the increased concentration of the body fluids, the osmotic gradient across the skin increases during dehydration, and water would be expected to move along this gradient more rapidly. Such effects must be relatively small, but could contribute to minor increases in water absorption as SPIGHT observed among urodeles. The increased osmotic pressure of the body fluids probably also exerts a direct action on the skin and increases its permeability to water, such as has been observed in the urinary bladders of toads and frogs (BENTLEY, 1964; PARISI, RIPOCHE, and BOURGUET, 1969). In 1936 NOVELLI found that neurohypophysial extracts could increase the permeability of the skin of anurans to water; this action has since been shown to be widespread within this group (Table 6.5). Mammalian neurohypophysial hormones, like vasopressin (ADH) can exert such an action on anuran skin but the amphibian peptide, vasotocin, is far more active. BOURGUET and MAETZ (1961) found that it increased the permeability of the skin of *Rana esculenta* even when present at the concentration of only 10^{-10}M. Vasotocin is released into the blood of dehydrated frogs and toads in which it is present at a concentration of 10^{-9} to 10^{-10} M, which should be adequate to mediate increases in the permeability of the skin of dehydrated anurans. Other hormones possibly could also be involved; thus adrenaline has been shown to increase *in vivo* the osmotic permeability of the skin of toads, *Bufo melanostictus,* (ELLIOTT, 1968) and frogs *in vitro* (JARD et al., 1968). If this catecholamine were released into the circulation during dehydration it could facilitate the action of vasotocin.

Vasotocin, and related peptides from the neurohypophysis, increase the permeability of the skin of anurans to water, the only known exceptions being, notably, *Xenopus laevis* and the crab-eating frog, *Rana cancrivora* (Table 6.5). The skins of a variety of urodeles have been examined for this response but none of

172

them have been found to react in this way to such peptides (Table 6.5). These observations are consistent with those showing that *Xenopus*, along with the urodeles, rehydrate more slowly than most anurans.

The entire body surface of amphibians may not be uniformly permeable to water or respond equally to neurohypophysial peptides. McClanahan and Baldwin (1969) have recently made the very exciting observation that dehydrated toads, *Bufo punctatus*, can absorb water across an area of ventral pelvic skin at the rate of 420 μ /cm^{2}h. In contrast water absorption across skin in the ventral pectoral area was too small to measure. These observations pose a number of interesting questions. Can the differences in permeability of the skin be related to its structure? To what extent does the variability in the speed of rehydration in different amphibians reflect the total area of such specialized regions in each species? The skin in the ventral pelvic area of many amphibians is an area that is readily apposed to damp surfaces, so that this could allow the animals to hydrate adequately without the necessity of it becoming more completely submerged or buried.

β) In Air. Amphibians exposed to the air can lose considerable water by evaporation through the skin. The rate of loss may be 50 to 100 times as rapid as in lizards of a comparable size (Claussen, 1969). Evaporation from amphibians occurs at a rate similar to that from a free water surface (Rey, 1937) and, as shown by Adolph (1933), is not altered by the presence of the skin. Small differences in the rates of evaporative water loss from different amphibians have been described, but as these often do not include an allowance for differences in the surface area, or rate of respiration, it is difficult to decide whether they really reflect variations in cutaneous loss (see Bentley, 1966b; Warburg, 1967; Claussen, 1969). Water loss from amphibian skin, like that from a free water surface, is inversely related to the content of water vapour in the air and increases considerably if air flows over the animal more rapidly. Amphibians have little physiological ability to limit evaporation from their integument but they can diminish these losses by appropriate changes in their behaviour. When they are exposed to a dehydrating atmosphere, amphibians often reduce their exposed surface by hunching themselves up or, as seen in urodeles, by coiling. I have observed that Australian leptodactylid frogs regularly make vigorous burrowing gestures under such conditions, even though they were confined to a glass beaker. Amphibians seek refuge from conditions that facilitate evaporation; they may hide away under the bark of trees or in cracks among rocks. A number of species, including the Australian leptodactylids and many of the Pelobatidae are morphologically well equipped for burrowing into the soil. We (Bentley *et al.*, 1958) have observed Australian frogs, *Heleioporus*, in burrows 80 cm below the surface and American spadefoot toads probably go to similar depths (Mayhew, 1965). During the heat of the day, the temperature in such burrows is considerably less than that at the surface, while the relative humidity of the air may be greater. Anurans can burrow down into damp layers of soil as spadefoot toads have been observed to do in the Colorado desert (Mayhew, 1965).

Amphibians may aestivate for many months in underground burrows. Mayhew (1965) observed that spadefoot toads, on emerging from such refuges after rain, were covered with a cocoon-like layer of dry skin. He suggested that this may help retard evaporation during aestivation. A number of Australian leptodactylids, which also aestivate during periods of drought, similarly have been shown to form such cocoons (LEE and MERCER, 1967). A careful examination showed that these sacs completely cover the animals, with the exception of small channels leading to the nares. These cocoons are composed of compacted layers of the shed epidermis. Evaporation across these membranes was found to be considerably less than across the integument of the living frogs. These epidermal structures may thus limit evaporation of water from the skin of aestivating frogs and are reminiscent of the cocoons that surround the African lungfishes, *Protopterus*, during similar periods of retirement.

b) The Kidney

Amphibians in fresh water excrete the water absorbed through their skin as a dilute urine. An anuran like the European frog, *Rana esculenta*, which weighs 100 g, forms about 60 ml of urine during a single day in fresh water (JARD and MOREL, 1963). Urodeles, in the same situation, secrete similar amounts of urine each day; the mudpuppy (200 g) forms about 50 ml while the alpine newt (10 g) produces 16 ml, 160% of the body weight (BENTLEY and HELLER, 1964). It can be seen that the daily volume of urine that an amphibian in fresh water may secrete is considerable, especially in small animals that have a relatively large surface area in relation to their body weight. When living in sea-water, *Rana cancrivora* forms far less urine; this frog, weighing 50 g, forms only about 2 ml of urine each day (SCHMIDT-NIELSEN and LEE, 1962). When amphibians are exposed to a drying atmosphere, urine formation is too small to measure and in leopard frogs amounts to an anuria (ADOLPH, 1933).

The relationship of the processes of glomerular filtration and tubular water reabsorption to the eventual urine volume has been examined in a number of anurans but in no urodeles. Both of these mechanisms play a prominent role in the regulation of the urine volume. The *Rana esculenta*, referred to above, have a GFR of 125 ml per day, corresponding to the 60 ml of urine per day, so that about 50% of the filtered water is reabsorbed under these conditions. In other circumstances tubular water reabsorption may be greater. The urine volume in *Rana esculenta* is almost directly related to the GFR, and in some individual frogs that form urine at twice the average rate, the GFR is also doubled (JARD, 1966). Crab-eating frogs (50 g) in sea-water filter 30 ml of water per day across their glomeruli but produce only 2 ml of urine, so that more than 90% of the filtrate is reabsorbed. The cessation of urine secretion, that is observed in dehydrated frogs probably results from the almost complete failure to form a glomerular filtrate.

Amphibians, in common with reptiles and fishes, cannot secrete urine that is osmotically more concentrated than their body fluids, so that the ability to prevent water loss in this way is limited. The urine can attain isotonicity with the plasma, as seen, for instance, in aestivating spadefoot toads (McCLANAHAN, 1967). In fresh water the urine is markedly hypoosmotic, less than 30 m-osmole/l (see for instance GORDON, 1962; JARD and MOREL, 1963). In sea-water the urine of crab-eating frogs is intermediate in concentration, being slightly hypoosmotic to the body fluids, 600 m-osmole/l compared to 830 m-osmole/l in the plasma (GORDON et al., 1961).

Table 6.6 *Distribution of antidiuretic action of neurohypophysial hormones in amphibians. (From data collected by* MOREL *and* JARD, 1968)

	Hormone	Antidiuretic Action	Nature of the effect
Anura			
Bufo arenarum	Mammalian neurohypo-physial extract	+	Tubular action and constriction of afferent glomerular vessels
Bufo carens	Vasopressin, oxytoxin	+	
Bufo regularis	Vasopressin, oxytocin	+	
Bufo marinus	Pituitrin, Pitressin Oxytocin	+	Tubular action and glomerular effect glomerular diuretic effect
Bufo bufo	AVT, AVP, LVP, Oxyt.	+	
Rana pipiens	Pitocin, Pituitrin	+ + —	Glomerular effect increase with large doses or decrease with small doses, of active glomeruli
	Pituitrin	+	inhibition of water diuresis
	Hypophysial extracts	+	Inhibition of salt diuresis
Rana catesbeiana	AVT, oxytocin, frog neuro-hypophysial extract	+ —	Glomerular and tubular antidiuresis glomerular diuretic effect
Rana temporaria	Oxytocin, vasopressin	+ +	
Rana esculenta	Pituitrin, Pitressin, AVT	+ +	Glomerular and tubular
Xenopus laevis	Arginine-vasotocin	0	antidiuresis
Urodela			
Ambystoma tigrinum	Arginine-vasotocin	+	Glomerular effect
Triturus alpestris	Arginine-vasotocin	+	
Triturus cristatus	Arginine-vasotocin	+	
Necturus maculosus	Arginine-vasotocin	+	
Salamandra maculosa	Arginine-vasotocin	0	

AVT = arginine-vasotocin, AVP and LVP = arginine or lysine-vasopressin

Neurohypophysial peptides, when injected, can reduce the urine volume in a large number of anurans and urodeles (Table 6.6). It is, however, notable that this response may be poor or undetectable in some amphibians, especially aquatic forms like *Xenopus laevis* and *Necturus maculosus*. Peptides extracted from either mammalian or amphibian neurohypophyses may produce an antidiuresis, but amphibian vasotocin is most active. In *Rana catesbeiana* and *R. esculenta* this peptide reduces the urine flow when injected to give concentrations estimated to be 10^{-10} to 10^{-11}M (URANGA and SAWYER, 1960; JARD, MAETZ, and MOREL, 1960). These are similar to the concentrations of vasotocin in the blood of dehydrated frogs and toads (BENTLEY, 1969a). Vasotocin has a dual action on the kidney; it depresses the GFR and increases the renal tubular reabsorption of water. The decreased GFR seems usually to result from a reduction in the number of active glomerule, as has also been observed among the reptiles. The early studies of WALKER *et al.* (1937) on the composition of the fluid in the renal tubules of frogs suggests that the tubular site of action of vasotocin is a distal one, like in the mammals. The other amphibian neurohypophysial peptides, oxytocin and mesotocin, when present at high concentrations, exert various effects on the kidney. In the bullfrog, oxytocin may decrease or even, on occasions, increase the flow of urine (URANGA and SAWYER, 1960). Mesotocin also has a diuretic effect when injected into *Rana esculenta* (JARD, 1966). Neither mesotocin nor oxytocin have a prominent effect on tubular water reabsorption, (acting mainly on the glomeruli) and indeed, as shown by MOREL and JARD (1963), may even antagonize this action of vasotocin. It seems likely that vasotocin performs the role of an antidiuretic hormone in many amphibians, especially the more terrestrial species. Nevertheless, it is also likely that other factors may act to mediate changes in urine flow.

c) The Urinary Bladder

As a tetrapod innovation, a urinary bladder derived embryologically from the cloaca, makes its phyletic debut in the Amphibia. It is a distensable sac into which the urine passes, and in some species, especially those that live in arid conditions, like the Australian leptodactylid frogs, it can hold fluid equivalent to 50% of the body weight (Table 6.7). TOWNSON in 1799 described how frogs 'have power of absorbing the fluids necessary for their support ... through the external skin ... a large part of them appearing to be retained in the so-called urinary bladder, though gradually thrown off again by the skin.'. The more recent experimental observations of EWER (1952a) and SAWYER and SCHISGALL (1956) have shown that toads and frogs reabsorb water from their urinary bladders when they are dehydrated (see also RUIBAL, 1962; SHOEMAKER, 1964). The urinary bladder of many terrestrial amphibians holds large quantities of water compared with more aquatic species (Table 6.7) and thus affords them a useful store of water, that can be used to replace body water at times when other supplies are limited.

Before the discovery of vasotocin, mammalian neurohypophysial peptides were shown, when injected, to increase water reabsorption from the bladders of toads and frogs (EWER, 1952a; SAWYER and SCHISGALL, 1956). Vasotocin has since been shown to be particularly active in promoting water reabsorption from this organ, being effective, *in vitro*, at concentrations of 10^{-11} to 10^{-12}M (JARD *et al.*, 1960; SAWYER, 1960). Mesotocin and oxytocin also increase the permeability of the anuran bladder to water but, in *Bufo marinus*, they are about 200 times less potent than vasotocin (Fig. 6.2).

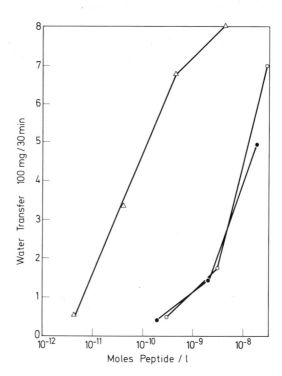

Fig. 6.2 Water transfer across the isolated urinary bladder of the toad *(Bufo marinus)* in the presence of vasotocin ($\triangle$), oxytocin ($\bullet$) and mesotocin ($\bigcirc$). Note: log-dose scale. (From Bentley, 1969b).

Neurohypophysial peptides increase the osmotic permeability of the bladder of a variety of anurans, including members of the families Ranidae, Bufonidae, Hylidae and Leptodactylidae, but not of the Pipidae *(Xenopus)* (Bentley, 1966a). They have been shown to exert such an action only in one species of urodele. The response of the bladder of the crab-eating frog is poor compared to other anurans (Dicker and Elliott, 1970b). The urinary bladders of the urodeles *Ambystoma*, *Necturus* and *Triturus* do not respond (Bentley and Heller, 1964), nor have I been able to demonstrate such an effect in *Siren* or *Amphiuma*. The fire-salamander, however, *Salamandra maculosa*, does reabsorb water from its bladder, *in vitro* or *in vivo*, in response to vasotocin (Bentley and Heller, 1965). It is interesting that vasotocin does not seem to act on the kidney of the latter.

Thyroid hormones have been reported to increase the permeability of the bladder (and skin) of toads, *Bufo bufo*, to water (Green and Matty, 1963) but a careful study by Taylor and Barker (1967) has failed to confirm these observations. The reason for the discrepancy is not clear.

d) Large Intestine (Colon)

The Amphibia have a well developed large intestine across which sodium is actively transported (Ussing and Andersen, 1955; Cooperstein and Hogben, 1959). Such transport can result in the production of osmotic gradients along which water

Table 6.7 *Capacity of the urinary bladders of different species of Amphibia living in various habitats. (From data collected by* BENTLEY, *1996 a)*

	Capacity of Bladder (% body wt. *)	Habitat
Anura		
Notaden nichollsi	50	Semi-arid tropical steppe, fossorial (Australia)
Neobatrachus wilsmorei	50	Arid desert, fossorial (Australia)
Cyclorana platycephalus	50	Tropical and arid desert, fossorial (Australia)
Bufo cognatus	44	Arid, semi-arid desert, fossorial (U. S. A. Mexico)
Bufo marinus	25	Wet tropics (central America)
Hyla moorei	30	Cool wet, aboreal (S. W. Australia)
Hyla leseuri	30	Summer rain forest, arboreal (N. Australia)
Xenopus laevis	1	Aquatic (African)
Urodela		
Salamandra maculosa	34	Terrestrial (viviparous S. Europe)
Ambystoma tigrinum	20	Temperate fossorial (N. America)
Triturus cristatus	2	Temperate, spends much time in water (Europe)
Necturus maculosus	5	Aquatic (eastern N. America)

* weight with bladder empty

can move. Mammalian antidiuretic hormone (ADH) can increase the osmotic flow of water across this organ *in vitro* (COFRE and CRABBE, 1967) but this effect is very small, an increase from 3.9 μl/cm^{2}h to 5 μl/cm^{2}h, and is probably not physiologically significant.

2. Salt Exchange

Aquatic amphibians accumulate salts from the environment. KROGH (1937; 1939) showed that frogs, *Rana esculenta,* can take up chloride and sodium from solutions that contain such ions only at a concentration of 10^{-5}M. Later studies by this Copenhagen school demonstrated that the skin of this frog, *in vitro,* could transport sodium actively from such external solutions while the chloride moves as a result of the electrical differences that attend the cation transfer (USSING and ZERAHN, 1951). Active chloride transport can, however, take place *in vivo* across the skin of frogs, *R. esculenta* and *R. temporaria,* and toads, *Bufo bufo* (JORGENSEN, LEVI, and ZERAHN, 1954). Frog skin, *in vitro,* has been shown to transport chloride actively in the South American frog, *Leptodactylus ocellatus* (ZADUNAISKY and CANDIA, 1962) but attempts to show this in other species have not been successful. It seems likely that chloride, as well as sodium, may be actively transported across the skin of at least some amphibians, but the former process may not readily survive *in vitro* procedures.

A Chilean leptodactylid frog, *Calyptocephalella gayi*, like *Rana esculenta*, can also accumulate sodium and chloride, independently, across its skin *in vivo* (GARCIA ROMEU, SALIBIAN, and PEZZANNI-HERNANDEZ, 1969). The Na^+ and Cl^- were found to be exchanged, respectively, for H^+ and HCO_3^- in the body. KROGH (1939) also found that European frogs exchanged Na^+ and Cl^- for endogenous solutes. Such exchanges cannot readily be demonstrated *in vitro*. The skin of amphibians plays an important respiratory role as it is an avenue for the excretion of carbon dioxide. In an aquatic environment this takes place as a diffusion of HCO_3^-, a function that would be expected to be more prominent *in vivo* than *in vitro* (GARCIA ROMEU *et al.*, 1969). As will be described later, fish in fresh water may exchange Na^+ for NH_4^+ across their gills (MAETZ and GARCIA ROMEU, 1964) and it has been suggested (GARCIA ROMEU *et al.*, 1969) the Na^+/H^+ substitution that is observed in these frogs is an adaptation to a terrestrial life, and reflects the adoption of ureotelism as opposed to ammoniotelism.

The exchange of sodium between intact amphibians and their aqueous environment responds to the prevailing osmotic conditions and to the injection of certain hormones. KROGH (1939) thus found that depleting frogs of salt by bathing them in solutions of distilled water for several weeks enhanced their ability to accumulate sodium and chloride. The French group at Saclay (see MAETZ, 1959) have also shown that maintaining anurans in distilled water increases their rate of sodium accumulation, while if they are kept for several weeks in 0.7% sodium chloride solutions, the rate is greatly reduced. Injections of neurohypophysial peptides increase the rate of accumulation of sodium from solutions bathing larval and adult *Ambystoma* (JORGENSEN, LEVI, and USSING, 1946; ALVARADO and JOHNSON, 1965), *Triturus alpestris* and *T. cristatus* (BENTLEY and HELLER, 1965), *Bufo regularis* (MAETZ, 1963) and *Rana catesbeiana* (ALVARADO and JOHNSON, 1966). The corticosteroid hormone aldosterone has been shown to increase the net uptake of sodium by *Rana esculenta* (MAETZ, 1959) and larval *Ambystoma tigrinum* (ALVARADO and KIRSCHNER, 1964). Such changes in the salt balance reflect actions at several distinct morphological sites.

a) Skin

The skin of amphibians can actively take up salt from the solutions that bathe their external surface. Sodium, and in some instances chloride, can be accumulated against an electro-chemical gradient. Potassium is not accumulated actively from the external solutions. The mechanism for the active sodium transport is very efficient, for, as shown by KROGH, this ion may be accumulated from solutions as dilute as 10^{-5}M. Active accumulation of sodium across the skin (*in vitro* and *in vivo*) has been demonstrated in a number of anurans from diverse families including the Ranidae, Bufonidae, Hylidae, Leptodactylidae and Pipidae. It has also been shown to occur across the integument of some urodeles including *Triturus alpestris* and *Ambystoma mexicanus*. I have been unable to find evidence from electrical measurements *in vitro* of any substantial active sodium transport across the skin of the aquatic urodeles, *Necturus maculosus, Siren lacertina* or *Amphiuma means* (BENTLEY, 1969b) but this could be due to damage that occurs during the removal of the skin from these animals. Such transport could also be obscured if sodium and an accompanying anion were transported in electrically equivalent quantities so that no electrical p.d. would be observed.

α) *Aldosterone.* The rate of sodium transport across anuran skin *in vitro* increases when the frogs or toads have been previously maintained for several weeks in distilled water, while it decreases when they are kept in 0.7% sodium chloride solution (MAETZ, 1959; see also CRABBE, 1966; HORNBY and THOMAS, 1969). MAETZ and his collaborators also found that the prior injection of aldosterone into frogs, *Rana esculenta* kept in saline greatly increased the rate of sodium transport that was subsequently observed across the skin *(in vitro)*, but it had a much smaller effect in frogs that had been kept in fresh water. This suggests that aldosterone is involved in the adaptational changes that are observed *in vivo.* Aldosterone has since been shown to increase sodium transport *in vitro* across the ventral skin of *Rana esculenta* and *Bufo marinus* (CRABBE, 1964). Previous to this TABENHAUS, FRITZ, and MORTON (1956), forecast such an action in what appears to be a relatively forgotten series of observations. They found that desoxycorticosterone and cortisol promoted fluid transfer across the isolated skin of *Rana pipiens.* Sodium was not actually determined but from measurements of the osmotic pressure of the transported fluid it clearly was transferred. The levels of aldosterone in the blood of *Bufo marinus* kept in distilled water are far higher than in those kept in dilute saline (CRABBE, 1961a). Aldosterone thus would seem to assist the regulation of sodium transport across the skin of anurans. There is, however, little information about its role in urodeles. As many species of this group live an almost exclusively aquatic existence, such a regulatory mechanism could be useful, though it is conceivably not vital. If, for instance, aquatic amphibians normally took up sodium at a constant (and maximal) rate through their skin, any excess salt, if modest, could be excreted through the kidneys just like excess water.

Frogs that enter, or live in, saline solutions would seem to have some special problems with regard to the permeability of their skin to salt. ADOLPH (1933) described how leopard frogs, *Rana pipiens,* when placed in hypoosmotic saline solutions gained weight due to the accumulation of salt. When anuran skin is exposed, *in vitro*, to external solutions of increasing sodium chloride concentration, the rate of sodium transport increases, due, presumably, to a depression of its chemical gradient across the membrane. It seems likely that the salt accumulation observed by ADOLPH results from such increased permeability to salt, in conjunction with an inability of the kidneys to excrete it. I have observed such salt accumulation in the anurans, *Rana catesbeiana, Bufo marinus* and *Xenopus laevis,* kept in 0.5% sodium chloride solutions (but not in the urodeles *Necturus, Amphiuma* or *Siren*). For these anurans a reduction in the permeability of the skin to sodium may be useful in such circumstances. When the green toad, *Bufo viridis,* adapts to living in 2% sodium chloride solutions, the electrical properties of its skin indicate a reduction of about 95% in the sodium transport (GORDON, 1962). Information about the skin of the crab-eating frog in sea-water is sparse, but the results of GORDON *et al.* (1961) suggest that there is also a considerable reduction in sodium transport. The ability to 'turn off' such cutaneous processes may be as important as 'turning (them) on' so that hormonal control could conceivably play an important role in such species. At present there is no information about this.

β) *Neurohypophysial Peptides.* Neurohypophysial peptides not only increase the permeability of the anuran skin to water but also promote active sodium transport. This also occurs in some urodeles. FUHRMAN and USSING (1951) showed that mammalian neurohypophysial peptides increase sodium transport across frog skin and vasotocin has since been shown to be even more effective, being effective at concentrations as low as 10^{-10}M (JARD *et al.*, 1960; MAETZ, 1963). This action on sodium transport ('natriferic effect') has been demonstrated in a number of anurans including *Rana esculenta, R. catesbeiana, Bufo bufo, B. marinus* and *Xenopus laevis* (MAETZ, 1963; BENTLEY, 1969 b). The isolated skin of the urodeles *Triturus alpestris* and *T. cristatus* (BENTLEY and HELLER, 1964) and *Ambystoma mexicanus* (ACEVES, ERLIJ, and EDWARDS, 1968) also respond in this manner. I have been unable to show such an effect in *Ambystoma tigrinum, in vitro,* but the results of ALVARADO and JOHNSON (1965) suggest that it occurs *in vivo.* The increased transport of sodium and water in response to neurohypophysial peptides does not represent a single general action of such peptides on the membrane; the two processes, water and sodium transfer, are quite distinct. This is indicated by several observations. Thus BOURGUET and MAETZ (1961) showed that the relative activities of different neurohypophysial peptides on water and sodium transfer are not parallel. In addition, some amphibians like *Triturus alpestris* and *Xenopus laevis,* exhibit the natriferic but not the hydrosmotic (water) response. Whether the natriferic effect of such peptides respresents a significant osmoregulatory response is uncertain. However, it should be remembered that this action of neurohypophysial peptides is less sustained than that of the corticosteroids, and that frogs and toads placed in salt solutions release large amounts of vasotocin into the circulation. The latter does not make homeostatic sense with respect to sodium regulation. In the Amphibia the natriferic effect of vasotocin is probably not osmoregulatory in its nature though it may conceivably have a more localized function, or even reflect some osmotic significance in the ancestors of this group.

Other hormones have also been reported to increase sodium transport across the amphibian skin. Adrenaline has this effect in *Rana esculenta;* this action is probably mediated by production of cyclic AMP, which is also a metabolic intermediate in the action of neurohypophysial peptides (BASTIDE and JARD, 1968). Adrenaline promotes loss of chloride from the frog skin by stimulating the secretion of skin glands (USSING, 1960). Thyroid hormones have also been reported to increase sodium transport across the skin of *Bufo bufo* (GREEN and MATTY, 1963) but others (TAYLOR and BARKER, 1967) have been unable to confirm this. The skin of *Bufo marinus* exhibits an increased active sodium transport when it is exposed to insulin and if this acts in conjunction with aldosterone, sodium transport is considerably higher than if each is present alone (ANDRE and CRABBE, 1966). Thus optimal rates of sodium transport *in vivo* may be the result of an interaction of several hormones.

b) Kidney

The urine of amphibians kept in fresh water contains little salt. Thus the ureteral urine from *Rana esculenta* living in fresh water contains only about 5 m-equiv/l sodium (JARD and MOREL, 1963) while in urodeles, such as *Triturus,* the sodium levels in the bladder urine are similar to this. Analysis of the glomerular filtrate obtained by micropuncture of the renal tubules indicates that sodium and chloride are progressively reabsorbed as they pass down the nephron, this occurring mainly in the proximal but also in the distal segment (WALKER *et al.*, 1937).

The role of hormones in regulating these processes in amphibians is not clear. While from the mammalian evidence it seems likely that corticosteroids influence sodium reabsorption and potassium excretion this has not been directly shown in frogs and toads. Injections of aldosterone into *Bufo marinus* have no consistent effect on renal sodium losses (MIDDLER *et al.*, 1969). MAYER (1963; 1969) also found this to be so in *Rana esculenta*. Information about the role of corticosteroids in controlling renal electrolyte excretion in the Amphibia is still sparse.

Vasotocin increases sodium reabsorption by the renal tubule of *Rana esculenta* (JARD and MOREL, 1963; JARD, 1966), an observation similar to that described previously in the snake, *Natrix sipedon*. Vasotocin also decreases the urinary sodium loss in *Triturus* (BENTLEY and HELLER, 1964). The injection of large amounts of mammalian neurohypophysial peptides, as well as mesotocin, increases the renal sodium excretion in *Rana pipiens* which have been 'loaded' with saline (BOYD and WHYTE, 1939; SAWYER and SAWYER, 1952). These effects probably result from an increased GFR, perhaps reflecting the vasoactive properties of these peptides when present at high concentration. The natriuretic effects of such peptides are only of pharmacological significance but their role in promoting renal sodium conservation is uncertain.

c) The Urinary Bladder

The urinary bladder of the Amphibia, in addition to performing the function of a water reservoir, can decrease the sodium content of the urine. Active sodium transport has been shown *(in vitro)* to take place across the urinary bladders of frogs, *Rana esculenta*, and toads, *Bufo marinus* (LEAF *et al.*, 1958).

Sodium transport across the urinary bladders of *Bufo marinus* and *Rana ridibunda* is increased by aldosterone, either when the animals are injected 24 h prior to measuring sodium transport *in vitro*, or after the tissue is directly exposed to the steroid's action (CRABBE, 1961b, a; 1963). Other steroids, including corticosterone, are not as active and at very high concentrations (100 times as great) may inhibit its action. Changes in the sodium concentration of the media bathing the toads influences the rate of sodium transport across the bladder in a manner that parallels that of aldosterone in the blood; a low external sodium level increases aldosterone in the blood and stimulates sodium transport.

LEAF *et al.* (1958) found that oxytocin increases sodium transport across the toad urinary bladder and vasotocin has been found to be very active in this process in frogs (JARD *et al.*, 1960). In *Bufo marinus* vasotocin is about 100 times more

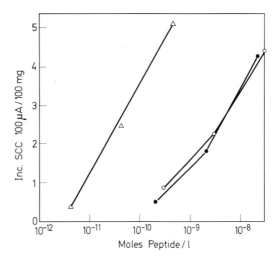

Fig. 6.3 Sodium transfer
(as short-circuit current) across
the isolated toad *(Bufo marinus)*
urinary bladder in the presence
of vasotocin (△), oxytocin (●)
and mesotocin (○).
Note: Log-dose scale.
(From BENTLEY, 1969b).

active in increasing sodium transfer across the bladder than either oxytocin or mesotocin (Fig. 6.3).

Like in the anuran skin, several other hormone preparations have been shown to promote sodium transport across the bladder. These include insulin (HERRERA, 1965) and adrenaline (JARD *et al.*, 1968). Various endocrine factors may interact in their effects on the bladder. Aldosterone can facilitate the effects of neurohypophysial peptides (FANESTIL *et al.*, 1967; HANDLER, PRESTON, and ORLOFF, 1969). Aldosterone and insulin, similarly, together can produce a far greater increase in sodium transport *in vitro* than either can produce alone (CRABBE and FRANCOIS, 1967). Such endocrine interactions probably assist all such tissues to attain optimal levels of sodium transport.

Sodium reabsorption from the urinary bladder probably occurs in the urodeles *Necturus maculosus* and *Ambystoma tigrinum* but vasotocin has not been found to increase this (BENTLEY and HELLER, 1964). When mudpuppies are preinjected with aldosterone their urinary bladders are subsequently found *(in vitro)* to have a facilitated rate of sodium transport, though this steroid could not be shown to have an effect when applied directly *in vitro* (BENTLEY, 1971a).

Sodium reabsorption from the urine stored in the bladder could contribute to the sodium conservation of some amphibians. If sufficient sodium were reabsorbed this could be particularly important in aquatic species. There is, however, no definitive information available as to the relative (to other organs and in different species) physiological significance of this. Such sodium transfer, at least in anurans, would appear to be influenced by changes in the circulating levels of aldosterone and could also increase on occasions when vasotocin is released.

d) The Large Intestine (Colon)

Sodium is actively transported from the mucosal to serosal side of the colon in anurans (USSING and ANDERSEN , 1955; COOPERSTEIN and HOGBEN, 1959). In the toad, *Bufo marinus,* this process can be increased by aldosterone and neurohypophysial peptides (COFRE and CRABBE, 1965; 1967). FERREIRA and SMITH (1968) found that when they placed these toads in dilute saline for two to three weeks, the sodium transport across the colon *(in vitro)* reversed so that it was occuring from the serosa to mucosa. This resulted in the sodium concentration at the mucosal surface being higher than it was in the plasma. It would thus seem that in such saline adapted toads, there is an extrarenal sodium excretion across the solon, but whether or not this plays any role in the life of anurans that live in salt water is unknown. This very interesting directional change in sodium transport across the colon is, as we shall see later, reminiscent of the reversal in ion transfer that may occur when euryhaline fishes are transferred from fresh water to sea-water.

3. Nitrogen Metabolism

The end-products of nitrogen metabolism require water for their excretion and the amount that is required depends mainly on whether ammonia, urea or uric acid is formed. The formation of uric acid is a reptilian and avian prerogative, while the Amphibia, like the lungfishes, can form both ammonia and urea.[1]

The formation of ammonia requires the presence of abundant water for its excretion, as it is very toxic; this occurs in larval amphibians and in the neotenous urodele, *Necturus maculosus.* In the latter about 90% of the total waste nitrogen is excreted as ammonia, not through the gills or kidneys, but through the skin (FANELLI and GOLDSTEIN, 1964). Even normal adult forms of amphibians can excrete large amounts of ammonia; the toad, *Xenopus laevis,* does this when it is in fresh water, but if kept out of water or in dilute saline solutions, it accumulates urea instead (BALINSKY, CRAGG, and BALDWIN, 1961). This is similar to what happens in the African lungfish, *Protopterus,* when it aestivates buried in mud. In *Xenopus* the increased formation and accumulation of urea is accompanied by a two-fold increase in the activity of the ornithine-urea cycle enzymes present in the liver (McBEAN and GOLDSTEIN, 1967). *Rana pipiens* can also be persuaded to accumulate extra urea when it is kept in dilute saline solutions, but this does not appear to be influenced by the pituitary hormones as it is not significantly changed after hypophysectomy (SCHEER and MARKEL, 1962).

As described earlier, some amphibians have a remarkable ability to accumulate and withstand high concentrations of urea in their body fluids. This is seen in aestivating spadefoot toads and in the marine crab-eating frog. In the latter the ability

---

[1] J. P. Loveridge (Arnoldia, Publication of National Museums of Rhodesia, Vol. 5, 1–6 (1970) has recently shown that a Rhodesian frog, *Chiromantis xerampelina,* excretes substantial amounts of uric acid in its urine. This fascinating xerophilous frog survives for prolonged periods in dry exposed situations. It loses water by evaporation very slowly (comparable with that in a local lizard, *Chamaeleo dilepis*) but can absorb water very rapidly through its ventral cutaneous surface.

to conserve this urea in the body fluids is vital for its osmoregulation as it is used to balance the body fluids osmotically with the external sea-water. The skin is presumably relatively impermeable to this solute, but precise measurements would be interesting as neurohypophysial peptides can increase the permeability of this anuran membrane to urea (see Ussing, 1960). The kidney is potentially an important avenue for urea loss, but as much as 99% of that filtered at the glomerulus is reabsorbed across the renal tubule of the crab-eating frog (Schmidt-Nielsen and Lee, 1962). It is unknown whether the actions of hormones on the kidney can interfere with this process.

4. Behaviour

The importance of behaviour in limiting the evaporation of water has already been stressed. Amphibians may hide in dark places under rocks or vegetation (termed cryptozoic) and so reduce their water losses. Others, provided with stout hind legs for digging, may burrow into the earth. Frogs like *Heleioporus* have been found 80 cm below the ground's surface while others like *Neobatrachus* only appear to dig down to 30 cm (Main *et al.*, 1959). The depth is probably largely determined by the friability of the soil in the area in which the animals live, rather than the industry of each group of frogs. Such anurans may burrow down to regions of damp earth where the humidity is elevated and they may even be able to absorb moisture. All burrowing frogs do not occupy arid areas, but most of the animals in such regions are efficient burrowers. There are exceptions like the small *Hyla rubella*, which is widespread throughout the dry areas of Australia and owes its survival largely to its cryptozoic habits (see Warburg, 1965 b).

During prolonged periods of drought frogs may aestivate in burrows. This process is common among desert species but little is known about its physiological basis. McClanahan (1967) has given an account of some of the physiological processes that occur during aestivation in spadefoot toads, *Scaphiopus couchi*. This work is, I think, the most complete account of the life of a single anuran living in a desert environment. I strongly recommend it to those interested in this subject. During aestivation McClanahan found that these toads sustain a large increase in the concentration of their body fluids due largely to the accumulation of urea. This is excreted when the toads emerge after rain has fallen. Such rain may fall only rarely and it is thought that these animals may aestivate for periods of longer than a year. Attempts to measure the oxygen consumption of the aestivating toads were not successful, as they were aroused by the experiment, but the rates of oxygen consumption of resting animals were recorded. From these values, and the fat content of the body, it was estimated that the aestivating toads could survive for at least one year with the available food stores in the body. If the metabolic rate of aestivating toads were reduced further they could survive even longer, providing that their ability to withstand the osmotic changes are also equal to the task. The endocrinology of such aestivating toads, both with respect to their osmoregulation and their general metabolism should be most interesting.

5. Reproduction

Alll amphibians undergo a period of larval development which usually takes place in pools of water. Some amphibians, especially certain urodeles, exhibit varying degrees of viviparity, but this is usually in species occupying cold alpine areas and seems to be an adaptation to low prevailing temperatures, rather than to a lack of water. Embryonic, and even larval, development can occur on the land, as seen in some Australian leptodactylid frogs. Such frogs, nevertheless, require a damp situation in which to lay their eggs and so do not inhabit arid areas (MAIN et al., 1959).

Frogs living in regions where water is only sporadically available, can breed only when suitable rain falls. In most desert areas, like those in the centre of Australia, rain is not only sparse but also unpredictable. Leptodactylid frogs in the genus *Heleioporus*, which live in more temperate areas where rain falls more consistently, have a precisely timed breeding season that coincides with the rain. However, other members of this family, like some species of *Neobatrachus*, which live in less favourable areas, only come into a reproductive condition when the rainfall and temperature are suitable. These species are called 'opportunistic breeders' (MAIN et al., 1959). This pattern of reproductive behaviour is the same as that observed in the budgerygahs and zebra finches which also live in central Australia. Reproduction is a special endocrine problem and it would be interesting to know something about the mechanisms which can so promptly initiate breeding.

When the eggs of amphibians are deposited and fertilized, they undergo a period of embryonic development and then on hatching, the larvae grow and eventually differentiate and metamorphose into the adult form. In most amphibians this process may take many months, such as in the bullfrog in which it takes more than a year. In desert areas in which such a prolonged embryonic and larval life is usually not possible it is far shorter, often only a few weeks. Young, but adult, *Scaphiopus couchi* have been observed in the field only four weeks after rain has fallen, while tadpoles, at an advanced stage of development (with fore and hind limbs), have been seen after only 10 days (MAYHEW, 1965). Just what are the factors, other than genetic, that influence the rate of development in such amphibians is not clear. The propensity of the thyroid to accelerate amphibian metamorphosis suggests that it may be involved. Environmental conditions such as temperature, food supplies and possibly the chemical (osmotic?) constitution of the pond undoubtedly influence the rate of larval growth.

When certain newts prepare to breed, they may undergo a morphological transition and return to an aquatic life. This has been likened to a second metamorphosis. The timing of these changes and the return of such newts to water is influenced by a hormone from the adenohypophysis. The life cycle of the eastern spotted newt, *Diemictylus* (also called *Triturus) viridescens*, takes place in three stages; an aquatic larval phase, followed by a terrestrial life which lasts from 2 to 4 years (the 'red eft' stage) and finally a return to water as sexual maturity approaches. REINKE and CHADWICK in 1939 showed that when pituitaries from the sexually mature newts were transplanted into the immature red efts they were induced to seek water ('water drive') and to assume their reproductive form. This effect also occurred after the injection of mammalian prolactin, even after removal of the thyroid and gonads (CHADWICK, 1940; 1941). Furthermore, pituitary trans-plants from a wide range of other vertebrates could also induce this 'water drive'. Prolactin is also thus effective in hypophysectomized newts, while other mam-malian pituitary hormones including growth hormone, thyrotrophin, luteinizing hormone, corticotrophin and neurohypophysial extracts are without such an action (GRANT and GRANT, 1958). Prolactin also can induce *Triturus cristatus* and *T. al-pestris* to seek water in order to breed (TUCHMANN-DUPLESSIS, 1948). Thus a pro-lactin-like hormone can initiate the change from a terrestrial to an aquatic life in some newts. Whether this hormone also confers any changes which directly in-fluence osmoregulation are unknown. Morphological changes take place in the skin of such newts and these conceivably could influence its permeability to water and salts. The responses of the three life stages of *Diemictylus* to vasotocin have been measured; all responded by accumulating water but, while this was less in the larval form, it was similar in magnitude in the terrestrial efts and aquatic 'adults' (OVER-ACK and DEROOS, 1967). Current interest in the possible osmoregulatory effects of prolactin are centered on its effects in certain euryhaline fishes. Injections of this hormone as we shall see, promote survival of such fish after they have been hypophysectomized and placed in fresh water.

6. Osmoregulation in Larval Amphibians

The larvae of Amphibia live an exclusively aquatic existence and even breathe with the aid of gills. They thus show more physiological affinities with their ancestors, the fishes, than do adult amphibians.

Anuran larvae, or tadpoles, usually live in fresh water, though those of *Rana cancrivora* live in the sea (GORDON et al., 1961). The larvae of the bullfrog, *Rana catesbeiana*, may attain a weight of nearly 20 g and hence have been especially fa-voured by the physiologists for the study of tadpole osmoregulation. The available information is, nevertheless, rather incomplete. Bullfrog tadpoles readily exchange sodium with the solutions that bathe them, but the influx and outflux of this ion are about equal and there is no evidence of an active accumulation by the animal other than by feeding (ALVARADO and JOHNSON, 1966). TAYLOR and BARKER (1965) measured the electrical p. d. across the isolated skin of these tadpoles and found it, in contrast to adult bullfrogs, to be zero in young larvae and did not begin to develop until metamorphosis was about to occur. This also suggests (but *not* con-clusively) that active ion transport does not take place across the skin of young tadpoles. This deficiency has also been correlated with the low levels of Na-K ac-tivated ATPase in the skin of the larvae (KAWADA, TAYLOR, and BARKER, 1969). The rate of water uptake *in vivo* is unknown, but the skin of tadpoles *(in vitro)* is osmotically permeable, water moving across it at the rate of about $3\mu l/cm^2 h$, which is only about 25% as fast as in bullfrogs (BENTLEY and GREENWALD, 1970). Presumably this water and any that is taken in through the gut, is normally excreted by the kidneys.

More recently ALVARADO and MOODY (1970) have shown that bullfrog tad-poles which have been salt-depleted (in distilled water) can actively accumulate sodium and chloride. This results mainly from a facilitated influx of sodium but there is also a small decrease in outflux (renal effect?). Experiments on the isolated skin confirm that the site is not a cutaneous one. Perfusion of the tadpoles branchial chambers suggests that the gills are the site of this active ion accumulation. Drinking in these tadpoles was considerable and was equivalent to about 33% of the body weight each day.

The neurohypophysis of bullfrog tadpoles contains an activity that increases the permeability of the toad bladder to water and this is presumably due to the vasotocin which is known to be present in adult anurans (BENTLEY and GREEN-WALD, 1970). The activity present, is only 25% as much (activity/kg body weight) as that in bullfrogs and, interestingly, is similar in quantity to that present in the neurohypophysis of a neotenous form, the mudpuppy (FOLLETT and HELLER, 1964a). When vasotocin is injected into these tadpoles it causes a water retention. This is rather less than in adults, especially in the younger tadpoles, but increases as metamorphosis approaches (ALVARADO and JOHNSON, 1966; BENTLEY and GREENWALD, 1970). We have found a small increase in the osmotic permeability of the skin (in vitro) when it is exposed to vasotocin, but the water retention in vivo is almost certainly mainly due to a reduced urine flow. Vasotocin also causes a net accumulation of sodium by bullfrog tadpoles, and this similarly results from a decreased renal excretion. The cutaneous effect of vasotocin on sodium movement in adults, is not apparent at this stage (ALVARADO and JOHNSON, 1966). The administration of thyroxine accelerates metamorphosis in amphibian larvae and when this hormone is given to bullfrog tadpoles their skin develops an electrical p. d. between its two sides, presumably reflecting the onset of active sodium transport (TAYLOR and BARKER, 1965). It would not be surprising to find that corticosteroids influence the electrolyte metabolism of the tadpole but I have no information about this.

The tadpoles of the marine frog, *Rana cancrivora*, live in the sea and so have rather special osmoregulatory problems. In contrast to the adults that live in this environment, they maintain their body fluids at a level that is hypoosmotic to the sea-water (GORDON and TUCKER, 1965). How they manage to do this is uncertain, but it was suggested that extrarenal salt excretion, possibly through the gills, may occur. Needless to say these tadpoles are particularly interesting as their pattern of osmoregulation shows a remarkable similarity to that of many marine fishes.

Larvae of the salamander, *Ambystoma tigrinum*, exchange water and sodium with the fresh water in which they live (ALVARADO and KIRSCHNER, 1963). Water uptake in these larvae is about 26% of their body weight in a day which is slightly less than that observed in the adults. This water is excreted as a dilute urine that also contains about 75% of the larva's total sodium efflux, compared with 55% in adults. Sodium is also exchanged across the skin with an active uptake of sodium and chloride.

ALVARADO and JOHNSON (1965) have compared the effects of vasotocin in larval and adult salamanders. They found that while this peptide caused water retention in both larvae and adults, the effect was less in the former. In both instances the effect was due solely to a reduction in urine flow resulting (in the larvae) from a decrease in the GFR. Vasotocin injections also resulted in the accumulation of sodium, a prolonged effect that persisted for 43 h in the adults. Such an effect has also been observed in the axolotl (JORGENSEN et al., 1946). Vasotocin usually only

188

acts for 3 to 6 h, when injected into amphibians, so that the extended period of action is rather surprising. Furthermore, I have been unable to detect any action of this peptide on sodium transfer *in vitro* across the skin of adult *Ambystoma tigrinum* (BENTLEY, 1969b). It seems likely that some other factor is involved *in vivo* and an endogenous release of aldosterone would be more consistent with the time course of the change in sodium metabolism. Injected aldosterone can increase the influx of sodium into larval *Ambystoma* (ALVARADO and KIRSCHNER, 1964).

The osmoregulation of larval amphibians seems to be well adapted to their aquatic life but they nevertheless still exhibit some similarities, such as a 'water balance effect', to their terrestrial adult forms. This response is poorly developed in the larvae, but in bullfrog tadpoles and in the axolotl (H. HELLER unpublished observations quoted by BENTLEY and HELLER, 1964) they can be increased by accelerating metamorphosis by administration of thyroid hormone. These animals thus offer a unique opportunity for studying the changes in the osmoregulatory processes that are associated with changing from an aquatic to a terrestrial mode of life.

7. Function of the Amphibian Adrenocortical Tissue in Relation to Osmoregulation

Amphibian adrenocortical tissue secretes principally corticosterone and aldosterone, and in some instances cortisol, (Table 6.4). Aldosterone, and corticosterone (less strongly) have been shown, both *in vivo* and *in vitro*, to increase the rate of sodium transport across the skin and urinary bladder of frogs and toads. In addition, the corticosteroid levels in the circulation change in response to alterations in the sodium content of the bathing media. This suggests that these steroids may be involved in assisting the regulation of amphibian salt metabolism.

The rates of secretion of corticosteroids are partly under the control of the adenohypophysis. CARSTENSEN *et al.* (1959; 1961) found that when adrenocortical tissue from bullfrogs was incubated with either mammalian corticotrophin or an extract of the frogs anterior pituitary, the rate of corticosteroid production increased; aldosterone being produced at 4 times the rate of corticosterone. The circulating levels of these corticosteroids have since been measured in bullfrogs, corticosterone being found to be predominant (JOHNSTON *et al.*, 1967). CRABBE (1961a) has also shown this to be so in *Bufo marinus*. Injection of mammalian corticotrophin into the bullfrogs produced a ten-fold increase in the aldosterone level in the blood, while corticosterone increased three-fold. However, in hypophysectomized bullfrogs the amount of aldosterone in the blood was similar to that in normal animals but the corticosterone level was reduced 5-fold. For the higher tetrapods there is evidence indicating that aldosterone secretion may be predominantly influenced by the release of renin from the kidneys. This enzyme initiates the formation of angiotensin in the blood and so stimulates steroid secretion from the gland. JOHNSTON *et al.* (1967) extracted a renin-like material from frog kidneys, which, when injected into hypophysectomized bullfrogs, increased the secretion of aldosterone, but left that of corticosterone unchanged. The renin activity is greater if the frogs are kept in distilled water instead of 0.02% sodium chloride solution (CAPELLI *et al.,* 1970). ULICK and FEINHOLTZ (1968) have shown that injection of mammalian corticotrophin also increases the rate of secretion of aldosterone into the blood of bullfrogs but that the injection of mammalian angiotensin fails to produce such an effect. This could, however, be due to differences in the structures of the mammalian and homologous bullfrog peptides. When the bullfrog were depleted of sodium by bathing them in distilled water the circulating levels of aldosterone increased. CRABBE (1961 a; 1966) previously found that while circulating levels of aldosterone were greatly increased in *Bufo marinus* depleted of sodium, those of corticosterone increased only slightly. The increased aldosterone secretion was accompanied by a parallel stimulation in the rate of sodium transport across the skin, urinary bladder and colon of the toads. CRABBE also found that the injection of both mammalian corticotrophin and angiotensin increased the rate of sodium transport across these membranes. He has suggested that, while corticotrophin injections may initiate secretion of aldosterone, the changes in the levels of hormone that normally accompany alterations in the available sodium are probably mediated by the formation of angiotensin.

The adrenocortical tissues in the Amphibia are closely associated with the surface of the kidney, so that adrenalectomy is a difficult procedure that may be accompanied by renal damage. Attempts to perform this operation are thus sparse. The toad, *Bufo arenarum,* has been shown to suffer an excessive loss of salt following adrenalectomy (MARENZI and FUSTINONI, 1938). FOWLER and CHESTER JONES (1955) destroyed the adrenocortical tissue in the frog, *Rana temporaria,* and found that while winter frogs survived this operation for prolonged periods, summer frogs died within 2 days. These frogs suffered large losses of sodium and an accumulation of potassium which is the classical mammalian reaction to adrenocortical insuffiency. When the summer frogs were placed in isoosmotic saline solution they survived (see CHESTER JONES *et al.,* 1959) suggesting that death was due to faulty electrolyte balance.

Since adrenocortical tissue in frogs is at least partly under the control of the adenohypophysis, further attempts to elucidate its role have been made by studying the effects of the simpler operation of hypophysectomy. It has become apparent recently that this will not completely extinguish corticosteroid secretion as aldosterone levels are little affected by this operation. Hypophysectomy in *Rana temporaria* results in a progressive loss of sodium and chloride (JORGENSEN, 1947) but this loss seems to be slower than that following adrenalectomy in this species. Sodium loss has also been observed in hypophysectomized *Rana esculenta* (JORGENSEN and ROSENKILDE, 1957) and *Bufo marinus* (MIDDLER *et al.,* 1969), though RIDLEY (1964) failed to observe such an effect in bullfrogs. MIDDLER *et al.* were able to prevent this loss of sodium by injecting mammalian corticotrophin into the toads and this hormone has also been shown to prolong the survival of hypophysectomized *Bufo bufo* (JORGENSEN and LARSEN, 1963). The sodium loss observed in hypophysectomized anurans seems to be extrarenal i.e. through the skin. MIDDLER *et al.* (1969) could detect no changes in renal sodium excretion following such an operation but this could be due to the continued presence of aldosterone. This steroid, however, could not be shown to alter renal sodium excretion in toads adapted to a bathing media of either distilled water or saline. Neither could MAYER (1963; 1969) detect such an effect in *Rana esculenta* prepared in this way. A renal site of corticosteroid action thus remains somewhat enigmatic in the Amphibia. Nevertheless, there remain three extrarenal sites where these steroids can act: the skin, the urinary bladder and the colon. Cutaneous losses of sodium chloride have been shown to increase in hypophysectomized frogs, *Rana temporaria* and *R. pipiens* (JORGENSEN, 1947; MYERS, FLEMING, and SCHEER, 1956). This is due to an accelerated outflux ('leakage') of sodium while the active transport mechanism *per se* appears to be little affected. It is interesting that the skin of *R. pipiens* also becomes more permeable to water following hypophysectomy (LEVINSKY and SAWYER, 1952). MIDDLER *et al.* were able partially to maintain hypophysectomized *Bufo marinus* with injections of cortisol or corticosterone, but aldosterone was completeley ineffective. The available evidence suggests that the excessive sodium loss experienced by hypophysectomized amphibians is due predominantly to a deficiency of 'glucocorticoids' (corticosterone) rather than the absence of a 'mineralocorticoid'. A combination of the two is probably optimal for maintaining such animals. This is well known to be the case in mammals which cannot be maintained during adrenal insufficiency by aldosterone alone, whereas cortisol is adequate. Corticosterone is usually considered to be a glucocorticoid but it also exerts mineralocorticoid effects in amphibians; it can, for instance, stimulate sodium transport across the toad urinary bladder (HANDLER *et al.*, 1969).

Little is known about the physiology of adrenocortical tissues in urodeles. Hypophysectomized newts, *Triturus cristatus*, suffer an excessive loss of both sodium and potassium (PEYROT *et al.*, 1963). The injection of aldosterone or cortisol into these newts results in a retention of sodium and a loss of potassium by the muscles (FERRERI *et al.*, 1967). When both of these steroids were given together their effects were greatly enhanced. Aldosterone, when injected into *Necturus maculosus*, increased sodium transport across their urinary bladder *in vitro* (BENTLEY, 1971a). The adrenocortical physiology of this interesting group of amphibians deserves much more attention.

The homologous corticosteroids of amphibians are doubtless involved in the regulation of electrolytes. The osmoregulation of the Amphibia is somewhat complicated compared to other tetrapods, as their skin provides a prominent pathway through which salts may travel in either direction. Such transfers may take place by passive diffusion or be influenced by metabolically controlled active transport processes. The integrity of this barrier is vital to osmotic equilibrium, especially in an aquatic environment. In other tetrapods renal processes are the principal determinant of salt conservation, but in the Amphibia the skin and urinary bladder are also involved. Corticosteroids affect the osmoregulation in amphibians in three main ways. Firstly, they influence the normal metabolic processes in the body (glucocorticoid action) and so probably play a role in maintaining the integrity of the skin to the passive transfers of molecules. Thus in the absence of such hormones the skin may become more 'leaky' resulting in a steady loss of solute. Secondly, they may play a role in providing suitable metabolic 'fuel' for the processes of active transport of ions, also a glucocorticoid effect. Thirdly, a mineralocorticoid effect (principally due to aldosterone?) may be more directly involved in controlling the day-to-day activity of the sodium 'pumps'. The sites of action of aldosterone in amphibians differ from those in mammals for while it has been shown unequivocally to increase sodium transport across the skin and urinary bladder, there is little or no evidence to indicate that it increases sodium conservation by the kidney. Physiologically this may be unnecessary owing to the presence of extrarenal sites where regulation may also be effected. Possibly the acquisition of a renal site of action represents a phyletic advance that evolved subsequently in the other tetrapods.

8. Physiology of the Amphibian Neurohypophysis in Relation to Osmoregulation

Vasotocin and mesotocin have been identified in a variety of anuran and urodele amphibians, and oxytocin has also been found in *Rana pipiens* (Table 6.3). Vasotocin is the predominant peptide in the neurohypophysis, usually being present at more than twice the concentration of mesotocin. The amounts of these materials found in the neurohypophysis are similar to those in other tetrapods, 10^{-8} to 10^{-9}M/kg body weight (Table 6.8). There are, however, some interesting exceptions, as in the aquatic urodele, *Necturus maculosus*, and the bullfrog tadpole in which the neurohypophysis contains only about 20% of the usual quantity of vasotocin. These amounts are similar to those usually observed in the fishes and it is tempting to suggest that they may reflect their aquatic habitats. Other aquatic amphibians, like *Xenopus laevis* and *Siren lacertina*, however, have neurohypophysial levels of vasotocin that are comparable with those of terrestrial species. As the tadpole is a larval form and *Necturus* is neotenous, the differences between these and other amphibians probably reflect the stage of their development.

Injections of vasotocin cause water retention in most amphibians and, as predicted by the results of STEGGERDA (1937), aquatic species respond poorly as compared with terrestrial ones (Table 6.9). We (HELLER and BENTLEY, 1965) also noted that urodeles generally do not respond as well as anurans even when they come from similar habitats. Vasotocin may affect the water (and sodium) metabolism of amphibians by an action at several distinct sites:

(i) The kidney, by reducing the flow of urine and sodium loss.

(ii) The urinary bladder, by promoting reabsorption of water and sodium.

(iii) The colon, where vasopressin has been shown to have a small effect in promoting reabsorption of water and sodium from the contents of the gut.

(iv) The skin, where it promotes uptake of water and sodium from the external solutions.

Not all of these effector sites are present in every amphibian and differences in the distribution of these account partly for the variations that are observed in the magnitude of the effects of injected vasotocin. Thus in the urodeles, neither the skin, nor the bladder (with one exception) undergo an increase in osmotic permeability in the presence of vasotocin. Indeed urodeles usually exhibit a renal response (antidiuresis) to such peptides; the only exception is the fire-salamander which, instead, increases the osmotic permeability of its urinary bladder. Most anurans, on the other hand, reduce their urine volume and also increase the osmotic permeability of their skin and bladders in response to vasotocin. The aquatic toad, *Xenopus laevis*, is the only known exception; water transfer in this animal is unaffected by vasotocin, though sodium transport across the skin is increased.

192

Table 6.8 *Storage of vasotocin in the neurohypophyses of various species of the Amphibia*

	$m\mu$-moles/kg body weight	Habitat
Anura		
Bufo bufo[1] (European toad)	4.9	Terrestrial
Bufo marinus[2] (Giant toad)	2.9	Terrestrial
Rana esculenta[1] (European frog)	2.2	Amphibious
Rana catesbeiana[2] (Bullfrog)	3.1	Amphibious
tadpole[3]	0.6	Aquatic
Xenopus laevis[1] (South African clawed toad)	4.4 (1.9)[2]	Aquatic
Rana cancrivora[5] (Crab-eating frog)	c. 0.5 / c. 1.0	Fresh water / Sea-water (2/3)
Urodela		
Necturus maculosus[1] (Mudpuppy)	0.6 (0.7)[2]	Aquatic
Triturus alpestris[1] (Alpine newt)	6.4	Terrestrial
Ambystoma tigrinum[2] (Tiger salamander)	6.7	Terrestrial
Salamandra maculosa[4] (Fire-salamander)	6.0	Terrestrial
Amphiuma means[2] (Congo eel)	0.6	Aquatic
Siren lacertina[2] (Mud eel)	2.4	Aquatic

Results are from or calculated from: [1]FOLLETT and HELLER (1964 a); [2]*BENTLEY (1969 b); [3]BENTLEY and GREENWALD (1970); [4]BENTLEY and HELLER (1965). [5]DICKER and ELLIOT (1970 a). *calculated from activity on toad bladder.

Differences in the morphological sites of action of vasotocin diminish or facilitate its overall effect. There are, in addition, variations in the sensitivity to this peptide; for instance, the toad, *Bufo bufo*, is ten times more reponsive ('water balance effect') than the frog, *Rana esculenta* (HELLER and BENTLEY, 1965). Both species can retain similar amounts of water following the injection of vasotocin but 10 times more of this peptide must be given to the frogs. As described earlier, the concentrations of vasotocin that are usually required to elicit responses at the various sites of its action, either *in vitro* or *in vivo*, vary from 10^{-10} to 10^{-13}M.

The neurohypophyses of a variety of amphibians thus contain a peptide, vasotocin, that can facilitate the retention and accumulation of water in a number of species. The effects are more prominent in terrestrial than aquatic species (Table 6.9). Vasotocin can act at low concentrations, which in the circulation of a frog

193

Table 6.9 *Maximal water uptake by amphibians injected with various neurohypophysial hormones. (From data collected by* MOREL *and* JARD, *1968)*

	Maximal water retention g/100 g initial BW	Hormone
Anura		
Xenopus laevis	0	Pitressin, Pitocin Arginine-vasotocin
Neobatrachus	10	Oxytocin
Leptodactylus	16	Pitressin, Pitocin
Discoglossus pictus	19	Arginine-vasotocin
Pelobates cultripes	13	Arginine-vasotocin
Rana clamitans	14	Pitocin⟩Pitressin
Rana pipiens	14—19	Pitocin⟩Pitressin
	18	Pituitrin
Rana esculenta	12—24	Pitocin⟩Pitressin AVT⟩AVP⟩Oxy⟩LVP
	20	Arginine-vasotocin
Rana temporaria	28	Pitocin⟩Pitressin
Bufo bufo		AVP⟩AVT⟩Oxy⟩LVP
	18	Arginine-vasotocin
	40	Pitressin⟩Pitocin
Bufo carens	22	Pitressin⟩Pitocin
Bufo regularis	22	Pitressin⟩Pitocin
	45	Pituitrin
Bufo americanus	45	Pitressin⟩Pitocin
Hyla hyla	53	Arginine-vasotocin
Urodela		
Necturus maculosus	4	Pituitrin
	2	Arginine-vasotocin
Ambystoma tigrinum	6—8	Arginine-vasotocin
Triturus alpestris	8	Arginine-vasotocin
Triturus cristatus	10	Arginine-vasotocin
Salamandra maculosa	10	Arginine-vasotocin

or toad would only represent about 10% or the total quantity stored in the neurohypophysis. Such information strongly suggests that this peptide may have a physiological role in influencing the water metabolism of some amphibians, but does not show this unequivocally.

Neurohypophysectomy has yielded some conflicting results about the role of the neurohypophysis. Removal of the neurohypophysial tissue is a difficult operation and the simplest way of doing it is to remove the entire pituitary. This procedure also results in the loss of a number of adenohypophysial hormones. A loss of these can result (in *Rana pipiens*) in an increased osmotic permeability of the skin and a greatly reduced response to injected neurohypophysial peptides (LEVINSKY and SAWYER, 1952). This is due mainly to the absence of corticotrophin as injections of this hormone substantially restore the animal to normal, while additional small doses of thyroxine almost complete the replacement. Such effects are only manifest after a period of several days and do not occur if the frogs are tested within a day or two of the operation. When the latter was done, LEVINSKY and SAWYER (1953) found that such animals could only rehydrate (after dehydration) at about 50% the rate of intact frogs. JORGENSEN, WINGSTRAND, and ROSENKILDE (1956) attempted to destroy the neurohypophysis, leaving the adenohypophysis intact, by cutting the praeoptico-hypophysial tract and allowing the peripheral neurohypophysial tissue to degenerate over a period of several weeks. Toads, *Bufo bufo*, treated in this way continued to form normal quantities of urine, while their rate of rehydration, after dehydration, was not significantly different from that of normal toads. A number of objections were made to these experiments as it seemed possible that release of neurohypophysial peptides from the cut rostral stump of the praeoptico-hypophysial tract could not be excluded. In order to overcome this difficulty two separate attempts to completely destroy the tissue, by directly coagulating the praeoptic nucleus, have been made. JORGENSEN, ROSENKILDE, and WINGSTRAND (1969) repeated their experiments on *Bufo bufo* and destroyed the praeoptic nucleus. They found evidence for a deficiency in the toads ability to reabsorb water from its urinary bladder, but again found that the operated toads rehydrated at the same rate as the normal controls. An examination of their results, however, shows that the latter control toads only rehydrated at a very slow rate, very much less than had been reported previously (JORGENSEN and ROSENKILDE, 1956a). The efficiency of the controls' neurohypophysial function is, therefore, also in doubt. The praeoptic nucleus has also been destroyed in the toad, *Bufo marinus* (SHOEMAKER and WARING, 1968), and it was found that in these animals, when dehydrated or injected with saline, cutaneous water uptake was less and urine volume greater than in sham-operated controls. Removal of the entire pituitary also resulted in a reduction of the responses of the skin and kidney, but was similar to the effect of adenohypophysectomy alone. This presumably reflects the absence of the adenohypophysial hormones, an effect that was also seen by LEVINSKY and SAWYER (1952). MIDDLER, KLEEMAN, and EDWARDS (1967) also compared the responses of adenohypophysectomized and totally hypophysectomized *Bufo marinus* to dehydration and injections of hypertonic saline solutions. They found that the totally hypophysectomized toads had little ability to reduce their urine flow or to reabsorb water from their urinary bladders as compared to animals that only had the adenohypophysis removed. This difference is presumably due to the additional absence of the neural lobe in the hypophysectomized toads. The variations in the results from different laboratories probably reflect mainly the physiological condition of the toads used, both that of the controls and of the operated animals. The evidence generally suggests that the neurohypophysis facilitates the osmoregulatory changes that occur in the skin, kidney and urinary bladder following dehydrational stresses. Nevertheless, it should be remembered that alterations in the functions of these tissues may still occur even in the absence of any endocrine stimulation. For example, changes in the circulation and osmotic concentration of the blood supplying these tissues may have direct effects on their function.

195

The final confirmatory evidence required to establish the endocrine role of the neurohypophysis in osmoregulation is the identification of the putative hormone in the circulation. It should be present in amounts consistent with the sensitivity of its supposed receptors and should be released in response to physiologically appropriate stimuli. Dehydration has been shown to reduce greatly the amounts of neurohypophysial peptides in the pituitaries of *Rana pipiens* and *Bufo bufo* (LEVINSKY and SAWYER, 1953; JORGENSEN *et al.*, 1956) and this probably reflects a release into the circulation. DICKER and ELLIOTT (1970a), on the other hand, found that the neurohypophysis of crab-eating frogs, adapted to living in dilute sea-water (2/3 concentration), contained twice as much peptide activity as when they were living in fresh water. It is uncertain whether this reflects a decreased release or an increased rate of synthesis, possibly combined with higher levels in the circulation. SHOEMAKER (1965) found that when he injected blood plasma from dehydrated toads, *Bufo marinus*, into toads kept in water, they hydrated at an increased rate, just as though vasotocin were injected. I have collected blood from several anurans, *Bufo marinus*, *Rana catesbeiana* and *Xenopus laevis*, when they were dehydrated or had spent an hour or so swimming in hypertonic saline. The plasma of all such anurans treated in this manner, contains activity identical in its pharmacological behaviour, and interactions with various enzymes, to vasotocin (BENTLEY, 1969a). It is present (if it is vasotocin) at a concentration of 10^{-9} to 10^{-10}M which would stimulate adequately the various receptor tissues. It is interesting that *Xenopus laevis*, which lacks such receptor tissues, still releases this peptide in response to a dehydrational stress. I could not find vasotocin-like activity in the blood of *Necturus maculosus*, even when the animals were placed in saline solutions. Indeed the role of the neurohypophysis in the physiology of aquatic amphibians remains an enigma. It has been formerly suggested that it may play a role in electrolyte metabolism (MAETZ, 1963; HELLER and BENTLEY, 1963), but this has not been substantiated. One cannot even venture, at present, to suggest a role for the other neurohypophysial peptide, mesotocin, in any of these tetrapods.

Despite some initial disadvantages, like a highly permeable skin and aquatic larvae, the amphibians have adapted to a number of different habitats with contrasting osmotic conditions. Their manner of doing this necessarily differs from that of other tetrapod vertebrates, but the result is often as effective. When living in fresh water the Amphibia depend more on extrarenal mechanisms in osmoregulation. This feature is shared with the fishes, though the principal sites in each, skin or gills, differ. When on land many amphibians have a remarkable ability to survive after being excessively dehydrated but they are generally more dependent on the proximity of fresh water than other tetrapods. Amphibians in contrast to the fishes, have a distensable cloacal urinary bladder that can be used to store large amounts of water. If dry conditions become extreme they may aestivate entombed under the ground for many months, and with the aid of this store of water, await the return of more favourable conditions.

The amphibian endocrine system is clearly involved in many osmotic adjustments. It has the usual complement of tetrapod hormones most of which have, apparently, been inherited from the fishes. Aldosterone, however, seems to make its initial hormonal appearance in this group. Corticosterone is also present in the fishes as well as in the phyletically subsequent reptiles and birds. The peptide hor-

196

mone vasotocin is also present in all of these groups. In the Amphibia corticosterone and aldosterone clearly play a role in regulating the conservation and accumulation of sodium. Vasotocin facilitates rehydration and the preservation of body
water, especially in terrestrial species. The kidney of amphibians, like that of all
vertebrates, plays a central role in osmoregulation, but additional processes at extrarenal sites are also very important in this group. It is at such non-renal sites that
the peptide and steroid hormones have a particularly prominent action. The Amphibia probably best illustrate the oft-quoted dictum that it is not the hormones
that have evolved but the uses to which they have been put. These uses, when compared with the fishes, would appear to include actions at sites like the skin, urinary
bladder, and the renal tubular epithelium where they have the novel effect of facilitating water conservation. It seems likely that, in addition, at least one new hormone, aldosterone, may have evolved; this is concerned with sodium conservation,
acting especially at extrarenal sites.

Chapter 7

The Fishes

The fishes, or Pisces, are aquatic vertebrates that breathe with the aid of gills. They are thought to have originated 500 million years ago, a time that predates the origin of tetrapods by some 200 million years. It is uncertain whether the transformation from an invertebrate ancestor took place in fresh water or in the sea (see Page 35), but subsequently fish have adopted both of these environments, as well as inland lakes and springs where the solute concentrations of the waters may even exceed those in the sea. The Pisces contains about 23 000 species, more than all of the tetrapods, which occur in several widely divergent evolutionary classes. The principal of these are: (see Table 7.1) the Agnatha (the jawless lampreys and the hagfishes), the Chondrichthyes (sharks and rays) and the Osteichthyes (bony fishes). The teleostean fishes, which are a group of the latter bony fish, are today the predominant group in both fresh water and the sea; there is more information about the physiology of these than for the other classes of fish. From the biological viewpoint, however, further information about osmoregulation in the sparser relict fishes would probably be even more interesting, especially as it could aid our understanding of the evolution of osmoregulation in vertebrates. Such archaic fish include the lungfishes, the coelacanth, the sturgeons, the bowfin and garpike, and the lampreys and hagfishes. At present physiological information about these fishes is somewhat limited.

The sharks and rays and the bony fishes probably originated in fresh water (see ROMER, 1955) but they have subsequently lived predominantly in the sea. While

Table 7.1 *A brief classification of the fishes emphasizing the principle lines of their evolution (based on* ROMER, *1955)*

Class Agnatha (jawless fishes)
    Petromyzontia (lampreys)
    Myxinoidea (hagfishes)
Class Chondrichthyes (also called Elasmobranchii)
    Selachii (sharks and rays)
    Holocephali (chimaeroids)
Class Osteichthyes (bony fishes)
    Choanichthyes
        Crossopterygii (coelacanth)
        Dipnoi (lungfishes)
    Branchiopterygii (the bichir)
    Actinopterygii (ray finned fishes)
        Chondrostei (sturgeons and paddlefish)
        Holostei (bowfin and garpike)
        Teleostei (most contemporary bony fish)

---

contemporary species of some of the main orders within these groups of fish may today live solely in fresh water (the lungfishes and bichir) or the sea (the coelacanth), in the course of geological time some representatives of all such evolutionary divisions have occupied both habitats (see DARLINGTON, 1957). The dominant teleosts are thought to have originated in the sea, but some subsequently moved into fresh water and a number of these have even returned to the sea. DARLINGTON (1957) gives a fascinating description of the dispersal of freshwater fishes throughout the world, and this has often involved several successive migrations between the sea and fresh water. The freshwater catfishes of Australia, for instance, moved into the antipodean rivers from the sea, but their immediate marine ancestors previously originated from freshwater ancestors that probably lived in Asia. When one considers the complexities of such transformations, it is not surprising to find considerable diversity in the detailed pattern of osmoregulation of fishes, though the basic blueprint remains remarkably uniform.

Most contemporary species of fish have only a limited ability to move between fresh water and salt solutions like sea-water. Such osmotically conservative fishes are 'stenohaline' while others which can adapt more readily to either type of solution are 'euryhaline'. As emphasized by FONTAINE and KOCH (1950), such a classification is not rigid, some fish being 'more or less' stenohaline while others are 'more or less' euryhaline. Thus the freshwater goldfish and the marine perch can both live in salt solutions about one-third the concentration of the sea (LAHLOU, HENDERSON, and SAWYER, 1969a; MOTAIS, GARCIA ROMEU, and MAETZ, 1966). Some euryhaline fishes can withstand direct transfer from fresh water to sea-water, while others such as the lamprey once having migrated into rivers, will die if they are then replaced in the sea (MORRIS, 1960). Euryhaline fishes are often migratory breeders, some like lampreys and salmon move from the sea into rivers where they produce their eggs, while others, like the eel, return to the sea to breed. Many euryhaline fishes occupy estuaries and certain inland lakes and streams where the salinity may vary considerably during different seasons. The osmotic adjustments, which must be made by such fishes, would seem to offer a situation for endocrine coordination. In natural conditions adaptation to solutions of differing osmotic concentration probably can take place gradually, as when such water is evaporated from an inland lake or fish migrate through estuarine areas between rivers and the sea (FONTAINE and KOCH, 1950). Such situations would seem to be particularly suitable to the relatively long term actions that are characteristic of many hormones.

Fish can live in solutions with a wide range of osmotic concentration. As described above, fish that normally live in the sea or fresh water can often tolerate solutions of intermediate concentration. Others can live in waters with a solute concentration considerably higher than that of sea-water; *Tilapia mossambica* may survive in a salinity of 6.9%, while the cyprinodont *Cyprinodon variegatus* has been found in saline waters with a concentration of 14.2% (see PARRY, 1966). Such osmotic tolerance presumably constitutes a rather special physiological adaptability about which we have little knowledge.

Fishes contain an amount of water that is equivalent to 70 to 75% of their body weight which is similar to the water content of most other vertebrates (Table 1.2). The distribution of this water in the body varies somewhat in different species (THORSON, 1961 and Table 1.2). The plasma volume of agnathans and

Table 7.2 *Osmotic constituents in the plasma of various fishes*

	Habitat	Sodium m-equiv/l	Potassium m-equiv/l	Urea mM	Osmolarity m-osmole/l
Sea-water		470	12		1070
Osteichthyes					
Latimeria chalumnae[1] (Coelacanth)	SW*	181	—	355	1181
Protopterus aethiopicus[2] (African lungfish)	FW*	99	8		238
Carassius auratus[3] (Goldfish)	FW	115	3.6		259
Opsanus tau[4] (Toadfish)	SW	160	5.2	—	392
Platichthys flesus[5] (Flounder)	FW	157	5		304
	SW	194	5		364
Anguilla anguilla[6] (European eel)	FW	155	2.7	—	323
	SW	177	2.8	—	371
Salmo salar[7] (Atlantic salmon)	FW	181	1.9	—	340
	SW	212	3.2	—	400
Chondrichthyes					
Raja clavata[8] (Thornback ray)	SW	289	4	444	1050
Squalus acanthias[9, 10] (Spiny dogfish)	SW	287	5.4	354	1000
Chimaera montrosa[9] (Rabbitfish)	SW	360	10	265	
Pristis microdon[11] (Freshwater sawfish)	FW	170 (Cl)	—	130	540
Agnatha					
Myxine glutinosa[12] (Hagfish)	SW	549	11	—	1152
Lampreta fluviatilis[13] (River lamprey)	FW	120	3	—	270

*SW = sea-water; FW = fresh water

References: [1]PICKFORD and GRANT (1967); [2]SMITH (1930 a)
[3]MAETZ (1963); [4]LAHLOU, HENDERSON, and SAWYER (1969 b); [5]LANGE and FUGELLI (1965);
[6]SHARRATT, CHESTER JONES, and BELLAMY (1964 a); [7]PARRY (1961); [8]MURRAY and POTTS
(1961); [9]FANGE and FUGELLI (1962); [10]BURGER (1965); [11]SMITH (1936); [12]BELLAMY and
CHESTER JONES (1961); [13]ROBERTSON (1954).

chondrichthyeans constitutes about 5% of their body weight, which is similar to that of the tetrapods, but in osteichthyeans it is, according to THORSON (1961), about half as large. However, as the number of species of bony fishes is vast and the measurements have been made on comparatively few, it remains to be seen whether such a difference is a general characteristic of these animals.

The solute content of the plasma in representatives of the main groups of fishes has been measured (Table 7.2). The bony fishes that live in the sea generally have higher solute levels than those of fish in fresh water. Euryhaline fish, like the flounder, salmon and eel, have a plasma osmolarity that is about 20% greater in the sea than in fresh water, and this is the result of an elevated sodium concentration. Such marine fish are considerably hypoosmotic, 2 to 3-fold less, to the sea-water that bathes them. The only exception among the Osteichthyes appears to be the relict crossopterygian (originally a predominantly freshwater group) *Latimeria* whose plasma is slightly hyperosmotic to sea-water, due mainly to a retention of urea in the fluids. It is interesting that the closest living phyletic relatives of this fish are the lungfishes, and one of these, *Protopterus*, can also tolerate large amounts of urea in its body fluids. The marine chondrichthyean fishes, which like the crossopterygians are thought to have originated in fresh water, also maintain their body fluids slightly hyperosmotic to sea-water by retaining urea as well as some sodium (H. SMITH, 1936). Chondrichthyeans that live in fresh water, like the sawfish *Pristis,* have a much lower plasma concentration due to a reduction in the levels of both urea and sodium (Table 7.2). The agnathans exhibit both types of osmotic constitution, the lampreys are hypoosmotic to sea-water, while the hagfishes are slightly hyperosmotic. The latter group, however, does not retain urea for this purpose, but has high salt concentrations in its body fluids.

We can predict the osmotic stresses on the various fishes from the osmotic concentrations in their body fluids. In fresh water all fishes tend to gain water by osmosis while at the same time they may be expected to lose solutes by diffusion. In the sea, chondrichthyeans, *Latimeria* and the hagfishes, all gain water by osmosis, and (except for the latter) may be expected to accumulate sodium by diffusion. The marine bony fishes and lampreys, which are hypoosmotic to sea-water, lose water osmotically and gain sodium. Measurements of water and solute balance in the different groups of fishes indicate that such osmotic changes are occurring, but they differ widely and are adequately compensated for by physiological adjustments.

1. The Piscine Endocrines

The fishes possess tissues that are homologous to the endocrine glands of tetrapods and that form similar biological products (see for instance: BERN and NANDI, 1964; BERN, 1967). Whether these all have an endocrine function is uncertain, but in many instances this seems likely. The precise molecular structures of the products of such tissues in fish often differ somewhat from those of the secretions from the homologous tissues in other groups of vertebrates. This is particularly apparent among the proteinaceous hormones which may exhibit different potencies, pharmacological activities and immunological behaviour from those in other vertebrates. An orderly pattern in their distribution may, nevertheless, be apparent. In addition, while the fish seem to have all the endocrine tissues that have been identified in higher vertebrates some have two additional putative ones; the urophysis and the corpuscles of STANNIUS (see page 78).

a) The Neurohypophysis

The neurohypophysis of fish is not as well developed as that of the tetrapods, it lacks a distinct neural lobe while the portal circulation between the median eminence and the adenohypophysis is not usually well defined. This tissue in the fishes nevertheless contains peptides that have a similar structure to those in the tetrapods. Vasotocin has a widespread distribution in the fishes (Table 7.3 and 7.4) ranging phyletically from the agnathans to the lungfishes and is subsequently perpetuated in the amphibians, reptiles and birds (Fig. 7.1) (SAWYER, 1969). It appears to be present, usually in small amounts, in some, but not all of the Chondrichthyes. Vasotocin may have been the original parent neurohypophysial hormone in vertebrates from which others have been derived as a result of successive mutations (see SAWYER, 1967a). Such mutations presumably gave rise to isotocin, mesotocin, oxytocin and glumitocin. Speculation regarding the genetic changes which may have occurred is very tempting, but has become successively more complicated as further such peptides have been found and chemically characterized (for a discussion see GESCHWIND, 1967; SAWYER, 1968). Vasotocin in the neurohypophysis is usually associated with the presence of another peptide with a similar amino acid make-up. This is not always apparent as vasotocin seems to be present by itself in the agnathan and possibly in the chondrostean fishes (paddlefish). Mesotocin and sometimes oxytocin, is present, along with vasotocin, in reptiles and the Amphibia and it is interesting that both these peptides have also been identified among their closest living piscine relatives, the Choanichthyes (lungfishes). The rest of the bony fishes lack mesotocin and oxytocin but possess instead, isotocin, a peptide that differs from mesotocin by the presence of serine at position 4 in the molecule (see

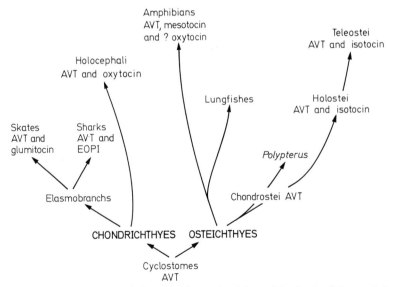

Fig. 7.1 Phyletic distribution of the neurohypophysial peptides in the fishes and the Amphibia. AVT = arginine vasotocin, EOP I = an unidentified oxytocin-like principle found in some elasmobranchs. (From SAWYER, 1969).

Table 7.3 *Neurohypophysial hormones in bony fishes (Osteichthyes)*

	Vasotocin	Mesotocin	Oxytocin	Isotocin
Choanichthyes				
Protopterus aethiopicus[1,2] (African lungfish)	+	+	+ ?*	—
Lepidosiren paradoxa[1,3] (South American lungfish)	+	?	+	—
Neoceratodus forsteri[1] (Australian lungfish)	+	+	—	—
Branchiopterygii				
Polypterus senegalis[4] (Bichir)	+	—	—	+
Actinopterygii				
Chondrostei				
Polyodon spathula[5] (Paddlefish)	+	—	—	—
Acipenser[5] (Sturgeon)	+	Little oxytocin-like		
Holostei				
Amia calva[5] (Bowfin)	+	—	—	+
Lepisosteus sp.[5] (Garpike)	+	—	—	+
Teleostei				
Salmo gairdneri[5,6] (Rainbow trout)	+	—	—	+
Cyprinus carpio[5,7] (Carp)	+	—	—	+
Anguilla anguilla[5] (Eel)	+	—	—	+
Esox lucius[5] (Pike)	+	—	—	+
Fundulus heteroclitus[8] (Killifish)	+	—	—	+
Pollachius virens[9,10]	+	—	—	+

and in many other fish (see SAWYER, 1968).

* see SAWYER (1969).

References: [1]FOLLETT and HELLER (1964 a); [2]SAWYER (1968); [3]PICKERING and McWATTERS (1966); [4]SAWYER (1969); [5]FOLLETT and HELLER (1964 b); [6]HELLER and PICKERING (1961); [7]ACHER et al. (1965 a); [8]SAWYER and PICKFORD (1963); [9]HELLER et al. (1961); [10]ACHER et al. (1962).

Table 7.4   *Neurohypophysial peptides in the Chondrichthyes and Agnatha*

	Vasotocin	Glumitocin	Other
Chondrichthyes			
Selachii			
Raja clavata[1] (Thornback ray)	+	+	
Raja ocellata[2] (winter skate)	—	+	?
Rhinobatos rhinobatos[3] (Guitar fish)	—	?	+ (EOP 1 ?)
Squalus acanthias[2, 4, 5] (Spiny dogfish)	+	—	+ EOP 1 (SAWYER)
Squatina squatina[3] (Monkfish)	+	—	+ (EOP 1 ?)
Scyliorhinus caniculus[3] (Spotted dogfish)	—	+	+ (EOP 1 ?)
Bradyodonti (or Holocephali)			
Hydrolagus colliei[3, 4, 6, 7] (Ratfish)	+	—	Oxytocin 7
Chimaera montrosa[3] (Rabbitfish)	?	—	Oxytocin?
Callorhynchus sp.[3]	?	—	Oxytocin ?
Agnatha			
Lampetra fluviatilis[8] (River lamprey)	+		—
Petromyzon marinus[4] (Lamprey)	+		—
Myxine glutinosa[8] (Hagfish)	+		?

[1]ACHER *et al.* (1965 b); [2]SAWYER *et al.* (1969); [3]ROY (1969); [4]SAWYER (1965); [5]SAWYER (1967); [6]SAWYER, FREER, and TSENG (1967); [7]PICKERING and HELLER (1969); [8]FOLLETT and HELLER (1964 b).

Fig. 2.4). The pattern of distribution of the neurohypophysial peptides within the Chondrichthyes is still not clear but there appears to be a distinct difference between those in the Holocephali (chimaeroid fish) and in the Selachii (sharks and rays) (Table 7.4). ACHER et al. (1965 b) have chemically characterised another such peptide in some of the Selachii. This contains glutamine at position 8 in the molecule and serine at position 4 so that it differs from isotocin by a single amino acid substitution. It has been called glumitocin. This peptide does not seem to be present in the Holocephali which may posses oxytocin. At least one other peptide, SAWYER's EOP 1, has been found in a number of the selachians (Table 7.4) but its chemical structure has not been worked out. The distribution of the various peptides among fishes and the Amphibia is shown in Fig. 7.1. While in non-mammalian tetrapods vasotocin is quantitatively the predominant neurohypophysial peptide, this does not always seem to be so in fishes. In addition, the concentrations of vasotocin in the neurohypophyses of fish (in relation to their body weight) is usually (though not invariably) more than 10-fold less than those in terrestrial tetrapods. This probably reflects differences in the development of the neural lobe and possibly also the physiological significance of the neurohypophysial peptides in the fishes which, as we shall see, is not at all clear.

b) Corticosteroids

Corticosteroids with the same chemical structures as those found in tetrapods, have been identified in the blood of a variety of fishes ranging from the agnathans to the lungfishes (Table 7.5). Cortisol and corticosterone usually predominate but their presence in the Agnatha is in doubt (WEISBART and IDLER, 1970). Earlier reports that aldosterone is present have not been confirmed, though it can be formed from steroid substrates during *in vitro* incubation of the interrenal tissues of a variety of fish. There is some doubt, however, as to whether such products represent a normal secretion in fish. Other steroids have been found in the blood of fishes; in the chondrichthyean, *Raja radiata*, 1α-hydroxycorticosterone is the principal corticosteroid present (IDLER and TRUSCOTT, 1966; 1968). TRUSCOTT and IDLER (1968b) have shown that the interrenal tissues of eleven species of skates, rays, dogfish and sharks (Chondrichthyes) produce mainly 1α-hydroxycorticosterone during *in vitro* incubation.

There is abundant evidence to indicate the presence of a corticotrophin-like hormone in the pituitary gland of fish. RASQUIN in 1951 showed that injection of mammalian corticotrophin into a teleost fish, *Astyanax mexicanus*, resulted in a hypertrophy of its interrenal tissues. When carp pituitaries were implanted into these fish they produced a similar response. The pituitaries of the salmon, *Oncorhynchus keta*, and the cod, *Gadus morhua*, have also been shown to contain activity which, like mammalian corticotrophin, decreases the ascorbic acid content of the hypophysectomized rat adrenal cortex (RINFRET and HANE, 1955; WOODHEAD, 1960). Evidence of a more direct nature, indicating that there is a relationship between the adenohypophysis and the secretion of corticosteroids in fish has been derived from several types of experiments. Hypophysectomy results in decreased levels of corticosteroids in eels (LELOUP-HATEY, 1961; BUTLER, DONALDSON, and CLARK, 1969) and rainbow trout (DONALDSON and McBRIDE, 1967). In addition, the injection of mammalian corticotrophin increases the circulating corticosteroid levels in eels (LELOUP-HATEY, 1961; BUTLER *et al.*, 1969) and salmon (HANE *et al.*, 1966). A negative feed-back control system between the adenohypophysis and the interrenals also seems to exist, as in mammals. The injection of large doses of very potent synthetic corticosteroids (for example dexamethasone) suppresses the activity of the interrenals of salmon (FAGERLUND and McBRIDE, 1969) and eels (BUTLER *et al.*, 1969). The corticotrophin cells in the adenohypophysis of eels hypertrophy after interrenalectomy (HANKE, BERGERHOFF, and CHAN, 1967; OLIVEREAU and OLIVEREAU, 1968) which is also consistent with the presence of a negative feed-back system of control. While a corticotrophin-like activity has been found in the pituitaries of agnathans (STRAHAN, 1959), the evidence for the presence of

Table 7.5 *Adrenocortical steroids in the blood of fishes*

	Cortisol	Corticosterone	Other
Osteichthyes			
Chondrichthyes			
Protopterus aethiopicus[1] (African lungfish)	+	+[2]*	
Actinopterygii			
Chondrostei	+*,**	?	
Holostei	?	?	
Teleostei			
Salmo gairdneri[3] (Rainbow trout)	+	+[4]*	Cortisone
Anguilla anguilla[5] (Eel)	+	+	Cortisone
Fundulus heteroclitus[6] (Killifish)	+		
Cyprinus carpio[6, 7] (Carp)	+	+	
Chondrichthyes			
Raja radiata[8, 9] (starry ray)	trace	trace	1α-hydroxy-corticosterone*
Squalus acanthias[10, 11] (Spiny dogfish)		+*	aldosterone* 1α-hydroxy-corticosterone
Hydrolagus colliei[10] (Ratfish)	+*		aldosterone*
Carcharhinus[11, 12] (Sharks)	+	+	1α-hydroxy-corticosterone*
Agnatha			
Petromyzon marinus[12, 13] (Sea lamprey)	? + or —	? + or —	
Eptatretus stouti[12] (Pacific hagfish)	+	+	
Myxine glutinosa[13, 14] (Atlantic hagfish)	? + or —	? + or —	

* *in vitro* tissue incubation ** see IDLER and SANGALANG (J. Endocrin. 48: 627, 1970).

TRUSCOTT and IDLER (1968) have shown that the interrenal tissues from eleven species of the Chondrichthyes mainly produced 1α-hydroxycorticosterone during *in vitro* incubation.

References: [1]PHILLIPS and CHESTER JONES (1957); [2]JANSSENS *et al.* (1965); [3]HANE and ROBERTSON (1959); [4]NANDI and BERN (1965); [5]LELOUP-HATEY (1964); [6]CHESTER JONES *et al.* (1959); [7]BONDY, UPTON, and PICKFORD (1957); [8]IDLER and TRUSCOTT (1968); [10]BERN, DEROOS, and BIGLIERI (1962); [11]TRUSCOTT and IDLER (1968); [12]PHILLIPS (1959); [13]WEISBART and IDLER (1970); PHILLIPS *et al.* (1962 b).

207

such a principle in chondrichthyean fish is confused. Reports on the effects of hypophysectomy on the interrenals conflict (see HOAR, 1966). Mammalian corticotrophin fails to stimulate corticosteroid production of chondrichthyean interrenal tissues *in vitro* (BERN, DEROOS, and BIGLIERI, 1962) but this could reflect the limitations of such a system, or differences in the chemical structure of the mammalian and chondrichthyean hormones. Corticotrophin-like activity has been directly demonstrated in the pituitary of a chondrichthyean fish (DEROOS and DEROOS, 1964). The fishes thus usually appear to possess a corticotrophin-like hormone(s) that influences the activity of the interrenals.

In 1942 FRIEDMAN *et al.* demonstrated the presence of renin in the kidneys of a number of freshwater teleost fish, though it was rather interesting that it was not then found in marine species. At that time the existence of a renin-angiotensin system for controlling interrenal secretion, as in mammals, was unknown. In the latter group renin is principally concerned with the regulation of aldosterone, though to a lesser extent it can also influence corticosterone secretion. Fish do not seem to possess aldosterone but corticosterone is prominent. The kidneys (and corpuscles of STANNIUS in eels) of a variety of marine as well as freshwater teleost fish have since been shown to contain a renin-like material (CHESTER JONES *et al.*, 1966; MIZOGAMI *et al.*, 1969, CAPELLI *et al.*, 1970). The renin-activity is generally higher in freshwater than in marine species and the levels increase two-fold when Japanese eels and toadfish are adapted to life in fresh water (SOKABE *et al.*, 1966; CAPELLI *et al.*, 1970). The latter changes would be consistent with a role in sodium conservation as seen in mammals. Attempts to identify renin in the kidneys of elasmobranchs (4 species) and cyclostomes (2 species) have not been successful (NISHIMURA *et al.*, 1970) so that such a mechanism may not be present in all of the fishes.

c) Adenohypophysis – Prolactin

The adenohypophysis of fishes contains a product which is homologous with mammalian prolactin. Its precise actions and properties differ somewhat from the tetrapod secretion, but in a manner that may be expected to result from differences in their amino acid constitution. 'Prolactin' from a variety of vertebrates can initiate the pre-breeding 'water drive' in the efts of certain newts (see page 186) and this includes most groups of fishes (GRANT and PICKFORD, 1959). The eft 'water drive' response is quite specific to this type of secretion and has provided a very useful biological method for detecting the presence of such substances. There are also some additional methods for characterizing 'prolactins' that are based on their effects in other vertebrates. In mammals prolactin promotes the growth and secretion of the mammary glands, while in columbiform birds, such as doves and pigeons, it initiates hypertrophy and secretion of the crop-sac glands. These three responses have been particularly useful tests in comparing the properties of the prolactin secretions from different groups of vertebrates (see BERN and NICOLL, 1968). Prolactin can also initiate a variety of other responses in different vertebrate preparations, and these include a luteotrophic action on the corpus luteum of rats and increased growth in amphibian tadpoles. However, the most interesting effect related to our present subject is its ability to enhance the survival of certain hypophysectomized fishes in fresh water (PICKFORD and PHILLIPS, 1959) and to elevate the plasma sodium concentrations under such conditions (BALL and ENSOR, 1965). 'Prolactins' derived from the pituitaries of different vertebrates exhibit a different spectrum of activity when tested in these various ways. BERN (1967) has summarized such effects of the pituitaries of tetrapods and fishes (Fig. 7.2). The eft 'water drive' response can be initiated by the 'prolactins' from the widest phyletic range of vertebrates and is prominent in both fish and tetrapod pituitaries. The ability to stimulate mammary gland tissues, however, appears to be an exclusive property of the tetrapod product and is absent in the fishes (NICOLL, BERN, and BROWN, 1966). The pigeon crop-sac glands have an almost identical sensitivity as the mammary tissue, but with the very interesting exception that they respond to 'prolactin' from the lungfish, *Protopterus*, pituitary but not to that of other fishes (NICOLL and BERN, 1968). This has phyletic appeal as the tetrapods have more ancestral affinities with the lungfishes than with other fish. CHADWICK (1966) found that fish pituitaries had modest effects on the mammalian and avian preparations but they were somewhat different in character from those exhibited by tetrapod pituitaries (BERN and NICOLL, 1968). The full phyletic distribution of the ability of vertebrate 'prolactins' to enhance survival of hypophysectomized freshwater fish and elevate their plasma sodium concentrations has not yet been described. Mammalian, amphibian and fish prolactin is effective (see ENSOR and BALL, 1968 a). It has usually been assumed that the observed differences in the spectrum of biological activities result from variations in the structure of the 'prolactin' molecule. This may have evolved in the course of time (BERN and NICOLL, 1968). The eft 'water drive' would seem at this time, to reflect a property of the molecule that is most basic or common,

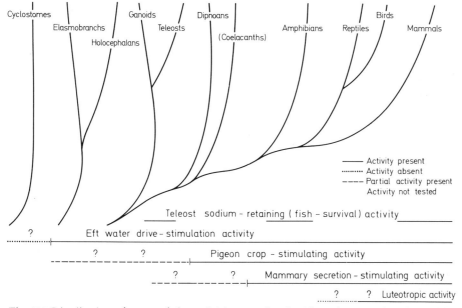

Fig. 7.2 Distribution of some of the activities associated with prolactin and prolactin-like hormones (including paralactin) in vertebrates. (H.A. BERN, personal communication).

while other properties, such as the ability to stimulate the mammary gland arose as a result of subsequent changes. That the various characteristics of vertebrate 'prolactins' are indeed due to a fundamentally similar molecule is seen when one considers their immunochemical behaviour. Anti-mammalian (ovine) prolactin serum combines with fish 'prolactin', while anti-fish 'prolactin' serum reacts with ovine prolactin (EMMART and WILHELMI, 1968). Fish 'prolactin' has been termed 'paralactin' by BALL (1965), a term that has received the blessing of others (BERN and NICOLL, 1968).

2. Water Exchanges

a) The Integument

The skin and gills of fishes are continuously bathed by the environmental solutions and are avenues for movements of water in or out of the animal. The direction of any net transfer depends on the direction of the osmotic gradient.

The relative surface areas of the skin and gills vary considerably in different fishes; in the toadfish, *Opsanus tau*, the gills have an area 10 times larger than that of the skin, while in the mackerel, *Scomber scombrus*, it is 60 times greater (see PARRY, 1966). The toadfish has recently been shown to have a very low permeability to water and sodium (LAHLOU and SAWYER, 1969) and this may partly reflect its relatively small branchial surface.

The skin of fishes is a complex structure, which is liberally supplied with mucous glands; it is often covered with scales, which give it the appearance of a substantial barrier to molecular transfer. Unfortunately, however, there are few direct measurements of this. KROGH (1939) after considering the evidence that was available at the time, concluded that the gills, rather than the skin, are the principal pathway for the movements of water and ions in and out of fish. There is no subsequent contradictory information. MOTAIS et al. (1969) have recently measured the movements of tritated water across the gills of several freshwater and marine fishes (*Anguilla, Carassius, Platichthys* and *Serranus*), and found that virtually all of the water that was exchanged could be accounted for by that which crossed the gills. The permeability of the gills of these fishes to water was found to be low compared with other epithelial membranes like amphibian skin and urinary bladder. It was also interesting to observe that the permeability to water was less when the fish were in sea-water than in fresh water, despite the larger osmotic gradient between the fish and the former solution. The reasons for this difference or change in the permeability of the membranes is not known. It was suggested that it may be an example of 'rectification of flow' such as I had observed (BENTLEY, 1961; 1965) in the urinary bladders of toads bathed on their external surface with hyperosmotic solutions. It is thus not necessary to postulate any hormonally mediated adjustments, but they could occur. BOYLAN (1967) has shown that the gills of chondrichthyeans are permeable to water and solutes but he also found that this was somewhat restricted as compared to amphibian membranes.

The agnathan fishes, like their more sophisticated relatives, accumulate water across their integument. The river lamprey, *Lampetra fluviatilis*, takes up water at about the same rate as the goldfish, equivalent to about 33% of the body weight

210

in a day (BENTLEY and FOLLETT, 1963). HARDISTY (1956) compared the rates of water uptake in the river lamprey, after it had undertaken its breeding migration, with that of brook lampreys (*L. planeri*) which live their entire lives in fresh water. After making adjustments for the differences in size of these two agnathans he found that the brook lamprey accumulated water through its integument at twice the rate of the river lamprey. Water moves across the skin of the river lamprey at the rate of 2.5 μl/cm^{2}h (*in vitro*, osmotic gradient about 210 m-osmole/l; BENTLEY, 1962a). This value is similar to that estimated for the entire integument (skin and gills) *in vivo* (WIKGREN, 1953). The marine myxinoid agnathans have an integument that is very permeable to water, as seen in the Pacific hagfish (McFARLAND and MUNZ, 1965), but as the osmotic gradient between the body fluids and the sea is small, little water is normally accumulated in this way, less than 0.5% of the body weight in a day in *Myxine glutinosa* (MORRIS, 1965).

The relative permeabilities of the integuments of an osteichthyean and an agnathan fish have been compared with those of the European frog (WIKGREN, 1953). The ratios of the rates of water accumulation across the integument of the eel: river lamprey: frog in fresh water are: 1:20:60.

While their skin is normally not very permeable to water, when fish are stressed, such as when they are handled, they excrete large volumes of urine. This is usually considered to result from damage to the skin, possibly by removing part of the coating of mucous on its surface. Exposure of fish, such as the eel, *Anguilla anguilla*, and killifish, *Fundulus kansae*, to external solutions that contain no calcium may also produce an increase in the permeability of the skin (SHARRATT *et al.*, 1964 b; PICKFORD *et al.*, 1966). Thus, despite its somewhat rugged appearance the skin of fishes is a delicate barrier whose integrity must be continually maintained.

b) The Gut

Fishes living in fresh water drink little water, but teleosts in the sea must swallow sea–water in order to compensate for their osmotic losses (H. SMITH, 1930b). Indeed, if marine teleost fish are prevented from drinking they rapidly die from dehydration. The rate of drinking appears to relate directly to the fishes' needs, for, as the concentration of the external media increases, so does the amount that the fish drinks; eels in sea-water drink 325 μl/100g h, but if they are placed in a solution twice as concentrated as the sea, they drink 800 μl/100 g h (MAETZ and SKADHAUGE, 1968). The quantity of sea-water swallowed by marine teleosts each day varies from volumes equivalent to 5% of the body weight in the flounder to 12% in the sea-water perch (Table 7.6). These quantities are presumably largely dictated by the osmotic water loss of the fishes, a factor influenced by their surface area, as well as the precise properties of the skin and gills. There are few determinations of drinking rates in non-teleost fishes; the lamprey drinks when placed in 50% sea-water and myxinoids have been observed to drink, but this water does not appear to be absorbed (MORRIS, 1965; McFARLAND and MUNZ, 1965). Marine chondrichthyeans do not appear to drink (H. SMITH, 1931).

About 75% of the sea-water that is imbibed by teleost fishes is absorbed into the body fluids. The intestine absorbs water, but as this is a consequence of the absorption of sodium, a large amount of salt is also absorbed. Most of the divalent ions appear to be retained in the gut, though some are absorbed and later excreted by the kidney. The bulk of the salt is sodium chloride, which is excreted through the gills (Keys, 1931) leaving osmotically-free water in the body. The amounts of sodium (and chloride) absorbed across the intestine of fishes kept in sea-water amounts to 100 μ-equiv/100 g h in the flounder, *Platichthys flesus* (Maetz, 1969b), 280 μ-equiv/100 g h in the sea perch, *Serranus scriba* (Motais and Maetz, 1965) and 170 μ-equiv/100 g h in the eel, *Anguilla anguilla* (Maetz and Skadhauge, 1968). This is equivalent in a day to 100 to 200% of the total sodium present in the fish.

Table 7.6   *Rates of drinking of various fish in fresh water (FW) and sea-water (SW)*

		ml/100 g body weight day	
Osteichthyes			
Carassius auratus (Goldfish)	FW	2	Lahlou, Henderson, and Sawyer (1969 a)
Tilapia mossambica	FW SW	6 27	Potts *et al.* (1967)
Anguilla anguilla (European eel)	FW SW	3 8	Maetz and Skadhauge (1968)
Anguilla japonica (Japanese eel)	FW SW	Not detectable 8	Oide and Utida (1968)
Platichthys flesus (Flounder)	SW	5	Motais *et al.* (1969)
Salmo gairdneri (Trout)	SW	8	Oide and Utida (1968)
Serranus scriba (Sea-water perch)	SW	12	Motais and Maetz (1965)
Chondrichthyes	SW	nil.	H. Smith (1931)
Agnatha			
Lampetra fluviatilis (River lamprey)	SW (50%)	15	Morris (1960)
Myxine glutinosa (Hagfish)	SW	6.5	Morris (1965)

The process of absorption of salt and water across the gut of teleost fish has been examined both *in vitro* and *in vivo*. There are some interesting differences in this process between fish adapted to sea-water and fresh water. Water and salt movement across isolated sacs of eel intestine were examined by Sharratt *et al.* (1964b). Their results indicated that when sea-water enters the gut, it is initially diluted with water from the body fluids, while sodium simultaneously moves back into the plasma. Sodium can be actively transported across the intestine of fish (HOUSE and GREEN, 1963) and this presumably accounts for the absorption of a large additional amount of salt, which takes water with it across to the plasma (see also SKADHAUGE, 1969b). SKADHAUGE and MAETZ (1967) devised an ingenious system for following the absorption of water and salt across the eel intestine *in vivo*. The gut of these fish was cannulated at both ends and a solution was slowly circulated through it. From the changes of phenol red and ^{24}Na in this fluid the process of its absorption could be seen. For the first 90 min following the admission of the sea-water into the gut, fluid moved into the lumen, but after this period absorption into the body fluids took place. There was a continual transport of sodium from the gut, the water absorption being coupled to this. No fluid transfer could be seen if a non-absorbable solute such as mannitol was used to replace the salts in the sea-water.

The rate of absorption of water from a salt solution perfused into the gut is greater in fish that are adapted to sea-water than those in fresh water. This has been observed in eels both *in vitro* (SHARRATT et al., 1964b; OIDE and UTIDA, 1967) and *in vivo* (SKADHAUGE and MAETZ, 1967; SKADHAUGE, 1969b). The latter investigators found that the rate of fluid absorption from dilute sea-water solutions by the intestine perfused *in vivo* was 350 μl/100 g h in eels adapted to fresh water, compared with 1100 μl/100 g h in eels kept in sea-water. The absorption of water also appears to occur more efficiently in sea-water eels, as nearly twice the quantity is moved for each mole of sodium transferred. This change in the ability to reabsorb fluid from the gut takes place during a period of several days which follow placing the eels in sea-water (OIDE and UTIDA, 1967). If the adenohypophysis is removed from Japanese eels, *Anguilla japonica*, this adaptation in sea-water fails to appear, but it can be promoted again by injections of corticotrophin or cortisol (HIRANO, 1967; HIRANO and UTIDA, 1968). Sodium transport across the isolated intestine of these eels increases if the fish are pre-treated with cortisol (but not corticosterone), and this presumably partly accounts for the facilitated rate of fluid absorption. However, as observed by SKADHAUGE and MAETZ more water is also transferred in relation to the sodium moved so that a direct increase in permeability to water may also be involved.

The enzyme Na-K activated ATPase has been identified in the gut of the goldfish (M. SMITH, 1964; 1967) and of the Japanese eel (OIDE, 1967) and inhibition by the drug ouabain blocks both the absorption of sodium and water. The activity of this enzyme in the gut changes in different conditions, and is higher in eels adapted to sea-water than in those in fresh water (OIDE, 1967). It is not clear if this is related to the activity of pituitary or interrenal glands, but it is interesting that increases in the level of this enzyme in the gills of the killifish do not occur after hypophysectomy (EPSTEIN et al., 1967). It is possible that the pituitary-interrenal axis is concerned with inducing long-term changes in the activity of Na-K activated ATPase but at present the evidence is inconclusive. However, cortisol injections have recently been shown to restore the Na-K activated ATPase levels in the intestine (as well as gills and kidneys) of hypophysectomized killifish kept in sea-water (PICKFORD et al., 1970).

213

c) The Kidney

The kidney structure of fishes conforms mainly to the general tetrapod pattern. There is usually a glomerulus across which the plasma can be filtered and the filtrate is then exposed to the modifying influences of a renal tubule where reabsorption and secretion of water and solutes take place. The arrangement of the different types of cells in the renal tubules of fish, however, is not the same as in tetrapods and a different segmental arrangement may be present. As many as four or five different segments may be distinguished in, for instance, the goldfish and lamprey (see HICKMAN, 1965; BENTLEY and FOLLETT, 1963) but the exact role of each of these is unknown. In the absence of micropuncture studies we have no direct information about the physiological role of various segments of the nephrons in fish.

The relative development of the glomeruli may vary considerably in different fishes, and in some, like the goosefish, *Lophius americanus,* and the toadfish, *Opsanus tau,* they may be completely absent. This is relevant to the life of fish in hypoosmotic environments and has been the subject of considerable discussion, especially in relation to the origins of the different groups of fishes (see page 35). A glomerulus appears to be particularly suited to the excretion of large volumes of water as are accumulated by fish living in fresh water. Most fishes, with the notable exception of certain marine species, have well developed glomeruli which prompted HOMER SMITH to suggest that this was additional evidence to indicate the freshwater origins of vertebrates. Glomeruli are present in the agnathans, chondrichthyeans and most osteichthyeans. Their absence from certain marine teleosts can be considered an adaptation to a life in hyperosmotic media where a limited glomerular filtration may be an advantage. Nevertheless, it should be remembered that many marine fish, including hypoosmotic teleosts, retain glomeruli while aglomerular fishes can also survive in hypoosmotic media. The toadfish thrives in 10% sea-water, a concentration that it normally experiences in coastal waters, and despite its lack of glomeruli, adequately secretes the water that it accumulates (LAHLOU et al., 1969 b). Whether or not the glomerulus arose as an adaptation to life in hypoosmotic solutions is still being debated, but when present it assists the excretion of water rather well. The vascular supplies to the glomeruli

FW = fresh water; SW = sea-water

References: [1]SAWYER (1966 a); [2]MAETZ (1963); [3]HICKMAN (1965); [4]LAHLOU et al. (1969 b); [5]LAHLOU (1967); [6]SHARRATT et al. (1964 a); [7]HOLMES and McBEAN (1963); [8]FROMM (1963); [9]R. M. HOLMES in W. N. HOLMES and McBEAN (1963); [10]see B. SCHMIDT-NIELSEN (1964); [11]BURGER and HESS (1960); [12]H. SMITH (1936); [13]BENTLEY and FOLLETT (1936); [14]MORRIS (1965).

Table 7.7 Composition of the urine of fishes

		vol. (ml/kg hr)	GFR	Tubular H₂O% reabsorption	Sodium conc. m-eq/l	total µeq/kg h	Potassium conc. m-eq/l	total µeq/kg h	osmolarity m-osmole/l
Osteichthyes									
Protopterus aethiopicus[1] (African lungfish)	FW	4.9	14	69	5.5	27	—	—	17
Carassius auratus[2] (Goldfish)	FW	13.7	20.4	33	11.5	158	1.4	19	36
Esox lucius[3] (Pike)	FW	0.6	3.1	77	<0.5	≈0.5	5.5	3.3	37
Opsanus tau[4] (Toadfish)	SW	0.18	aglom.	—	73	13	9	16	356
Platichthys flesus[5] (Flounder)	FW	1.8	4.2	57	30	43	1.2	2.2	90
	SW	0.6	2.4	75	60	36	2.5	1.5	275
Anguilla anguilla[6] (Eel)	FW	3.5	4.7	25	19	56	0.7	2.5	
	SW	0.6	1	40	6.5	4	2.1	1.3	
Salmo gairdneri[7,8,9] (Trout)	FW	4.0	6.5	38	9.3	37	1.3	5.2	
	SW	0.03	0.4	93	220 (Cl)	—	—	—	
Chondrichthyes									
Squalus acanthias[10,11] (Spiny dogfish)	SW	0.18	3.0	94	339	61	2.0	0.4	780
Pristis microdon[12] (Freshwater sawfish)	FW	10.4	19	45	'Traces'				54
Agnatha									
Lampetra fluviatilis[13] (River lamprey)	FW	13.7	21	34	15	106	2.7	37	—
Myxine glutinosa[14] (Atlantic hagfish)	SW	0.22	—	—	480	72	12.8	3.8	1000

215

and renal tubules are relevant to our ideas about the regulation of renal function. A renal portal system, such as occurs in non-mammalian tetrapods, allows for the separate functioning of glomeruli and tubules. Many fish have a renal portal blood supply, but this may be absent in some, even within groups like the Teleostei (see GERARD, 1954).

The rate of urine flow is very much greater in fish living in fresh water than in those in the sea and this reflects the excessive accumulation of water in the former environment (Table 7.7). The high rate of urine flow in freshwater fishes is accompanied by an elevated GFR and a relatively restricted rate of water reabsorption from the renal tubule. This usually amounts to less than 50% of that filtered across the glomerulus, but may vary from about 25% in the eel living in fresh water, to 69% in the African lungfish. In fishes, alterations in the urine volume are mainly determined by changes in the GFR, but some changes in the rate of water reabsorption across the tubules are also apparent (Table 7.7). This is in direct contrast to mammals, where tubular water reabsorption predominates, though in other vertebrates, like the reptiles and amphibians, changes in the GFR are also prominent. The rainbow trout, when kept in fresh water, reabsorbs 38% of the water filtered at the glomerulus, but in sea-water this rises to 93% (Table 7.7) indicating that the glomerulus is not in exclusive control.

Euryhaline fishes in nature experience the contrast of the osmotic circumstances of fresh water and sea-water. When such fish are transferred from fresh water to sea-water the urine flow decreases. This is not an immediate effect but takes place slowly and in the flounder is not complete for two days (LAHLOU, 1967). This suggests that physiological changes are occurring and that the altered urine flow is not simply the direct effect of a decrease in the amount of filterable water being delivered to the kidney. Killifish, *Fundulus kansae*, when transferred from fresh water to sea-water show, over a period of several hours, a gradual drop in the rate of urine formation (FLEMING and STANLEY, 1965). Eels, *Anguilla anguilla*, placed in sea-water also show a gradual decline in urine flow that does not stabilize for about 6 h (CHESTER JONES, CHAN, and RANKIN, 1969 a). In the first 20 min there is an increased GFR rate in eels and this presumably reflects the initial increase in arterial blood pressure. Thereafter the GFR and the blood pressure gradually decrease until a stable level is reached about 5 h later. Using Japanese eels, *Anguilla japonica*, OIDE and UTIDA (1968) also observed such a decline in urine flow GFR, but after 10 days the GFR, but not the urine flow, of these fish returned to the levels seen in fresh water. The reabsorption of water across the renal tubule increased so that the urine flow remained low.

The physiological control of GFR in fishes is not clear, but several factors could be influencing it. The rate of filtration across individual glomeruli could be influenced by changes in the osmotic pressure of the plasma proteins as would be expected if water were added to, or subtracted from, the plasma and this has been shown in eels (KEYS, 1933). As described above, the blood pressure of eels placed in sea-water decreases and this could also reduce filtration. Apart from changes in such balances of hydrostatic force across the glomerular membrane, alterations in the number of functioning glomeruli would be reflected by changes in the total GFR of the fish. MAETZ (1963) suggested that such 'glomerular recruitment' occurs in the goldfish and LAHLOU (1966) has also demonstrated it in the flounder. HICK-

216

MAN (1965) also considers it likely that such a mechanism is present in the pike, *Esox lucius*. There is evidence to indicate that 'glomerular recruitment' also occurs in amphibians and reptiles.

In tetrapods, changes in the GFR (and tubular water reabsorption) may result from the action of the neurohypophysial peptide, vasotocin. This hormone is released during dehydration in tetrapods and decreases the urine flow by increasing tubular water reabsorption and also often decreases the GFR. However, such effects have not been observed in the fishes, even though vasotocin is present in their neurohypophyses. Injections of neurohypophysial peptides do not have an antidiuretic effect in fishes, though in certain circumstances excessive doses can produce such an effect transiently. MAETZ and his collaborators (1964) showed that small

Table 7.8   *The effects of neurohypophysial peptides on water and sodium metabolism of fishes*

| | | Kidney | | Gills | |
		Water	Sodium	Sodium	Peptide
Osteichthyes					
Protopterus aethiopicus[1]	FW	Loss	Loss		Vasotocin
(African lungfish)		0*	0		Mesotocin
Ameiurus nebulosa[2]	FW	0			Vasopressin
(Catfish)					
Carassius auratus[3]	FW	Loss	Loss	Gain	Vasotocin
(Goldfish)					
		Weak antidiuresis		Gain	Isotocin
Opsanus tau[4]	SW	0			Vasotocin
(Toadfish, aglomerular)					
		Loss			Mesotocin (v. high dose)
Anguilla anguilla[6]	FW	Loss	Loss	0 SW[5]	(5) vasopressin
(Eel)					Isotocin
					Vasotocin
Salmo gairdneri[7,8]	FW	Inc. GFR	Retention	0	Vasopressin
(Rainbow trout salt-loaded)					
Fundulus kansae[9]	FW		0	Gain	Oxytocin
(Plains killifish)					
Platichthys flesus[10]	FW	0			Vasotocin or
(Flounder)					Isotocin
	SW	0			
Chondrichthyes		?	?	?	
Agnatha					
Lampetra fluviatilis[11]	FW	0	Loss	0	Vasotocin
(River lamprey)					

* 0 = no change

References: [1]SAWYER (1966 a); [2]BURGESS *et al.* (1933); [3]MAETZ *et al.* (1964); [4]LAHLOU, HENDERSON, and SAWYER (1969 b); [5]KEYS and BATEMAN (1932); [6]CHESTER JONES, CHAN, and RANKIN (1969 b); [7]HOLMES (1959); [8]HOLMES and MCBEAN (1963); [9]MEIER and FLEMING (1962); [10]LAHLOU quoted by MAETZ (1968); [11]BENTLEY and FOLLETT (1963).

doses of vasotocin (but not isotocin) when injected, increase the urine flow of the goldfish. Vasotocin (but not mesotocin) also has such a diuretic effect in the African lungfish (SAWYER, 1966a). This action has also been demonstrated in the eel but is absent in the flounder and lamprey (Table 7.8). The diuretic effect of neurohypophysial peptides is the direct result of an increased GFR, the rate of tubular water reabsorption being unchanged. The reason for the elevated GFR is uncertain, but, at least in eels, it cannot be always accounted for by changes in the blood pressure (CHESTER JONES, CHAN, and RANKIN, 1969b). However, a local vasodilatatory action in the kidney could mediate its effect especially if it resulted in a 'recruitment' of further glomeruli into activity. In the African lungfish the diuresis is initially associated with an increased arterial pressure (SAWYER, 1966b) but this does not persist for the duration of the effect (SAWYER, 1970). There are many difficulties in considering the possible physiological significance of the diuretic action of neurohypophysial peptides in fishes. The effect is not seen in all species, while the sparse (and flimsy!) available evidence suggests that if the hormone is released at all it may be in response to dehydration. This is not a physiologically appropriate stimulus for a diuretic response. At present, judgement about such a possible role should be reserved. From the phyletic standpoint it is, however, very interesting that while the neurohypophysial peptides are present in both fishes and tetrapods they only exert an antidiuretic action in the terrestrial group.

The control of urine volume in fishes may not depend on neurohypophysial peptides, but could be dependent on an, as yet, unidentified hormone. It conceivably could also be directly controlled by regional changes in the blood supply to the kidney that could result from haemodynamic effects dictated by changes in circulating catecholamines and the volume and salt content of the fishes' body fluids.

3. Salt Exchanges

Fishes may gain salts from, or lose them into, the fluids that bathe them. The principal exchanges involve sodium and chloride and, to a much lesser extent, potassium.

On simple physico-chemical grounds a net loss of sodium and chloride would be expected to occur across the integument of fishes into fresh water, while a gain of these ions may be expected to take place from sea-water. In addition, accumulation of salts occurs as a result of drinking and feeding. Earlier experiments, especially those of KROGH (1939), indicated that such movements of sodium chloride do indeed take place, but the technical methods then available did not allow a prompt assessment of the relative magnitude of the various exchanges to be made. The subsequent ready availability of radio-isotopes considerably facilitated such measurements. In 1950 MULLINS used ^{24}Na to measure the total rate of salt exchange in sticklebacks, *Gasterosteus aculeatus*. The isotopes were placed in the external bathing fluids (alternatively they can be injected into the fish), and the amounts accumulated by the fish were measured at intervals. If the fish are in salt balance the total influx and outflux of the ion will be similar and can be taken to indicate the total rate of its turnover by the fish. This has been called the rate constant (K) for the ion and is often expressed as the percentage of the total (exchangeable) amount of that ion in the body that moves in (K_i) or out (K_o) of the fish each hour. This provides a relatively simple and accurate method for comparing the permeability of different fish in various circumstances. Sticklebacks in sea-water were found to exchange 20% of their total exchangeable sodium each hour, while in fresh water only about 1% was moved. Potassium was exchanged far less rapidly; in sea-water only 0.4 m-equiv/kg h was transferred (compared to 11 m-equiv/kg h for sodium), but this increased about 3-fold in fresh water.

Table 7.9 *Total sodium fluxes* in fish bathed in various media*

		Total Na flux µ-equiv/100 g h	K (% of total exchangeable Na/h)	Weight grams
Osteichthyes				
Carassius auratus[1, 2]	FW	27	$\langle 1$	90—220
(Goldfish)	190 mM Na Cl	553	8	
Opsanus tau[3]	10% SW	52	$\langle 1$	200—700
(Toadfish)	SW	805	16	
Blennius pholis[4]	10% SW	50	8	2.5—7
(Blenny)	SW	2700	45	
Platichthys flesus[5]	FW	43	$\langle 1$	60—390
(Flounder)	SW	2600	45	
Anguilla anguilla[6]	FW	4	$\langle 0.1$	60—120
(Eel)	SW	1321	33	
Fundulus heteroclitus[7, 8]	FW	60	$\langle 0.1$	9—20
(Killifish)	SW	2020	35	
Gasterosteus aculeatus[9]	FW	60	1	1—2
(Stickleback)	SW	1000	20	
Tilapia mossambica[10]	FW	200	3	0.5—3
	SW	6000	66	
Serranus scriba[11]	SW	3100	Aprox. 60	30—80
(Sea-water perch)				
Chondrichthyes				
Scyliorhinus caniculus[12]	SW	59	0.5	40—380
(Spotted dogfish)				
Squalus acanthias[13]	SW	90	0.9	4000
(Spiny dogfish)				
Hemiscyllium plagiosum[14]	SW	75	0.74	155—1100
(Lip shark)				

* given from measurements of outflux.

References: [1]BOURGUET, LAHLOU, and MAETZ (1964); [2]LAHLOU *et al.* (1969 a); [3]LAHLOU and SAWYER (1969); [4]HOUSE (1963); [5]MOTAIS and MAETZ (1965); MOTAIS *et al.* (1966); [6]MAETZ, MAYER and CHARTIER-BARADUC (1967 a); [7]MAETZ, SAWYER, PICKFORD, and MAYER (1967 b); [8]POTTS and EVANS (1967); [9]MULLINS (1950); [10]POTTS *et al.* (1967); [11]MOTAIS and MAETZ (1964); [12]MAETZ and LAHLOU (1966); [13]BURGER and TOSTESON (1966); [14]CHAN, PHILLIPS, and CHESTER JONES (1967 b).

When the rate constants for sodium were measured in other fishes in sea-water they were found to vary greatly, though the movement of this ion in fresh water is always small (Table 7.9). In sea-water, *Tilapia mossambica* exchange 66% of their total sodium each hour, but in the toadfish, *Opsanus tau*, only 16% is moved in this time. Marine chondrichthyeans exchange sodium far less rapidly than teleosts: less than 1% each hour.

Differences in the total ion exchange of different fish are due to several factors, probably the most prominent being their relative surface areas. Nevertheless other factors also can result in differences in the rates of exchange and these include the specific permeability of the gills and skin, the amounts of salt that may be accumulated through the gut as a result of feeding and drinking, and that which is unavoidably lost in the urine. As will be seen these latter factors are usually of minor importance.

a) The Skin and Gills

As described in the previous section, the skin is generally considered to be relatively impermeable to the movements of solutes, as well as water, the principal exchanges taking place through the gills. We must, however, await precise measurements of cutaneous permeability in different species of fish.

α) *Gills in Sea-Water.* The flounder, *Platichthys flesus*, in sea-water exchanges about 2600 μ-equiv of sodium/100 g body weight each hour (Table 7.9). This represents 40% of its total exchangeable sodium. About 25% of this sodium is absorbed through the gut (MOTAIS and MAETZ, 1965) so that 75% of the total sodium accumulated by these fish takes place through the gills. In the sea perch, *Serranus scriba*, 90% of the sodium taken up from the sea-water takes place through the gills. The extrabranchial gains of sodium, which occur as a result of drinking, are extruded through the gills, urinary losses making up less than 0.1% of the total. The gills of such fish in sea-water are thus the site of considerable sodium exchange, with the outflux exceeding the influx by an amount which is about equivalent to that gained through the gut.

KEYS (1931) used a perfused heart-gill preparation of the eel to demonstrate an active extrusion of chloride by the branchae into the external sea-water. Isolated gills of eels have also been shown to secrete chloride actively into the sea-water that bathes them, and this is dependent on oxidative metabolism (BELLAMY, 1961). Such an extrarenal mechanism for excretion of sodium chloride provided the channel that HOMER SMITH concluded must be present in marine teleost fish in order that they can drink sea-water and maintain a positive water balance. Such active salt excretion appears to be characteristic of marine teleost fish and probably the lampreys (see MORRIS, 1960) but has not been demonstrated in chondrichthyeans. Some of the latter, however, possesss an alternative channel for extrarenal salt excretion, the rectal gland.

Immediately following the original demonstration, by KEYS, of chloride secretion by fish gills, a histological search was made for a structural element which could be involved in this process. KEYS and WILLMER (1932) found some large and prominent epithelial cells at the base of the gill leaflets in the eel (Fig. 7.3). As these

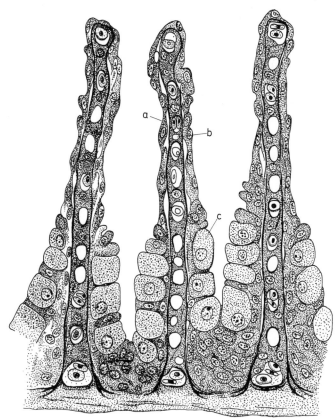

Fig. 7.3 Diagramatic representation of the gill leaflets of the eel showing the 'chloride se-cretory' cells. *a* respiratory epithelium; *b* blood vessels; *c* 'chloride secretory' cell. (Drawn after KEYS and WILLMER, 1932).

were not seen in freshwater teleosts or marine chondrichthyeans it was concluded that they may be concerned with the extrusion of chloride and were termed *'chloride secreting cells'*. This type of cell has been identified in a large variety of marine teleosts, and in lampreys during their migration from the sea into rivers (MORRIS, 1960). They are absent in the myxinoid agnathans (MORRIS, 1965; McFARLAND and MUNZ, 1965). These 'chloride secreting cells' contain, rather notably, large numbers of mitochondria and have also been called 'mitochondria-rich cells' (MORRIS, 1957; CONTE and LIN, 1967). Unequivocal evidence that it is these particular cells that secrete salt from the gills is lacking, (see PARRY, HOLIDAY, and BLAXTER, 1959; FLEMING and KAMEMOTO, 1963), but it is generally considered to be likely that they have such a role.

Epithelial cells in the basal regions of the gill lamellae of young salmon have been shown to undergo rapid renewal (increase of ³H-thymidine into DNA) in sea-water (CONTE and LIN, 1967). The activity of a mitochondrial enzyme, glutamic dehydrogenase, also increased in sea-water suggesting that increased formation of 'mitochondria-rich cells' may be occurring. The enzyme, Na-K activated AtPase, has also been identified in the gills of the killifish (EPSTEIN *et al.*, 1967) and Japanese, European and American eels (UTIDA *et al.*, 1966; MOTAIS, 1970; JAMPOL and EPSTEIN, 1970). The activity of this enzyme also increases considerably when these fish are transferred from fresh water to sea-water. The levels of Na-K activated ATPase in the gills are also 4 to 5 times greater in teleosts that normally are exclusively confined to sea-water than in those which are limited to fresh water (KAMIYA and UTIDA, 1969; JAMPOL and EPSTEIN, 1970). Marine chondrichthyean fishes had similar branchial enzyme levels to the fresh water teleosts. In other tissues, this enzyme is known to be intimately concerned with active sodium transport and a specific inhibitor of its activity, ouabain, has been shown to abolish the ability of the isolated gills of the Japanese eel to secrete sodium (KAMIYA, 1967; KAMIYA and UTIDA, 1968). This enzyme also requires potassium for its activation and the flounder fails to extrude sodium from its gills when potassium is excluded from the surrounding sea-water solution (MAETZ, 1969b). It is not known whether the Na-K activated ATPase is present in 'chloride secreting' (or 'mitochondria-rich') cells, but the pseudobranch of the killifish is particularly rich in these cells as well as the enzyme (EPSTEIN *et al.*, 1967). The increased active extrusion of sodium that occurs from the gills of eels after they are adapted to sea-water can be prevented when they are treated with actinomycin D (MAETZ *et al.*, 1969b). This antibiotic specifically inhibits nuclear RNA polymerase and so prevents the formation of RNA and protein synthesis. The increase in the activity of Na-K ATPase is also prevented (MOTAIS, 1970). It seems likely that the antibiotic is acting to reduce protein synthesis (including Na-K ATPase) and the increased branchial cell differentiation that accompanies the adaptation of eels to sea-water.

When killifish are hypophysectomized the increase in activity of the branchial Na-K activated ATPase in sea-water does not occur (EPSTEIN *et al.*, 1967). This effect was, however, not observed in the Japanese eel (UTIDA *et al.*, 1966). Unequivocal evidence for a hormonal role in the increased induction of this enzyme is thus lacking. Nevertheless, endocrines can influence the process of sodium transport across the gills in a more direct manner, but just how they interact with Na-K activated ATPase is unknown. Injections of neurohypophysial peptides, vasotocin and oxytocin, facilitate the branchial outflux of sodium that occurs when the flounder is transferred from fresh water to sea-water (MOTAIS and MAETZ, 1967). The mechanism of this effect is not clear but it could reflect the vasoactive properties of such peptides in changing the circulation of blood in the gills (MAETZ, 1968). Corticotrophin, when injected, also increases the rate of sodium exchange across the gills of eels in sea-water and cortisol and aldosterone have a similar effect (MAYER and MAETZ, 1967). The mammalian adenohypophysial hormone prolactin reduces branchial sodium exchanges in killifish, *Fundulus heteroclitus*, in sea-water (MAETZ, MOTAIS, and MAYER, 1969). The physiological significance of such observations is not clear at present and will be discussed later.

β) Gills in Fresh Water. Fishes in fresh water exchange sodium with the fluid that bathes them at less than 1% of the rate in sea-water (Table 7.9). In the goldfish nearly all of the sodium loss is accounted for by that in the urine, while in the eel and flounder extrarenal losses make up about 10% of the total (Tables 7.9 and 7.7). In fish that are not feeding the accumulation of sodium is almost entirely accounted for by the active uptake of sodium across the gills. KROGH (1939) depleted goldfish of sodium chloride by keeping them for several weeks in distilled water; at the end of this time they were able to accumulate sodium chloride actively from solutions as dilute as 10^{-5} to 10^{-4}M. The isolated gills of the eel can also actively absorb sodium from dilute solutions (0.04 mM) while the intact fish can accumulate sodium from fresh water containing only 0.05 mM sodium chloride (BELLAMY, 1961; GARCIA ROMEU and MOTAIS, 1966). Chloride is apparently not taken up actively by the gills of the eel (KROGH, 1939; GARCIA ROMEU and MOTAIS, 1966). When eels are kept in distilled water for three weeks, and then replaced in tap water, the rate of branchial accumulation of sodium increases three-fold (HENDERSON and CHESTER JONES, 1967). Such experiments are performed on fasting fish and it should be remembered that feeding results in an accumulation of sodium through the gut and that this normally contributes to the animals' needs. However, many fish do normally undergo periods of prolonged fasting, and in these circumstances the ability to accumulate sodium chloride actively may well be vital to their survival.

The mechanism of sodium and chloride accumulation by the gills of fish in fresh water involves exchanges with ions in the body. In the goldfish sodium enters the gills in exchange for NH_4^+ while external chloride exchanges with bicarbonate (MAETZ and GARCIA ROMEU, 1964; GARCIA ROMEU and MAETZ, 1964). A similar $Na^+ - NH_4^+$ exchange occurs across the gills of eels (GARCIA ROMEU and MOTAIS, 1966). It is not known if Na-K activated ATPase is concerned in the mechanism of active accumulation of sodium in fresh water as it is with extrusion of this ion in the sea. This enzyme has been identified in the gills of killifish and European eels adapted to fresh water (EPSTEIN et al., 1967; MOTAIS, 1970) though the level of activity is less than half as great as in sea-water. However, the rate of accumulation of sodium in fresh water is far less than its rate of secretion in sea-water, so the requirements in the former media are probably less. MOTAIS (1970) found that while the levels of Na-K ATPase activity in the gills of eels in sea-water could be reduced by treatment with actinomycin D, the freshwater fish was unaffected by this antibiotic. He has suggested that there are two forms of Na-K ATPase present. One, which is situated at the external border of the epithelial cells in the gills, mediates sodium extrusion in sea-water, while the other, at the opposite side of these cells, is involved with sodium absorption. As yet there is no direct evidence to indicate that Na-K ATPase is concerned with active branchial sodium accumulation in fresh water but it seems likely that it may perform this role, just as it does in so many other tissues.

Sodium-chloride depletion, as we have seen, stimulates the rate of sodium uptake by the gills of goldfish and eels. Injection of solutions that dilute the plasma electrolytes (but not its osmotic concentration) increases absorption of sodium and chloride by the gills of goldfish (BOURGUET, LAHLOU, and MAETZ, 1964). Whether such effects are mediated by hormones is unknown but the injection of several 'putative' ones has a similar effect (Table 7.8). Isotocin, and to a lesser extent vasotocin, when so administered, stimulate the rate of sodium uptake across the gills of the goldfish. It is possible that such effects are mediated by changes in the branchial circulation (MAETZ and RANKIN, 1969). We (BENTLEY and FOLLETT, 1962; 1963) could find no evidence that vasotocin altered sodium exchange across the gills of lampreys. The effects of such peptides in chondrichthyeans do not seem to have

Table 7.10   *Effects of adrenocordicosteroids on sodium metbolism of fish*

		Kidneys	Gills	Corticosteroid
Osteichthyes				
Carassius auratus (Goldfish)	FW	.	Gain[1] or loss[2]	(1) Aldosterone (2) Deoxycorticosterone
Anguilla anguilla[3, 4, 5] (Eel)	FW	0[*]	Gain or loss	Gain aldosterone or cortisol small dose
	SW	.	Loss	Cortisol
Salmo gairdneri[6] (Rainbow trout salt-loaded)	FW	Retention (slight)	Loss	Cortisol
Agnatha				
Lampetra fluviatilis[7] (River lamprey)	FW	0	Gain	Aldosterone
		0	0	Cortisol

[*]0 = no change

[1]Favre (1960); [2]Sexton (1955); [3]Chester Jones *et al.* (1962 b); [4]Henderson and Chester Jones (1967); [5]Mayer *et al.* (1967); [6]Holmes (1959); [7]Bentley and Follett (1962; 1963).

Table 7.11   *Effects of hypophysectomy on sodium metabolism in various teleost fish in fresh water*

	Sodium loss	Site of effect	Replacement hormone
Fundulus heteroclitus[1] (Killifish)	+	Gills	Prolactin
Fundulus kansae[2] (Plains killifish)	+	Kidney	Prolactin and ACTH *not* effective
Poecilia latipinna[3]	+	?	Prolactin
Tilapia mossambica[4]	+	?	Prolactin partly
Carassius auratus[5] (Goldfish)	+	Gills	Prolactin partly
Anguilla anguilla[6] (Eel)	+	Gills	Prolactin partly but ACTH more effective

Note: Schreibman and Kallman (1966) also found four species of hypophysectomized platty-fish (*Xiphophorus sp.*) died in fresh water but could be maintained with prolactin.
References: [1]Burden (1956); Pickford and Phillips (1959); Potts and Evans (1966); Maetz *et al.* (1967 b); [2]Stanley and Fleming (1966 a, b; 1967 a, b); [3]Ball and Olivereau (1964); Ball and Ensor (1967); Ensor and Ball (1968 b); [4]Dharmamba *et al.* (1967); [5]Lahlou and Sawyer (1969); [6]Butler (1966); Olivereau and Chartier-Baraduc (1966); Maetz *et al.* (1967 a); Chan, Chester Jones, and Moseley (1968).

been tested. Aldosterone has not been found in the circulation of fish but when this steroid is injected it, nevertheless, promotes accumulation of sodium across the gills of eels (HENDERSON and CHESTER JONES, 1967) and goldfish (FAVRE, 1960) and reduces the rate of branchial sodium loss in lampreys (BENTLEY and FOLLETT, 1962; 1963) (Table 7.10). Cortisol, which appears in the circulation of fish, when injected, increases sodium loss from the gills of eels but in small doses can promote sodium accumulation in the hypophysectomized fish (HENDERSON and CHESTER JONES, 1967). Hypophysectomy results in an increased rate of sodium loss in fish and in many cases this can be prevented by the injection of prolactin (Table 7.11).

b) The Kidney

The kidney is not a major site for the exchange of osmotically important ions in fish.

In sea-water, the flounder and sea perch excrete only about 0.1% of the total accumulated sodium in their urine (MOTAIS and MAETZ, 1965). When euryhaline fish such as the flounder and eel are transferred from fresh water to sea-water the total renal sodium and potassium excretion changes little, if anything it may decrease in the latter medium (Table 7.7). In fresh water, sodium and potassium are lost in the copious dilute urine that is formed. In goldfish this daily sodium loss equals about 8% of the total sodium in the body, an amount similar in magnitude to the total accumulated by the fish (MAETZ, 1963). Lampreys lose a similar proportion of their sodium in this manner. The pike, *Esox lucius*, loses as little as 0.02% of its body sodium each day (HICKMAN, 1965). These renal losses can be replaced by active branchial uptake of sodium, but in some feeding fish this may not be necessary. Potassium is also excreted in the urine of fish but in smaller quantities than sodium. Eels and flounder in fresh water lose less than 0.1% of their total body potassium in this way each day, and this declines in sea-water (Table 7.7). Urinary potassium losses in fasting fish probably largely reflect those which arise in the body as a result of tissue catabolism.

Fish that are feeding in fresh water may gain an excess of potassium in their food, and the kidneys, most likely, have some role in its excretion. The kidneys of fish do not appear to respond dramatically to excesses of sodium chloride but this is unlikely to occur in fresh water. HOLMES (1959) observed that sodium loads were mainly excreted extrarenally in rainbow trout. The urinary losses of sodium are little affected when sodium chloride solutions are injected into goldfish, indeed if these are hypertonic there is a decreased renal loss (BOURGUET et al., 1964). In lampreys we (BENTLEY and FOLLETT, 1963) found that only 10% of a dose of injected sodium chloride was excreted in the urine after 24 hours.

Sodium is absorbed from the glomerular filtrate of fish. In the goldfish and lamprey 93% of this is reabsorbed (MAETZ, 1963; BENTLEY and FOLLETT, 1963) while in the pike 99.95% may pass back into the plasma (HICKMAN, 1965). Potassium is also reabsorbed from the glomerular filtrate of fish and as observed, for instance, in the pike and the white sucker, *Catostomus commersoni*, it may be secreted into the urine across the wall of renal tubules (HICKMAN, 1965). The sites of such transfers along the renal tubules of fish are unknown and must await collection of this fluid with the aid of micropipettes such as are used in experiments with mammals and amphibians.

α) *The Urinary Bladder.* Many fish have a urinary bladder that originates as an expansion of the mesonephric ducts. When the composition of the urine stored in such receptacles is compared with that collected directly from the ureters, differences in composition are apparent. In the flounder the sodium chloride levels are lower in the bladder urine of fish in either fresh water or sea-water (LAHLOU, 1967). This seems to result from reabsorption of these ions, which in sea-water leads to additional water conservation and in fresh water to a further saving of sodium chloride. LAHLOU *et. al.* (1969b) have also found evidence for such ionic reabsorption from the urinary bladder of the toadfish.

β) *Neurohypophysial Peptides and Adrenocorticosteroids.* The injection of various neurohypophysial peptides can influence renal sodium excretion in some fish. Vasotocin has been shown to increase renal sodium losses in goldfish, eels, the African lungfish and lampreys (Table 7.8). Isotocin also has this action in eels, but in goldfish it sometimes produces a renal sodium retention. Such peptides do not, however, invariably alter renal sodium losses in fish while the physiological significance of such an action, when it does occur, is unknown. The increases in renal sodium excretion result (except in the lamprey) from increases in the GFR. The adrenocorticosteroids appear, in contrast to their effects on the gills, to be without an action on the sodium and potassium content of the urine (Table 7.10). This is in contrast to the actions of such steroids in the higher tetrapods, but is reminiscent of the failure to demonstrate such an action in amphibians.

c) The Gut

The intestine of fish living in fresh water actively transports sodium from the lumen to the blood. SMITH (1964; 1966) found that sodium and potassium moved from the fluid bathing the mucosal surface of isolated sacs of the goldfish intestine. Sodium transport greatly exceeded that of potassium and resulted in the generation of an electrical p.d., serosal side positive, of about 7 mV across the membrane. The process is inhibited by a Na-K ATPase inhibitor, ouabain, and the enzyme has been isolated and histologically identified in the mucosal epithelial cells lining the intestine (M. SMITH, 1967; HOLLANDS and SMITH, 1964). The level of the Na-K activated ATPase in the goldfish intestine is labile, as also seen in marine teleosts. When the fish were kept in water at 8° the activity of the enzyme increased to about twice the level seen when they were maintained at 30°. This may reflect the operation of a mechanism for controlling the rate of sodium absorption at different temperatures, similar to that influencing the levels of this enzyme in fish kept in sea-water.

The relative importance of sodium chloride absorption from the gut and gills of freshwater fish probably varies in different conditions. Fasting fish in fresh water are expected to gain little sodium from the water that they drink as this usually contains less than 1 m-equiv/l. Less than 0.5% of the total sodium accumulated by the

226

fasting eel kept in fresh water is absorbed through the gut (MAETZ and SKADHAUGE, 1968). However, fish that are feeding actively will be expected to gain more salt, due to its absorption from the food. The magnitude of this varies with the nature of the diet. In fresh water an herbivorous or detrital diet contains less sodium chloride than a carnivorous one. JORGENSEN and ROSENKILDE (1956b) calculated that the expected accumulation of chloride by a goldfish from its normal diet would be less than 25% the quantity that is normally absorbed by the gills. Carnivorous freshwater fish, like the pike, probably gain adequate salt from their diet.

4. Nitrogen Metabolism

In fishes the metabolic breakdown of amino acids may lead to the formation of either ammonia or, by way of the ornithine cycle, to urea. The particular product formed is closely related to the fishes' pattern of osmoregulation. Some additional urea may be formed as the result of the breakdown of uric acid which is the end-product of purine metabolism. This requires the presence of certain enzymes, notably uricase. While this process has been shown to occur in fishes, including the African lungfish (BROWN et al., 1966), it is probably not a major source of urea (FORSTER and GOLDSTEIN, 1966) in ureotelic fishes.

When adequate water is available ammonia can be readily lost by passage across the gills. In the trout, *Salmo gairdneri,* in fresh water, 60% of the fishes' total nitrogen excretion takes place as ammonia, and less than 2% of this is in the urine (FROMM, 1963). The balance is excreted across the gills, which may also be its principal site of formation (GOLDSTEIN and FOSTER, 1961). The loss of ammonia (as NH_4^+) across the gills of freshwater fish may be coupled with the active uptake of sodium (MAETZ and GARCIA ROMEU, 1964).

Some fishes can alter the relative quantities of ammonia and urea that they form. The African lungfish is enzymatically equipped to make either ammonia or urea, but when it is in fresh water the former predominates (H. SMITH, 1930a). However, if water is restricted, as when these fish aestivate, they only form urea. The Australian lungfish, *Neoceratodus forsteri,* is normally ammoniotelic, though it also possesses the ornithine-urea cycle enzymes, but these are only present at low levels of activity (GOLDSTEIN, JANSSENS, and FORSTER, 1967).

Urea is relatively non-toxic and can be accumulated at high concentrations in the bodies of many vertebrates. The chondrichthyean fish are ureotelic and have utilized retention of this metabolite to maintain the osmotic concentration of their body fluids hyperosmotic to the surrounding sea-water (see H. SMITH, 1936). Such a strategem has been used on other occasions by distinctly different groups of vertebrates, including the marine crab-eating frog and the coelacanth, *Latimeria* (BROWN and BROWN, 1967; PICKFORD and GRANT, 1967). The likely phyletic history and distribution of the ornithine-urea cycle in vertebrates is shown in Fig. 7.4.

Trimethylamine oxide is another nitrogen compound that is found in the body fluids of many fishes, especially in marine species. Its origin is uncertain and there is little evidence to indicate that it is formed metabolically in the Chondrichthyes (GOLDSTEIN, HARTMAN, and FORSTER, 1967). It has been suggested that it is ac-

227

cumulated from the diet, especially if this consists of marine invertebrates that contain large amounts of this compound (BENOIT and NORRIS, 1945). Whatever its origins, it is, like urea, retained in the body as a form of osmotic 'stuffing'. In the Chondrichthyes it makes up about 25% of the non-ionic osmolytes (SMITH, 1936). It is also present in the intracellular fluids of teleosts, like the flounder and stickleback, where it accumulates in higher concentrations when the fish are in sea-water, than when they are in fresh water (LANGE and FUGELLI, 1965). In the agnathan, *Myxine glutinosa*, it is almost absent from the plasma, but is contained in the cells at a concentration of about 200 mM (BELLAMY and CHESTER JONES, 1961) and so makes a substantial contribution to their osmotic balance.

Although both urea and trimethylamine oxide can be filtered across the glomerulus, only small quantities are excreted in the urine. Reabsorption of these com-

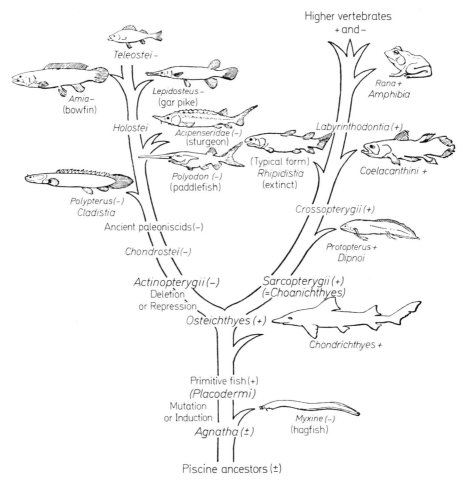

Fig. 7.4 Occurrence of the ornithine-urea cycle in different groups of vertebrates. Presence of cycle, +; absence, −; presumed presence, (+); presumed absence, (−). (From BROWN and BROWN, 1967; copyright 1967 by AAAS).

pounds from the renal tubule is a very efficient process (SMITH, 1936). Similarly the gills of chondrichthyean fish are also relatively impermeable to urea and trimethylamine oxide. The possible actions of hormones, particularly the neurohypophysial peptides, in changing renal urea secretion in chondrichthyeans have been suggested to me by HANS HELLER, but this interesting possibility has not been investigated.

5. Adaptations in the Osmoregulation of Teleost Fish during Transfers between Fresh Water and Sea-water

Many fish can adapt to life in either fresh water or the sea; they are euryhaline. Migrations, such as are often associated with breeding, involve the movement of the fish from the sea into rivers and back into the sea again. Salmon, trout and lampreys breed in fresh water where their young undergo initial growth and development before returning to the sea. Others, like eels, breed in the ocean but subsequently spend a large part of their life in fresh water. Some species, like the trout, *Salmo gairdneri*, and the marine lamprey, *Petromyzon marinus*, have landlocked populations that have discarded such migratory habits, and undergo their entire life cycle in fresh water. Certain other species like the flounder, killifish and toadfish occupy coastal waters near the mouths of rivers and streams, where they experience variations in the salinity of the water in which they live. There is not only a variety of euryhaline fishes, but they exhibit diverse morphological development and physiological conditions associated with different stages of their life period. These include the juvenile freshwater 'parr' and premigatory 'silver' eels, as well as the adult fishes in various stages of their breeding cycle. Many such fish cannot withstand direct transfers from fresh water to sea-water during all stages of their development, or even at any season of the year. When such a migration is made, several days are required to adjust completely to the new osmotic conditions. Our knowledge of the physiological changes that accompany the development of euryhalinity and the acute changes that occur at the time of the transition are still incomplete, but it is often considered likely that hormones may have a role to play.

CONTE, WAGNER and their collaborators (CONTE and WAGNER, 1965; CONTE *et al.*, 1966) have made some interesting comparisons of the migratory behaviour, and ability to adapt to sea-water, between the steelhead trout, *Salmo gairdneri*, and the coho salmon, *Oncorhynchus kisutch*. Populations of these fish breed and undergo their initial development in the rivers of the northwest of the United States. When very young juveniles of these salmonids are placed in sea-water they die, but subsequently as they grow larger they can adapt to such solutions. It is well known that this often occurs at about the time when they metamorphose from a 'parr' to a 'smolt', and this also corresponds to their seaward migration. It was found that it is not this metamorphosis *per se* that results in their ability to adapt to sea-water, but rather their size. When they attain a length of about 15 cm they can adapt to sea-water and indeed in the juvenile 'parr' coho salmon studied, this was usually seen 6 or 7 months *before* they metamorphosed into 'smolts'. This suggests that surface area relative to the body weight may be important, and that physiological and morphological features may develop that are associated with the general growth rate, rather than any sudden metamorphic transformation. HOAR (1951) found that the branchial 'chloride secreting cells' in Pacific salmon underwent a period of rapid development before migration. This is consistent with the recent observations of CONTE and LIN (1967), who found an increased rate of branchial cell renewal during salt water adaptation in young salmonids (see page 221). The levels of Na-K activated ATPase in the gills of developing salmonids has not been reported, but such measurements would be most interesting.

Transformation of the salmonid parr into a smolt, prior to its seaward migration, is associated with many body changes, including endocrine ones. Thus, there is an increased activity of the thyroid gland (FONTAINE, 1954; CONTE and WAGNER, 1965); the 17-hydroxysteroids in the plasma have an elevated concentration (FONTAINE and HATEY, 1954) and morphological changes take place in the hypothalamo-hypophysial tract (ARVY, FONTAINE, and GABE, 1959). The ability to osmoregulate in sea-water, however, does not reflect any single factor. As HOAR (1963) has warned, any effects of endocrines are probably rather general and it is unlikely that there is any specific hormone concerned with triggering such migratory behaviour and the associated adaptations to life in the sea.

The population of steelhead trout studied by CONTE and WAGNER (1965) attain the ability to regulate in sea-water in March, but if they fail to migrate at this time, they lose their ability to osmoregulate in sea-water and this does not return until the following year. Pacific salmon, on the other hand, retain their ability to adapt to sea-water (CONTE et al., 1966). Why such an ability should be kept by one salmonid and not another is unknown, but such cyclical changes in the ability to withstand transfers from fresh water to sea-water are noteworthy. On their terminal migration to breed in fresh water lampreys and salmon lose their ability to regulate in sea-water, the physiological condition of the fish apparently being inadequate to such a task. Clearly the ability, and predisposition, to osmoregulate in either fresh water or sea-water is at least partly subject to physiological changes which could involve the endocrine system.

As described above, migration into the sea and the ability to osmoregulate in such a solution may temporarily coincide, but these two factors are evidently not causally related. Thus physiological and endocrine changes that are associated with migration of fish do not necessarily directly influence osmoregulation. The increased thyroid activity that occurs in salmonids prior to migration cannot be strictly correlated with their ability to osmoregulate in the sea (D. SMITH, 1956; CONTE and WAGNER, 1965). The starry flounder, *Platichthys stellatus*, has an increased rate of oxygen consumption when it is placed in sea-water and this may be related to the added osmoregulatory needs of such fish (HICKMAN, 1959). The thyroid gland of the flounder also has an increased activity at this time, but the relationship of this to oxygen consumption (or osmoregulation) is not clear, as attempts to demonstrate such a causal relationship have not usually been successful. Removal of the thyroid gland thus, does not alter the metabolic rate in the parrot fish (MATTY, 1957). Changes in the hypothalamo-hypophysial tract of salmonids during their transformation to a smolt most probably occur subsequently to the ability to osmoregulate in sea-water.

Numerous attempts have been made to demonstrate changes in the ability of fish to adapt to altered salinity by injecting them with various thyroid hormone

230

preparations. Koch and Heuts in 1942 (quoted by Fontaine, 1956) found that feeding thyroid gland to one species of stickleback, *Gasterosteus aculeatus,* decreased its ability to withstand transfer into fresh water, while in another species, *Pygosteus pungitius,* the salinity tolerance was increased. Smith (1956) observed that administration of thyroid hormone (thyroxine) increased the ability of trout to survive in salt solutions, while antithyroid drugs decreased this ability. Fontaine and his collaborators (see Fontaine, 1956) have also found that antithyroid drugs decrease the salinity tolerance of various marine fish. On the other hand, Hoar, Black, and Black (1951) (quoted by Fontaine, 1956) found that keeping juvenile salmon in thyroxine solutions did not alter their ability to osmoregulate in salt water. The physiological significance of the various observations on the thyroid gland is not clearcut and the evidence that it is involved in osmoregulation has been treated with some suspicion. Smith (1956) commented that the doses of thyroxine that he used were excessive, and found that in nature there was no correlation between the thyroid activity of the trout and their ability to osmoregulate. In addition the administration of thyrotrophin did not alter the trout's salinity tolerance. As commented upon by others, the actions of antithyroid drugs may have actions other than on the thyroid gland for such substances may exhibit non-specific toxicities, as are readily apparent when they are given to mammals. Although the thyroid probably does not have a direct effect on osmoregulation in fish, it could interact with other physiological factors, which together could influence euryhalinity in the fish.

The *pituitary,* as will be described in more detail later, can influence osmoregulation in fish and the whole gamut of pituitary hormones has been tested on fish placed in solutions of differing salt concentration. The most dramatic effect found is the action of mammalian *prolactin* in restoring the ability of hypophysectomized killifish, *Fundulus heteroclitus,* to survive in fresh water (Pickford and Phillips, 1959). Other pituitary hormones, including corticotrophin, are ineffective. Injections of large doses of *corticosteroids,* including cortisol and aldosterone, were not found to affect survival, or electrolyte composition, of several species of teleosts after they were placed in solutions of increased salt concentration (Edelman, Young, and Harris, 1960). Smith (1956) demonstrated that pituitary *growth hormone,* when injected into trout, produced a dramatic increase in the ability to survive in salt solutions. The effective dose was quite small, but these interesting observations do not appear to have received the subsequent consideration that they deserve. As growth and differentiation appear to play an important role in the ability of salmonids to osmoregulate in the sea, the action of this hormone may be the result of an increased rate of development, possibly an accelerated maturation of the branchial 'chloride secreting cells'.

The frequent failure to demonstrate any dramatic changes in the salinity tolerances of various fishes after they have been injected with large doses of endocrine extracts does not necessarily indicate that these hormones are not involved in such changes. The hormone preparations are sometimes those that are only found in tetrapods and may not be identical in their structure to the piscine secretions. Even if such a hormone is present in fishes, its endogenous levels may already be maximal at the time of the injection, so that the extra hormone will have no added effect. In addition, physiological adaptations to solutions of different osmotic con-

231

centration almost certainly result from the interaction of a large number of factors, so that a solo role for one such hormone would be rather unexpected.

When fish are *transferred from fresh water to sea-water,* or from sea to fresh water, a number of changes take place in the osmotic composition of their body fluids. Initially the changes are usually more pronounced than are subsequently maintained. This period of equilibration lasts for about 48 h in eels and flounder (KEYS, 1933; LAHLOU, 1967) transferred from fresh water to sea-water, but may be as long as 170 h in the steelhead trout (HOUSTON, 1959). Some of the changes

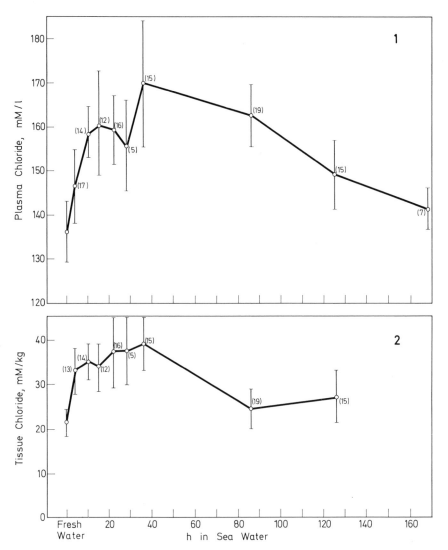

Fig. 7.5 Changes in the plasma and muscle chloride concentrations in trout, *Salmo gairdneri,* after transfer from fresh water to sea-water. (From HOUSTON, 1959; by permission of the National Research Council of Canada).

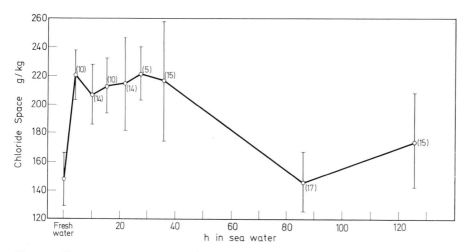

Fig. 7.6 Changes in the chloride space of trout, *Salmo gairdneri*, after transfer from fresh water to sea-water. (From HOUSTON, 1959; by permission of the National Research Council of Canada).

which occur in the steelhead trout during this time are shown in Fig. 7.5 and 7.6. Initially there is a marked increase in the concentration of chloride in the plasma and the tissues, while the overall chloride space increases. After the rather abrupt initial increases these levels start to decline after about 24 h, and eventually reach equilibrium 4 or 5 days later. The electrolyte concentration in such sea-water adapted fish, however, remains somewhat elevated, compared to fish in fresh water.

a) Changes in Permeability that Accompany Variations in Osmotic Environment

The reasons why certain, euryhaline, fishes can survive moving between fresh water and saline solutions, like sea water, were until recently very poorly understood. Some recent studies utilizing radio-isotopes to measure the permeability of fish to electrolytes and water have indicated the essential processes that are involved.

α) *Salt*. It has been rightly assumed that stenohaline marine and freshwater fish die, respectively, in fresh water and sea water as a result of a rapid loss or accumulation of salt. The pathways for such transfers have, however, until recently been poorly understood and this has severely hampered the search for the physiological mechanism involved. RENE MOTAIS and JEAN MAETZ, working on the French Riviera, have carried out an extensive series of studies on the adaptation of fish to different osmotic environments. Their work is principally responsible for the relatively healthy state of our knowledge in this field. MOTAIS and MAETZ have utilized radio-isotopes to study the permeability of fish to salts and, more recently, water in order to define the relative importance of the various pathways involved in molecular transfers between fish and their environment. It had earlier been assumed that almost the entire quantity of sodium and chloride accumulated by fish in sea-

233

water occurred, as a result of drinking, through the gut. However, as described in the previous section, although this is important it rarely contributes more than about 10% of the total salt taken up by such fish (MOTAIS and MAETZ, 1965; POTTS et al., 1967). The kidney was also found to be only a minor avenue for salt loss in fish living in the sea. It thus became apparent that the major part of the very large amount of salt which is exchanged by marine fish takes place through their gills. It had previously been recognized that the gills of marine fish were the principal avenue for active extrusion of sodium chloride, but the fact that large amounts of

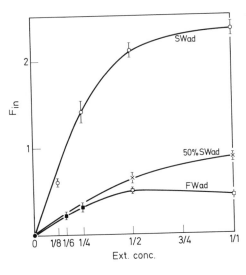

Fig. 7.7 Sodium *fluxes* of the flounder as a function of the external sodium concentration.
a) F_{in}, *influx* in m-equiv/h/100g + S.E. External sodium concentration given in terms of dilutions of sea-water (SW). Influxes measured during the first 30 min after transfer from the adaptation media (sea-water, 50% sea-water and fresh water). *ad*, adapted.

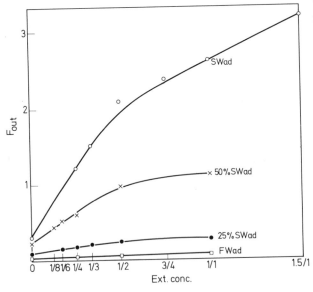

b) *outfluxes* (F_{out}) of the flounder as a function of the external sodium concentration. Details the same as above except that some fish were also adapted to 25% sea-water. (Modified from MOTAIS et al., 1966).

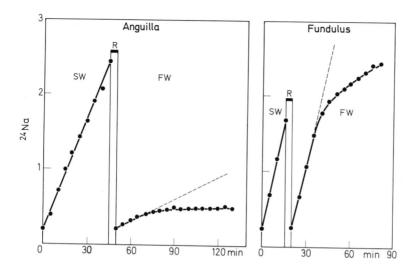

Fig. 7.8 Relative outflux of sodium upon transfer from sea-water (SW) to fresh water (FW) in eels, *Anguilla anguilla,* and killifish, *Fundulus heteroclitus.* External ^{24}Na concentration (in CPM $\times$ 10^{-4}) as a function of time in min. Observe the immediate decline in sodium loss from eels placed in fresh water, but this is not apparent in the killifish. The onset of 'delayed regulation' (derived from the broken line) can be seen about 20 min after both species have been placed in the fresh water. (Modified from MOTAIS *et al.,* 1966).

sodium also entered by diffusion through these structures was not generally appreciated. This indicated that the gills may be of major importance in the adaptation of fish to either fresh water or sea-water.

MOTAIS (1961 a, b) found that the euryhaline flounder, *Platichthys flesus,* exchanges about 25% of its total body sodium each hour when it is in sea-water, but only 0.7% of this when it is adapted to life in fresh water. When these flounder are transferred from sea-water to fresh water they reduce their rate of sodium outflux very rapidly (almost ceasing in about one hour). In contrast, when flounder, adapted to life in fresh water, are transferred to sea-water, the rate of sodium exchange only increases slowly to the more usual marine levels, the whole process taking about 30 h. However, if marine-adapted flounder are placed in fresh water for 30 min, during which time their rate of sodium exchange drops by about 90%, and they are then returned to sea-water, the sodium fluxes immediately returned to levels normal in this solution. The delayed increase in sodium extrusion, which is characteristic of freshwater adapted flounder placed in sea-water, is no longer apparent. These observations suggested that the permeability of the fish to sodium can be changed, and that a physiological adaptation takes place in the different media which takes some time to transpire. It later became apparent that these adjustments are occurring in the gills of the fish.

MOTAIS, GARCIA ROMEU, and MAETZ (1965; 1966) compared the influx and outflux of sodium from the gills of the flounder and a stenohaline marine fish, the sea-water perch, *Serranus scriba.* Sea-water adapted flounder and *Serranus* rapidly

235

take up and extrude sodium from their gills. When they are transferred from sea-water to fresh water they respond differently, for while the rate of sodium loss from the gills immediately decreases by about 90% in the flounder, it only declines by 40% in *Serranus*. This stenohaline fish continues to lose salt at this rate through its gills and dies from salt depletion in about 3 hours. The euryhaline flounder, on the other hand, can rapidly 'shut down' its branchial permeability to sodium and chloride to a level which the stenohaline species cannot do.

In the flounder, the rapid exchange of sodium across the gills is directly dependent on the external salt concentration and not on the osmotic concentration of the medium. If the salt concentration in the external solution is decreased, but its osmotic concentration is maintained by the addition of mannitol, the decline in sodium exchange is the same as in flounders placed in the ordinary dilute salt solution (without mannitol). In contrast, the sodium exchange in *Serranus scriba* in such sodium plus mannitol solutions is similar to that in ordinary sea-water, the 40% decline that is observed in fresh water being directly related to the change in the osmotic concentration of the solution. When flounder are transferred from sea-water to various dilutions of this fluid, the rate of sodium extrusion and accumulation across the gills is closely related to the decline of the external sodium concentration, proportionately until, in fresh water the loss is negligible (Fig. 7.7). The uptake and output of sodium and chloride are linked to the concentration of sodium (and chloride) in the external media. This is reminiscent of the process of *'exchange diffusion'*, which USSING (1947) suggested accompanies the rapid exchanges of sodium observed in frog muscle; one ion moving into the fish exchanges with one that is leaving. This process is not directly dependent on the expenditure of energy. Such 'exchange diffusion' accounts for 85% of the sodium transferred across the gills of the sea-water adapted flounder but only 3% in *Serranus*. It is notable that the relative magnitude of this process decreases in flounders adapted to solutions with a lower sodium chloride concentration, until in freshwater adapted fish it cannot be seen (see Fig. 7.7b).

An additional *'delayed regulation'* of branchial permeability to sodium and chloride is observed to commence about 30 min after the flounder is transferred from sea-water to fresh water. The onset of this secondary phase can be seen in Fig. 7.8 which shows the changes which occur after eels and killifish are transferred from sea-water to fresh water. This reduction in the permeability of the gills almost completely restricts their permeability to the outflux of sodium. The nature of this change is at present unknown. MOTAIS and his collaborators examined four more euryhaline teleosts and six stenohaline ones. A significant exchange diffusion effect was seen in the stenohaline marine fish, *Scorpaena porcus*, while a euryhaline fish, *Fundulus heteroclitus*, apparently lacked such an effect. The latter species, nevertheless, still reduced the permeability of its gills in fresh water but this was related to the decreased osmotic pressure (see also POTTS and EVANS, 1967). The intertidal and partly euryhaline stachaeid blenny, *Xiphister atropurpureus*, also exhibits exchange diffusion (EVANS, 1967b). The presence or absence of branchial ion exchange diffusion is not strictly confined to, or present in, all the euryhaline fishes examined. The secondary 'delayed' phase, whereby gill permeability is gradually restricted in fresh water, was however, seen in all euryhaline fish, but never in a stenohaline one.

The extent of the exchange diffusion of ions across the gills of the flounder, and so the magnitude of the 'instantaneous' response of the gills to transfer to fresh water, can be adjusted; it is less in fish previously adapted to low salinities than to sea-water. Such an adaptation takes several days to develop fully or to reverse. MOTAIS et al., 1966) have suggested that a 'carrier' mediates the exchange diffusion and the processes of such 'long term' adaptation involve changes in the amounts of such a 'carrier', or alternatively the synthesis of a substance which inhibits its action.

The exchange diffusion of sodium and chloride across the gills of fish can be considered as an adaptation to euryhalinity (MOTAIS et al., 1966). In sea-water the outflux of sodium normally exceeds the influx, the difference being the result of the activity of an active transport mechanism for these ions. However, this latter process is accompanied by an unavoidable 'leakiness' of the gills to passive diffusion of these ions, the magnitude of which is 5 to 10 times greater than the active transport. In most stenohaline fish this passive diffusion, in either direction, proceeds in a random manner. However in other fish this may be an 'organized leakiness' that results in a direct exchange of one ion moving in one direction, for another moving in the opposite way. When present, the advantage of this to fish in marine environments is not clear though it is possible that such a linkage, by its increased efficiency, may conserve some energy that would otherwise be used for active transport of these ions. When such marine fish are transferred to fresh water, however, they fare better than fish lacking such a linkage, as they possess a mechanism that immediately, and automatically, reduces the sodium loss from the gills. This, followed by the 'delayed' reduction in gill permeability, seals off the gills to diffusional salt losses and allows the fish to live in fresh water. Teleost fish are thought to have originated in sea-water and the appearance of such a system for controlling the permeability of the gills may well have arisen, and been utilized, as a pre-adaptation for migration into fresh water.

The lability of these processes of ion transfer across the gills indicates possibilities for their control and possibly the intervention of the endocrines. Treatment of the flounder with an inhibitor of RNA depolymerase, actinomycin D, results in a reduction of the active extrusion of sodium from the gills of fish in sea-water, so that they only survive a short time (MAETZ et al., 1969a). If fish, so treated, are placed in fresh water they show no evidence of exchange diffusion, so that there is a rapid and fatal loss of sodium from the gills as seen normally in *Serranus scriba*. In contrast, when flounders in fresh water are treated with actinomycin D they live for at least a week, the process of active sodium uptake, in contrast to active sodium extrusion, being unaffected by the antibiotic. These results suggest (see MAETZ, 1970) that a nuclear controlled synthesis of proteins is involved in the regulation of such processes. It will be recalled that MOTAIS (1970) has shown that the synthesis of a Na-K activated ATPase present in the gills of eels in sea-water can be prevented by actinomycin D, while the enzyme present in freshwater fish is unaffected.

β) *Water*. Like the movements of salts, the net transfers of water in euryhaline fishes in fresh water and sea-water are opposite in direction. When in the sea such teleost fish lose water by osmosis, while in fresh water they gain water. Comparisons of

237

the permeability of such fish in different environmental solutions have recently been undertaken with the aid of tritiated water. POTTS et al. (1967) found that in *Tilapia mossambica* the turnover of body water amounted in each hour to 115% of the total present in the body when the fish are in fresh water, and this declined to 84% in sea-water. The eel and the flounder in fresh water exchange an amount equivalent, respectively, to 42% and 31% of their total body water in an hour (MOTAIS et al., 1969). When these fish are transfered to sea-water this fraction decreases to 29% in the eel and 20% in the flounder. It was further found that nearly all of this water exchange primarily takes place through the gill. It is replaced by drinking when the fish are in sea-water while it is excreted in the urine when they are in fresh water. The osmotic gradients between the fish and its bathing fluids are greater when they are in sea-water than in fresh water, but, despite this, the transfer of water is less in the former medium. The permeability of the gills is thus apparently least when the fish are in sea-water. This, as discussed earlier, could be due to a rectifying property of the gills whereby net transfers of water take place less readily outwards than inwards. Alternatively there may be an adjustable mechanism that can reduce the branchial permeability to water when the fish are in the sea. LAHLOU and GIORDAN (1970) have shown that hypophysectomized goldfish have a reduced permeability to water and that both prolactin and cortisol can restore this to normal. This may reflect an endocrine mechanism for controlling the permeability of fish to water.

The water that is accumulated by teleost fish in fresh water is excreted by the kidney, while the losses in sea-water are replaced by drinking. When euryhaline fish move between fresh water and sea-water these processes are adjusted according to their physiological needs. In the flounder (LAHLOU, 1967) and eel (CHESTER JONES el al., 1969a) the flow of urine slowly declines when they are transferred from fresh water to sea-water and this is mainly due to a reduction in the GFR. The blood pressure of the eel also drops in sea-water but the immediate reason for the change in kidney function is not yet known. The rate of drinking changes in media of different osmotic concentrations. *Tilapia mossambica* in fresh water only drink fluid equivalent to 0.26% of their body weight each hour, but this increases to 1.11% when they are in sea-water (POTTS et al., 1967). Similarly eels in fresh water drink fluid equivalent to 0.14% of their body weight in an hour, while in sea-water they consume 0.33% in this time (MAETZ and SKADHAUGE, 1968). The efficiency of the mechanism for the absorption of the sodium (and consequently water) from the gut of eels adapted to sea-water is greater than those in fresh water (see page 213) and this may be related to the activity of adrenocorticosteroids. Factors that control the rate of drinking in fish are unknown, perhaps they just feel thirsty.

b) Role of the Endocrines

The movement of fish between fresh water and sea water is accompanied by a number of endocrine changes but the relationship of these to the altered osmoregulatory requirements is not always clear. When salmon and trout migrate up rivers in order to breed, the levels of 17-hydroxycorticosteroids in their plasma increases (see for instance ROBERTSON et al., 1961). Such increases in the circulating corticosteroids

238

probably reflect changes in the general metabolism of the fish, rather than their osmoregulation (CHESTER JONES and PHILLIPS, 1960). Exercise and various non-specific stresses can result in elevated corticosteroid levels in the blood of trout and salmon (HILL and FROMM, 1968; FAGERLUND, 1967). When smolts of salmon, *Salmo salar,* are transferred from fresh water to sea-water the interrenals exhibit histological signs of increased activity (OLIVEREAU, 1966). ARVY *et al.* (1959) found a reduction in the neurosecretory products of the neurohypophysis in various fishes transferred into solutions of increased salt concentration. When rainbow trout, *Salmo gairdneri,* are shifted from fresh water into sea-water there is a decrease in the amount of peptides stored in the neurohypophysis and this returns to normal after about 3 h (CARLSON and HOLMES, 1962). LEDERIS (1964) also found such a change in trout treated in this manner, but found that the isotocin levels in the gland were unchanged while the amount of vasotocin decreased by 50%. Neurohypophysial peptides, like corticosteroids, can be released in response to non-specific stresses so that the significance of such changes in relation to os-moregulation should be interpreted with caution. We (BENTLEY and FOLLETT, 1963) could not detect any change in the vasotocin content of the neurohypophysis of lampreys transferred from fresh water to sea-water. When North American eels are moved from fresh water to sea-water there is also no significant change of the vasotocin content in the pituitary after 3 days adaptation (BENTLEY, 1971b). Changes in the histological appearance of the adenohypophysial *eta* cells, which secrete prolactin (or 'paralactin') are also prominent when euryhaline fish are transferred between sea-water and fresh water. These cells are more active when the fish are in fresh water than when they are in sea-water. This has been observed in *Fundulus heteroclitus* (BALL and PICKFORD, 1964), *Poecilia latipinna* (OLIVEREAU and BALL, 1964), *Tilapia mossambica* (DHARMAMBA and NISHIOKA, 1968) and *Anguilla anguilla* (OLIVEREAU and OLIVEREAU, 1968). The 'paralactin' content of the pituitary of *Poecilia* is more than doubled when the fish are moved from fresh water to sea-water (ENSOR and BALL, 1968a). On the other hand, the teleost fish, *Mugil cephalus,* caught in freshwater ponds in Israel, had a much greater content of 'paralactin' in their pituitary than those collected from the sea or 'hypersaline' lagoons (BLANC-LIVNI and ABRAHAM, 1970). The differences possibly reflect the relative degrees of adaptation of the two species to their environments or seasonal variations in their condition. As we shall see, 'paralactin' has a dramatic effect on the adaptation of such euryhaline fish to fresh water. The adenohypophysial somatotrophic cells of eels kept in fresh water are also strongly stimulated (OLIVEREAU, 1967) but it is unknown whether their secretion is involved in osmotic adaptation.

α) Effects of Extirpation of the Pituitary and Interrenals. A number of experiments have been carried out on the effects of extirpation of various endocrine tissues on the ability of teleost fish to adapt to fresh water and sea-water. FONTAINE, CALLAMAND, and OLIVEREAU (1949) found that the euryhalinity of the eel was not affected following hypophysectomy and this has since been confirmed. However, the ability of a number of teleosts to survive transfer from sea-water into fresh water is abolished by this operation. Even in the fresh water eel a closer examination has shown that hypophysectomy results in an accelerated rate of sodium loss (see Table

239

7.11). Extirpation of the pituitary gland from killifish, *Fundulus heteroclitus*, results in their death within 10 days if they are kept in fresh water, though they survive when in sea-water (BURDEN, 1956). Death in fresh water is accompanied by an excessive loss of salt from the body. In another species of killifish, *Fundulus kansae*, hypophysectomy is not fatal but these fish, like eels, suffer a decline in their plasma sodium levels when they are kept in fresh water (STANLEY and FLEMING, 1967a). In other fishes including *Poecilia* (BALL and OLIVEREAU, 1964; BALL and ENSOR, 1967) and *Tilapia* (DHARMAMBA *et al.*, 1967), hypophysectomy is fatal if the fish are kept in fresh water and this is due to a loss of sodium. PICKFORD and PHILLIPS in 1959 showed that the effects of hypophysectomy in *Fundulus heteroclitus* could be overcome by injecting the fish with prolactin; no other pituitary hormones have been found to be effective. This effect has since been confirmed in the other species described above, but in the eel complete replacement is dependent on the additional administration of corticotrophin (CHAN, CHESTER JONES, and MOSLEY, 1968). Transplantation of the rostral parts of the fish adenohypophysis, rich in *eta* cells but poor in other endocrine cells, ensures the survival of hypophysectomized *Poecilia latipinna* (BALL, 1965). Thus the adenohypophysial secretion prolactin ('paralactin') of teleost fish plays a vital role in their ability to adapt to fresh water, but corticotrophin may also be involved. SCHREIBMAN and KALLIMAN (1966, 1969) have demonstrated the fatal effects of hypophysectomy in more than a dozen species of teleosts kept in fresh water, indicating that the importance of this gland in osmoregulation is widespread in this group. In contrast, hypophysectomy in the cyclostome fish, *Lampetra fluviatilis*, has no adverse effect on its survival in fresh water (LARSEN, 1969). These lampreys were in the fasting condition which they exhibit after their migration to breed in fresh water and their survival was *prolonged* after removal of the pituitary. This is due to a decrease in the rate of tissue catabolism.

In spring, sticklebacks, *Gasterosteus aculeatus*, migrate from the sea into rivers in order to breed. The following autumn they swim back into the sea where they spend the winter. Such marine winter fish die if they are placed in fresh water. These fish thus apparently undergo a cyclical change in their ability to tolerate salt solutions, just as observed in the steelhead trout described earlier. However, when winter sticklebacks are injected with prolactin they can survive when placed in fresh water (LAM and HOAR, 1967; LAM and LEATHERLAND, 1969). This fascinating observation suggests that such cyclical changes in salinity tolerance may reflect the ability of these fish to synthesize or secrete 'paralactin'. In winter they may be deficient in this hormone, but with the onset of spring it can be secreted and allow the fish to make their annual nuptial migration into the rivers.

As the adrenocortical tissues influence the regulation of electrolytes in mammals, the possible significance of the homologous *interrenal tissues* in osmoregulation of fish has been the source of speculation for some time. The interrenal tissues of teleost fish are rather diffuse in their distribution so that surgical removal is very difficult. This operation has, nevertheless, been performed in eels (CHESTER JONES, HENDERSON, and MOSLEY, 1964b). The procedure involves the removal of certain blood vessels to which the tissues are attached and replacement of these with plastic tubing. This operation has since been slightly modified in an effort to remove the last vestiges of the interrenal tissues (CHAN *el al.*, 1967 a;

BUTLER *et al.*, 1969). As judged by the marked decline in circulating cortisol levels in eels, this operation removes most of the interrenal tissue (BUTLER *et al.*, 1969). Interrenalectomy results in a decline of the plasma sodium levels in European and North American eels, *Anguilla rostrata*, kept in fresh water. There were no consistent changes in the potassium concentrations (CHAN *et al.*, 1967 a; BUTLER *et al.*, 1969). MAETZ (1969 a) measured the influx and outflux of sodium in European eels and found that when they were in fresh water the influx of sodium across the gills declined following interrenalectomy. The decreased plasma sodium concentration observed in these freshwater eels is due to the failure of the gills to absorb adequate sodium from the external media, but this can be restored by the injection of aldosterone or small doses of cortisol (HENDERSON and CHESTER JONES, 1967). BUTLER *et al.* also noted profound changes in the carbohydrate metabolism of such eels indicating that corticosteroids, as in mammals, are also involved in the control of general metabolism. It is interesting that larger doses of cortisol enhance the loss of sodium from the gills. Thus, MAYER *et al.* (1967) found that interrenalectomy in sea-water eels resulted in an accumulation of sodium in the body, but the injection of cortisol could restore this to normal. The rate of the turnover of the sodium following interrenalectomy in eels kept in sea-water is reduced in each hour from 33% to only 6% of the total present in the body (MAETZ, 1969 a). This can also be restored by injecting cortisol. Cortisol thus appears to be able to stimulate active sodium transport in eels whether it is occurring from the external solution into the fish, or from the fish into the surrounding sea-water.

Hypophysectomy clearly disturbs the sodium metabolism of fish in fresh water though they do perfectly well in sea-water. In fresh water this extirpation is fatal to some, such as *Fundulus heteroclitus*, *Poecilia latipinna* and *Tilapia mossambica* and results in a depletion of the body sodium in others, like *Carassius auratus*, *Fundulus kansae* and *Anguilla anguilla* (Table 7.11). This is due to an excessive leakage (outflux) of sodium across the gills of *Fundulus heteroclitus*, the eel and the goldfish, while in *F. kansae* urinary losses are increased. In all fish, except the latter, the outflux can be reduced by the injection of mammalian *prolactin*. Another adenohypophysial hormone suspected of influencing the osmoregulation of fish is corticotrophin, which acts by initiating release of corticosteroids from the interrenals. Corticotrophin, however, does not generally have a dramatic effect on the sodium metabolism of hypophysectomized freshwater fish. Nevertheless it is effective in hypophysectomized eels, where, in conjunction with prolactin, it restores the osmotic balance (CHAN *et al.*, 1968; MAETZ, MAYER, and CHARTIER-BARADUC, 1967a). Tetrapod adrenocortical tissues are not completely controlled by corticotrophin, and if this is also so in fishes (if a renin-angiotensin system is present) sufficient corticosteroids may still be available to meet their needs following hypophysectomy. Prolactin reduces the branchial outflux of sodium in most hypophysectomized fish and this would appear to be its principal site of action. Interrenalectomized eels cannot take up adequate sodium through their gills, but this can be facilitated by the injection of corticotrophin or cortisol. Therefore, corticosteroids probably act in freshwater fish to facilitate branchial influx of sodium, the simultaneous action of these steroids and prolactin at this site regulating the sodium content of the fish.

The precise role and metabolic *site of action of prolactin and corticosteroids* during adaption of fish to fresh water or sea-water is unknown. Unequivocal evidence of their physiological action is indeed lacking. Measurements of their circulating levels during such transitional adjustments would help substantiate such a role. EN-SOR and BALL (1968a) have promised to measure circulating prolactin levels in fish, and it seems more reasonable to think that this should also be possible for corticosteroids. This evidence must be awaited. In the meantime we can speculate as to their roles in the physiological adjustments that are known to occur when fish are transferred from sea-water to fresh water. The prompt decrease in branchial permeability to sodium, which is observed on such occasions, probably does not depend on hormones. Endocrines may, however, play a role in the subsequent 'delayed' secondary decrease in branchial permeability (MOTAIS et al., 1966). The long-term adaptation of the Na-K activated ATPase levels and the sodium-exchange diffusion 'carrier' (or its inhibitor) could also be influenced by the endocrines. Hypophysectomy has, for instance, been shown to interfere with alterations in the branchial Na-K ATPase levels in killifish (EPSTEIN et al., 1967); a deficiency that is restored by the injection of cortisol (PICKFORD et al., 1970). The action of prolactin could be more banal and involve the integrity and secretion of the branchial mucous glands that have been observed to degenerate after hypophysectomy (BURDEN, 1956; PICKFORD, PANG, and SAWYER, 1966). Other hormones could also be released during transitions between fresh water and sea-water. It is notable that injections of vasotocin, corticotrophin and cortisol can facilitate the exchange of sodium that occurs across the gills of the flounder when it is transferred from fresh water to sea-water (MOTAIS and MAETZ, 1967; MAYER and MAETZ, 1967). Cortisol probably acts to stimulate ion extrusion from the 'chloride secreting cells' while vasotocin may increase the blood supply to these cells (MAETZ and RANKIN, 1969). A final judgement of the role of these hormones must however, await measurement of their circulating levels in the various osmotic circumstances. Preliminary results in Japanese eels show that plasma cortisol levels double when they are transferred from fresh water to sea-water (HIRANO, 1969). This effect may be mediated through the pituitary as it was not seen when hypophysectomized eels were treated in this manner. After 2 months in sea-water the circulating corticoid levels declined and did not change significantly when the eels were replaced in fresh water.

While *the kidneys* appear to play little role in the adjustments of electrolyte metabolism that accompany movements between fresh and salt water, they are vitally concerned in regulating body water. In fresh water urine flow is high, in order to excrete excess water, while in the sea it is low, reflecting its relative sparsity. Although, as discussed earlier, vasotocin can in some circumstances increase the flow of urine in freshwater fish this property is shared by other substances such as adrenaline and angiotensin which also increase the blood pressure (CHESTER JONES et al., 1969 b). As described below, the control of urine volume in fish may well be mediated haemodynamically, and while vasoactive substances in the circulation could be involved in this, neural mechanisms would be equally feasible.

There has recently been a revival of interest in the role of vascular changes and vasoactive substances in the osmoregulation of fish. KEYS and BATEMAN (1932) found that adrenaline produced a vasodilatatory action in the perfused gills of the eel, while vasopressin (from mammals) had little effect. When the gills were bathed

242

with sea-water, this vasodilatation was accompanied by a cessation of chloride extrusion. As already noted, increases in urine flow (and renal sodium loss) resulting from the injection of vasotocin are usually accompanied by an elevation in the blood pressure. Adrenaline, angiotensin II, and extracts of eel urophyses and corpuscles of STANNIUS also have parallel actions on both processes (CHESTER JONES et al., 1969b). The role of vascular changes in the normal regulation of piscine kidney function is unknown, but the circulatory supply to this gland is complex as it usually has both a venous renal portal (tubular) and an arterial (glomerular) supply. The kidneys of fish also can exhibit the phenomenon of 'glomerular recruitment' which suggests a lability in their vascular arrangements. Recently more information has become available on the pattern of circulation in the gills in which different regions are supplied by distinct blood vessels. The blood supply to the (respiratory) gill lamellae is increased by the action of adrenaline while that to the intermediate regions between these structures ('central compartments') is boosted by large doses of acetylcholine (STEEN and KRUYSSE, 1964; RICHARDS and FROMM, 1969). MAETZ and RANKIN (1969) found that small doses of vasotocin and isotocin have an effect similar to that of acetylcholine on the gills of eels. The 'chloride secreting cells' are situated at the base of the gill lamellae and are supplied by blood flowing through the 'central compartments'. Thus, in the presence of adrenaline, they have a reduced blood supply consistent with the cessation of their activity. Vasotocin, on the other hand, increases their blood supply and this may be the reason for the facilitated rate of branchial sodium exchange which is seen in the presence of such peptides (MAETZ and RANKIN, 1969). The concentrations of the neurohypophysial peptides required to elicit changes in the blood supply to the gills are 10^{-9} to 10^{-11} M for vasotocin and as little as 10^{-13} M for isotocin. There is no information available about the presence of such peptides in fish blood, but in mammals and amphibians their concentration is about 10^{-10} to 10^{-11} M. The effective levels of adrenaline are comparable with those known to be present in the blood of fish (FONTAINE, MAZEAUD, and MAZEAUD, 1963). Whether such vasoactive substances normally have a role to play in osmoregulation is unknown but the blood pressure of eels is quite labile. When, for instance, these fish are transferred from sea-water to fresh water the blood pressure increases (CHESTER JONES et al., 1969 a). If this reflects a release of adrenaline, such a hormone could produce a reduced secretion from the 'chloride secreting cells' and a facilitated water excretion by the kidney. On the other hand, if vasotocin were released in sea-water it would have the opposite effect on salt loss, stimulating its secretion from the gills. However, if present in sufficient quantities it would, like adrenaline, also be expected to increase the blood pressure and renal water loss. The potential actions of these hormones and the observed physiological responses of the fish in sea and fresh water thus cannot always be reconciled. The possibilities are, however, interesting and will no doubt be the object of further research.

I have recently attempted to demonstrate the presence of vasotocin in the blood of the North American eel, *Anguilla rostrata* (BENTLEY, 1971 b). The limits of sensitivity of the assay (toad urinary bladder) were 10^{-11} M (in eels blood). The concentrations required to alter the branchial blood flow, increase the blood pressure and promote a diuresis in European eels are usually greater than this. However, when the fish were transferred from fresh to sea-water, sea to fresh water or subjected to haemorrhage, vasotocin could not be detected in the blood. Possibly isotocin appears under these conditions but this would not have been detectable by the methods used.

c) Putative Endocrine Tissues in Fishes Possibly Involved in Osmoregulation

The caudal neurosecretory system (urophysis) and corpuscles of STANNIUS may produce endocrine secretions that influence the electrolyte metabolism of some fish. These tissues were introduced earlier (page 78).

α) *Urophysis.* The caudal urophysial, or DAHLGREN cells of teleost, chondrostean and chondrichthyean fishes react to changes in the osmotic environment (see FRIDBERG and BERN, 1968; and page 79). The evidence that they produce a secretion(s) that influences osmoregulation is equivocal.

Some fish die more readily if they are transferred to strange osmotic situations after their urophysis has been removed. Certain caudal neurosecretory fibres have also been shown to respond to changes in the external salt concentration (YAGI and BERN, 1963; 1965). Whether such responses reflect a reaction to specific stimuli associated with osmoregulation is, nevertheless, not clear. The cichlid fish, *Tilapia mossambica,* and the stickleback, *Gasterosteus aculeatus,* suffer increased rates of mortality when they are transferred from fresh water to saline solutions after extirpation of their urophyses (TAKASUGI and BERN, 1962; IRELAND, 1969). This effect in *Tilapia* was not, however, found by STANLEY and FLEMING (1964) but it is considered likely that the tissue may have regenerated in these fish. Recently CHAN et al. (1969a) removed the urophysis from freshwater eels, but this did not alter the renal losses of water or sodium, though urinary potassium, calcium and magnesium excretion initially increased, but later subsided. Some interesting observations of the histological appearance of the urophysis of fish living in different osmotic conditions have been made by FRIDBERG, BERN, and NISHIOKA (1966). The Hawaiian 'o'io', *Albula vulpes,* is often kept for extended periods of time in salt-water ponds adjoining the sea. The salinity in these pools varies considerably, due to evaporation and periodic flooding with fresh water. The fish kept under these conditions were found to have a hyperactive caudal neurosecretory system when compared with those taken directly from the sea. FRIDBERG et al. consider that this tissue may be involved in osmotic adaptations to the *changes* in salinity which are continually taking place in the ponds.

Extracts of the urophysis have yielded materials that have been characterized by a variety of biological actions. MAETZ et al. (1963) showed that extracts of the goldfish urophysis, when injected into goldfish, promoted a (net) accumulation of sodium by the fish. This was mainly due to an increased rate of uptake across the gills, but there was also a decrease in the renal losses. This effect is not to be confused with that of neurohypophysial peptides, which, although they stimulate branchial sodium uptake in these fish, produce a net loss of this ion due to prominent urinary losses (MAETZ et al., 1964). LEDERIS (1969) has also demonstrated an effect of urophysial extract on a piscine preparation; it contracts the urinary bladder of the rainbow trout *in vitro.* Urophysial extracts from mullet, *Mugil,* and eels, when injected into eels, elevate their blood pressure and produce an increased urinary loss of water and sodium (CHESTER JONES *et al.*, 1969 b). The urophysis, when extracted, can thus yield a substance (or more likely substances!) that has divers effects on teleostean preparations, but it is not known if this reflects a physiological role. Extracts from the urophysis of the mudsucker, *Gillichthys mirabilis*, can stimulate an amphibian tissue, the toad urinary bladder (*in vitro*) and increase the osmotic transfer of water from its mucosal to serosal side (LACANILAO, 1969). The exact chemical nature of the substance, or substances, which have these various actions, is not known. LEDERIS (1969) found that the ability to contract the trout bladder was not due to histamine, 5-hydroxytryptamine, acetylcholine, bradykinin or catecholamines in the extract and neither can these materials account for the facilitated osmotic permeability of the toad bladder. GESCHWIND *et al.* (1968) in a preliminary survey have described the material that acts on the trout bladder as being polypeptide in its nature, with a molecular weight of about 1000. It is emphasized, however, that while this material contracts the trout urinary bladder, it is not the same as that examined by MAETZ *et al.* (1964). The urophysial extract that elevates the blood pressure of the eel is also a polypeptide and is inactivated by trypsin but is unaffected by incubation with thioglycollate (CHAN, CHESTER JONES, and PONNIAH, 1969). The absence of an effect of the latter indicates that disulphide bonds are not essential for the material's activity, thus distinguishing it from the neurohypophysial peptides.

β) Corpuscles of Stannius. These are islets of tissue present in the caudal regions of teleostean and holostean fishes (see page 78). RASQUIN (1956) found histological changes in the appearance of these tissues in the teleost, *Astyanax mexicanus*, after it had been transferred to saline solutions. This suggested that they may be involved in osmoregulatory adjustments. When these tissues are extirpated from the eel there is a marked decline in plasma sodium concentration of freshwater fish and an elevation of the levels of potassium and calcium (M. FONTAINE, 1964). CHAN *et al.* (1967 a) have confirmed these observations. It is also interesting that the blood pressure of freshwater eels drops after removal of the corpuscle of STANNIUS (CHESTER JONES *et al.*, 1966). It seemed possible that such changes in blood pressure could influence osmoregulation by altering the GFR. However, BUTLER (1969) was unable to demonstrate any change in the GFR of stanniectomized North American eels, *Anguilla rostrata*. The decline in plasma sodium levels can be corrected by injecting the eels with aldosterone, but the calcium and potassium concentration are unaffected, though corpuscular extracts restore these also (LELOUP-HATEY, 1964; CHAN *et al.*, 1967a; FONTAINE, 1965).

The search for the chemical identity of the secretion(s) that may be produced by the corpuscles of STANNIUS has run along a somewhat confusing path. Substances with some properties similar to corticosteroids have been extracted from corpuscular tissues and may be produced during their *in vitro* incubation with various steroid substrates. The corticosteroid nature of such substances is in doubt as these results have not been generally confirmed (see for instance CHESTER JONES *et al.*, 1965; ARAI, TAJIMA, and TAMAOKI, 1969), though the possibility of species' differences cannot be ruled out. NANDI (1967) considered the available information and concluded that 'these data are not inconsistent with the notion that the corpuscles of STANNIUS are steroid storage organs and/or that they actively metabolize at least some steroid hormones'. She also considers it possible that they could 'convert plasma steroids, synthesized elsewhere, into other active hormones a process which might be called "cooperative steroidogenesis". CHESTER JONES et al. (1966; 1969 b) have found that corpuscular extracts from eels, when injected, raise the blood pressure of rats or eels. The extracts contain material that behaves in many respects like mammalian renin and they have suggested that it may function like the tetrapod renin-angiotensin system does in controlling the secretion of aldosterone. The effects of ablation of the corpuscles of STANNIUS on plasma sodium levels and their partial correction with aldosterone would also be consistent with this suggestion. Eels do not produce aldosterone but another steroid could be involved. The corpuscular material that influences calcium metabolism in eels is presumably a separate principle.

6. Osmoregulation and the Endocrines in Some Relict Fishes

a) The Lungfishes (Dipnoi)

These bony fishes originated in fresh water during Devonian times, about 320 million years ago. Although in the past a few species have lived in the sea, they have remained an essentially freshwater group. Their ability to withstand transfer to seawater does not, however, appear to have been reported. Today they are represented by two phyletic groups: the Ceratodontidae and the Lepidosirenidae. The former contains one species, *Neoceratodus forsteri*, which is confined to rivers in a small area of Queensland in Australia. The Lepidosirenidae contain two genera; *Protopterus* which lives in Africa and *Lepidosiren* in South America. Information about the water and salt metabolism of this group is, as far as I can ascertain, entirely confined to *Protopterus*. HOMER SMITH made some fascinating observations on the osmoregulation of these fish and an enjoyable account of his travels in search of them are described in *'Kamongo or the Lungfish and the Padre'* (H. SMITH, 1956).

The lungfishes appear to have a normal complement of endocrine glands, though these have not been exhaustively examined. The neurohypophyses of all three genera contain vasotocin, but, in contrast to other bony fishes, mesotocin and sometimes oxytocin, instead of isotocin. These peptides also occur in the Amphibia and the neurohypophysis of lungfish bears some morphological similarity to the amphibian gland (WINGSTRAND, 1966). Cortisol has been identified in the circulation of *Protopterus* (PHILLIPS and CHESTER JONES, 1957), while corticosterone can be formed by incubates of its adrenocortical tissues *in vitro* (JANSSENS et al., 1965). As aldosterone has not been found, this arrangement is more fishlike, though the tissues that form these steroids are morphologically like those 'of a primitive amphibian' (JANSSENS et al., 1965). MOORHOUSE (1956) found that the 'adrenocortical' tissues of *Protopterus* regress following hypophysectomy while the injection of mammalian corticotrophin stimulated them. JANSSENS et al. also found that these tissues hypertrophied in response to injections of corticotrophin and regressed following administration of cortisol. They thus appear to behave like the adrenocortical tissues of other vertebrates and, at least partly, are under the control of the adenohypophysis. The pituitary also contains a prolactin-like hormone which, as we have seen, like tetrapod prolactin but in contrast to the fish 'prolactin', can stimulate the pigeon crop-sac.

The African lungfish lives an aquatic life in regions of Africa that are subjected to seasonal drought, during which time the rivers and lakes it inhabits dry up. When this begins to happen the lungfishes dig burrows in the bed of the lake and retreat into them when the water disappears. While so entombed they surround themselves (except for the nares) with a cocoon made of secreted mucous and fine particles of mud, and this limits their evaporative water loss. They aestivate in this manner until the next wet season, which usually comes in 4 to 6 months, though they can survive 2 or 3 years and perhaps even longer (see SMITH, 1961). When in this condition they reduce their metabolic rate to a very low level and subsist principally by catabolizing tissue proteins. The carbohydrate reserves of these fish may even increase during aestivation which contrasts with the decrease that is observed when the aquatic fish are fasted (JANSSENS, 1964). Urine formation is too small to measure during aestivation and urea derived from the protein catabolism is accumulated in high concentrations in the body. HOMER SMITH found that after prolonged aestivation urea made up as much as 3% of the entire body weight. When water returns to the lakes in which the lungfish live, it submerges the burrows and the fish break out of their cocoons and rise to the surface. According to SMITH and JANSSENS they then excrete accumulated urea and sodium, most of this being lost in the first two days.

The role of hormones in the various phases of the life cycle of lungfishes is largely conjectural. When in their aquatic phase they take up water (the contribution of the gills and skin is not known) at a rate similar to that of other freshwater fishes and excrete this as a dilute urine (Table 7.7). While fishes do not respond to neurohypophysial peptides with an antidiuresis as seen in tetrapods, the phyletic relationship of the lungfishes and the Amphibia makes it seem possible that these two groups may both exhibit a 'water balance effect'. However, the injection of vasotocin was not found to produce such an effect in *Protopterus* (HELLER and BENTLEY, 1965) but instead it results in a diuresis (SAWYER, 1966 a). We also found that the injection of vasotocin into *Protopterus* resulted in a net loss of sodium by the fish and that this is the result of a facilitated urinary loss of this ion (SAWYER, 1966 a). HELLER and I, in some preliminary experiments, failed to show any effect of injected aldosterone on the total sodium balance of the lungfishes, but this needs to be explored in more detail, preferably with the aid of isotopes. Hypophysectomized lungfishes appear to live quite happily in fresh water (MOORHOUSE, 1956; GODET, 1961), but a detailed study of their osmoregulation has not been made. GODET, however, found that the urine volumes of hypophysectomized *Protopterus* are reduced, but as the rate of water accumulation is also less, there is no net change in their water balance. The primary site of such change in water balance is probably extrarenal rather than renal.

It seems likely that the endocrine glands play a role during the process of aestivation particularly during the period of its onset and during 'awakening'. The alterations in general metabolism probably involve the corticosteroids which facilitate protein catabolism in other vertebrates. The reason for the abolition of urine formation during aestivation is not known. If *Protopterus* is hypophysectomized while in its aestivating condition it appears to survive (GODET, 1961). There are

247

a number of acute osmotic problems involved in the 'awakening' process. During their long sojourn in the burrows the lungfishes suffer a considerable increase in the osmotic concentration in their body fluids, so that when exposed to water they accumulate it rapidly. As their kidney function is relatively deficient at this time they become swollen and oedematous in appearance. Subsequently the solutes and water are excreted and kidney function returns to normal. If aestivating lungfish are returned to water following hypophysectomy they die, apparently from over-hydration, as the kidneys cannot adjust and overcome the 'aestivation anuria' (GO-DET, 1961). SAWYER (1966 a) has tentatively suggested that the diuretic action of vasotocin could be utilized on such occasions, but other hormones are probably also involved.

While some fascinating information about the life of lungfishes has been disclosed the physiological information is still far from complete. The Australian and South American lungfish, as well as the African fish should provide material for further fruitful and enjoyable physiological and geographical excursions.

b) The Sharks and Rays (Chondrichthyes)

It may be objected that this group of fish is not really relict, especially as it contains a large number of species with a world wide distribution in the sea and even, occasionally, in rivers and lakes. However, the Chondrichthyes diverged from the main vertebrate stock about 300 million years ago and have evolved separately since that time. Their osmoregulatory pattern is different from that of the bony fishes (except probably for the coelacanth).

By retaining urea and an extra amount of sodium chloride in their body fluids they have, in contrast to the marine teleosts, attained a slightly hyperosmotic edge over the sea-water where most of them live. Water thus tends to move into the body by osmosis, probably mainly across the gills. The sharks and rays, unlike the marine teleosts, do not appear to drink sea-water (H. SMITH, 1931) but further species should be examined for such behaviour. The ability of these fish to excrete salt does not appear to be as great as in the marine teleosts, so that substantial drinking would be unexpected. A steady exchange of sodium takes place between the chondrichthyeans and the sea-water in which they live, the influx of which appears nearly all to take place across the gills. However, this is small compared to marine teleosts (but similar to freshwater ones!) making up less than 1% of the total body sodium each hour, compared to as much as 60 to 70% in some bony fishes (Table 7.9). About 7 to 10% of the total salt lost is excreted in the urine and a similar quantity is secreted from the rectal 'salt' gland. The balance appears to pass out through the gills, but whether this involves active transport is uncertain. The chondrichthyean gills appear to lack 'chloride secreting cells' and, probably as a result of this observation, it has been widely assumed that active salt secretion does not occur from their gills.

Sodium-potassium activated ATPase has been identified in the gills of a marine chondrichthyean, the dogfish, *Squalus acanthias* (JAMPOL and EPSTEIN, 1970). The concentration is low compared to marine teleosts but similar to that of freshwater species. Its role in the dogfish is unknown but it could possibly suffice for the smal-

Table 7.12 *Comparison of the composition of the rectal gland secretion of Squalus acanthius with that of its plasma, urine and sea-water*

	Osmolarity (m-osmole/l)	Sodium	Potassium (m-equiv/l)	Chloride
Sea-water	1000—1100	470	12	550
Rectal gland secretion	1018	540	7	533
Plasma	1018	289	5	246
Urine	780	339	2	203

From Burger and Hess (1960)

ler need for sodium excretion experienced by these fish. The evidence for active sodium excretion from the gills of marine chondrichthyeans is equivocal.

The sodium and chloride concentrations in the urine of sharks are slightly higher than those in the plasma, while the rectal gland can secrete fluid in which these ions are present at even higher levels than in sea-water (Table 7.12). The rectal gland is characteristically chondrichthyean and is a small gland, which, even in marine sharks weighing more than 10 kg, weighs only 1 to 3 grams (Bonting, 1966). Its duct opens into the rectal region of the gut. Histologically it consists of masses of tubules like in the avian salt gland (Doyle, 1962). These tubules may be arranged either radially and open to central ducts or longitudinally in lobules. The rectal gland contains high concentrations of Na-K activated ATPase (Bonting, 1966; Jampol and Epstein, 1970). Burger and Hess (1960) showed that in the spiny dogfish, *Squalus acanthias*, this gland secretes a fluid that is isoosmotic with the body fluids and contains sodium chloride at about twice the concentration seen in the plasma (Table 7.12). The total amounts of sodium chloride secreted from the salt gland are similar to those that appear in the urine; in the spiny dogfish it was found to be twice as great (Burger, 1962), while in *Scyliorhinus caniculus* it was half as great (Maetz and Lahlou, 1966) and in the lip shark, *Hemiscyllium plagiosum*, it may be greater or less (Chan, Phillips, and Chester Jones, 1967 b). The rectal gland appears to be able to direct the outflux of about 5 to 10% of the total sodium accumulated by the fish. Removal or incapacitation of this gland does not have a very dramatic effect on the levels of sodium and chloride in the plasma; they tend to rise, but only slightly (Burger, 1962; 1965; Chan et al., 1967 b). Renal salt excretion increases in such circumstances. This results from an increase in the urine volume rather than a change in concentration so that such a process is more wasteful of water. Burger has suggested that on these occasions the accumulation of salt is decreased due to a reduced branchial influx that may be controlled by 'humors not yet defined'. In the presence of additional salt loads the absence of the salt gland may be felt more severely while in normal circumstances it leaves the kidney free to carry out other tasks, such as the excretion of divalent ions.

Little is known about the control of secretion from the salt gland. In the spiny dogfish the injection of 1 M sodium chloride solutions initiates secretion (Burger, 1962; 1965). Immersion in dilute solutions, which have a hydrating effect and dilute

249

the plasma electrolytes, also stimulates secretion. It was thus suggested that secretion from the gland may be responsive to changes in volume, rather than concentration of the body fluids. In further contrast to the avian salt gland, no evidence of a neural control mechanism has been found, and the injection of vasopressin is without effect (BURGER, 1962). CHAN *et al.* (1967 b) also found that vasopressin, as well as oxytocin, did not alter secretion from the rectal gland of the lip shark, but cortisol injections decreased both the flow and concentration of this fluid. It is unknown whether this latter effect reflects a physiological or only a pharmacological action. BURGER (1962; 1965) suggested that 'volume receptors' may control the gland's secretion through the intervention of some 'undefined hormone'. At present we do not know what this is.

The hormones present in chondrichthyeans, and which are potentially concerned with osmoregulation, have been described earlier in this chapter (Tables 7.4 and 7.5); peptides are present in the neurohypophysis, cortisol, corticosterone and, at least in one species, large quantities of 1 -hydroxycorticosterone are present in the plasma. This steroid can stimulate sodium transport across membranes and when tested on the toad urinary bladder has 80% of the activity of aldosterone (GRIMM, O'HALLORAN, and IDLER, 1969). The interrenal tissues of at least eleven species of chondrichthyeans can produce 1 α-hydroxycorticosterone *in vitro* suggesting that it may also have a widespread distribution in the plasma of these fishes. Nevertheless, its role in the chondrichthyeans is unknown. The adenohypophysis also contains a 'prolactin-like' product and (probably) corticotrophin. There is, however, little information about how these secretions affect the osmoregulation of such fish.

Hypophysectomy has been performed in a number of chondrichthyeans, the object most often being to disturb the fishes' chromatic behaviour rather than their osmoregulation (see HOAR, 1966), but such fish appear to survive quite well in seawater. A closer examination of such deficient fish may, nevertheless, divulge some changes. Sharks and rays do not survive interrenalectomy for long but no prominent changes in their electrolyte metabolism have been reported (HARTMAN *et al.*, 1944). Interrenalectomy has been performed more recently on the skate, *Raja erinacea* (IDLER, O'HALLORAN, and HORNE, 1969). These fish survived at least 3 to 4 weeks and suffered no significant adverse changes in their liver glycogen levels. An account of the osmoregulation of these fish will be eagerly awaited.

Extracts of the neurohypophysis of chondrichthyeans do not alter overall water balance (weight change) when injected into nurse sharks, *Ginglymostoma cirratum* (HELLER and BENTLEY, 1965). CHAN *et al.* (1967b) found that cortisol did not change the plasma electrolyte concentrations of lip sharks, but that there was a rise in the muscle potassium levels due to loss of cell water. Such an effect could indirectly result from a sodium retention resulting in the expansion of the extracellular space at the expense of some cell water. As 1 α-hydroxycorticosterone appears to be a major interrenal steroid in many species of sharks, skates and rays it will be most interesting to see if it affects the salt and water metabolism of such fishes. The role of endocrines in the osmoregulation of sharks is a field that has been little investigated and only requires the facilities and courage to handle such beasts.

Adaptation of sharks and rays to fresh water. Most chondrichthyeans live in the sea but some frequent and may even live their entire lives, in fresh water. Such

250

migrations from the sea could involve the endocrine system, but I should honestly state at the outset that there is no information about this. Sharks and rays regularly move into brackish estuarine waters but the limiting dilution usually seems to be equivalent to about 50% sea-water (see PRICE, 1967). Attempts to adapt marine sharks and rays to lower concentrations in laboratory aquaria have not been successful (SMITH, 1936; PRICE and CREASER, 1967). PRICE (1967) measured the plasma concentrations of skates, *Raja eglanteria,* caught in different parts of Delaware and Chesapeake bays where the salinity varies from 1.8% to 3.3%. He found little difference in the concentration of electrolytes in the blood of these skates and none which could be correlated with the salinity of the water where the fish were caught. In the laboratory (PRICE, 1967) adaptation to such salinities was found to take place slowly over a period of two days, so that such fish could conceivably make rapid excursions into less saline water and return without suffering any marked changes in the concentrations of their body fluids.

When chondrichthyeans like the sawfish, *Pristis microdon,* live for prolonged periods in fresh water, the concentration of their body fluids is reduced by about 50% due to loss of both urea and salt (SMITH, 1936). Such fish produce a far more copious urine than they do in sea-water (Table 7.7) and this is hypoosmotic to their plasma and contains little salt. A rapid transfer from sea-water to fresh water would be expected to produce dramatic changes in hydration as these beasts are highly hyperosmotic to fresh water and would be expected to accumulate it at a far higher rate than adapted fish. PRICE (1967) has suggested that in such circumstances the increased accumulation of water would lead to an elevated GFR that exceeds the capacity of the renal tubules to absorb 'certain significant metabolities' (and sodium?) leading to the death of the fish.

SMITH (1936) suggested that the movements of sharks and rays between fresh and salt water may be a gradual process during which excretion or accumulation of urea and salts could slowly take place. This may be self regulating, thus when the fish move from the sea into more dilute solutions they may accumulate water which results in an elevated GFR and produces an increased loss of urea. Additional leakage of urea and salts across the gills would also be expected.

Apart from the problem of hyperhydration in fresh water it is also possible that sharks and rays suffer an excessive loss of sodium chloride, as observed in stenohaline marine teleosts that are placed in such media. This could occur in the urine as well as across the gill surfaces. The rectal gland regresses in freshwater sharks and presumably its secretions follow suit (OGURI, 1964). It should be simple to measure the overall sodium balance on such occasions. When chondrichthyeans live in fresh water they may be expected to face the additional problem of maintaining a positive salt balance, which in teleosts is assisted by an active uptake of sodium chloride from the bathing media. This process has not been demonstrated in chondrichthyeans but as they are carnivorous they probably obtain adequate salt from their food.

c) The Hagfishes (Myxinoidea)

The hagfishes, along with the lampreys (Petromyzontidae), are thought to be the closest contemporary relatives of the first vertebrates. It is uncertain whether the lampreys or hagfishes are more similar to the original progenitor but both groups have their adherents. It should be recalled that many generations of agnathans have come and gone since those nascent times.

Osmotically the hagfishes are the most exceptional and bear some interesting osmotic similarities to marine invertebrates; the extracellular fluids have an ionic composition similar to that of sea-water. The differences from the latter solution (Table 7.13) consist of (possibly) a slight hyperosmotic concentration in the serum that also has a sodium level about 20% greater, while the calcium and magnesium concentrations are lower. Potassium and chloride levels are similar in both solutions. As in other vertebrates, there are large differences between the concentrations of solutes in the serum and intracellular fluid. The ratio of concentration in the cell water/serum water is: 12/1 for potassium and 0.25/1 for sodium. The chloride levels are 5 times lower in the cell water, which also contains substantial amounts (200 mM) of trimethylamine oxide. The principal solute gradients in hagfishes are, thus, at the junction of the cells and extracellular fluids. This is a simpler situation than in other fishes, including the lampreys, in which there is also a large ionic gradient between the serum and the external sea-water.

Table 7.13 *Concentration (mM / kg water) of solutes in the body fluid of Myxinoidea*

		Myxine glutinosa			Eptatretus stouti
	sea-water[1]	serum[1]	muscle[1]	urine[2]	mucous secretion[3]
Sodium	470	549	132	481	95
Potassium	12	11	144	13	207
Chloride	550	563	107	536	—
Calcium	8	5	3	10	11
Magnesium	50	20	18	28	39
Trimethylamine oxide	—	nil	211	—	—

The concentrations of sea-water in all these observations was not as in 1 but the differences in composition are consistent on the various occasions.
References: [1]Bellamy and Chester Jones (1961); [2]Morris (1965); [3]Munz and McFarland (1964).

The small differences between the ionic composition of the serum and sea-water are no doubt vital to the fish and seem to be maintained as a result of the actions of the kidney, and, possibly, the secretion of the numerous mucous glands in the skin and gills. This is assisted by an integument which, while being readily permeable to water, restricts the movements of ions (Mc Farland and Munz, 1965). Morris (1965) considers that the Atlantic hagfish is slightly hyperosmotic to the sea-water and that this allows it to accumulate small quantities of water. Others

(MCFARLAND and MUNZ, 1965) consider the Pacific hagfish to be isoosmotic to sea-water. Whatever the source, water is secreted in the urine of both species and this is isoosmotic (or very nearly so) with the plasma, and contains slightly lower levels of sodium but higher concentrations of divalent ions. Sodium and water reabsorption do not appear to take place from the renal tubule, but glucose is conserved and potassium and divalent ions may be secreted by the tubules (MUNZ and MCFARLAND, 1964). The latter workers have also pointed out the considerable differences that exist between the ionic composition of the plasma and the secretions of the integumental mucous glands in hagfish (see Table 7.13). They have suggested that these may afford an additional avenue for ionic excretion. Thus the kidneys, and possibly the mucous glands, contribute to the maintenance of the composition of the extracellular fluids, slight though the differences from sea-water may seem.

Despite their low position on the phyletic scale the hagfishes possess endocrine tissues and 'hormones' similar to those present in other vertebrates. Cortisol and corticosterone, but not aldosterone, have been found in the blood of both the Atlantic and Pacific hagfishes and in concentrations similar to those which occur in mammals (CHESTER JONES and PHILLIPS, 1960; PHILLIPS et al., 1962 b). However, using more recently developed techniques WEISBART and IDLER (1970) have been unable to detect these steroids in Atlantic hagfishes or sea lampreys so that a reevaluation of the potential role of corticosteroids in cyclostomes may be necessary.

The pituitary gland of Myxine glutinosa contains an activity that, like corticotrophin, stimulates the growth of adrenocortical tissues in immature mice (STRAHAN, 1959). In addition, this gland contains a peptide that behaves in many ways like vasotocin (FOLLETT and HELLER, 1964 b). The endocrine products present in the Myxinoidea are thus distinctly similar to those in higher vertebrates, but this appears to be the end of their phyletic existence; thus corticosteroids have not been found in the protochordate Amphioxus (CHESTER JONES and PHILLIPS, 1960), while despite earlier reports, neurohypophysial peptides are not present in ascidians (SAWYER, 1959).

The hagfishes show evidence of osmoregulatory activity, but this seems to exist principally at the barrier between the cell and the extracellular fluid. The integument is relatively impermeable to sodium and there is no evidence to suggest that active uptake or extrusion of ions can occur across the skin or gills of hagfishes. Some osmoregulation occurs as a result of the activities of the kidneys and mucous glands but there is no information about the action of hormones on their secretions. When hagfishes are placed in dilute sea-water sodium loss is restricted but they accumulate large amounts of water. This is subsequently excreted indicating that 'volume regulation' is occurring. As MCFARLAND and MUNZ (1965) indicate, this may be an 'automatic' process due to an increased GFR resulting from the dilution of plasma proteins.

CHESTER JONES, PHILLIPS, and BELLAMY (1962) measured the effects of various hormone preparations on the water and electrolyte composition of Myxine glutinosa. In order to promote the need for regulatory action they placed the fish in dilute (60%) sea-water and then injected them, over a period of several days, with the hormones. Aldosterone, deoxycorticosterone and corticotrophin decreased the levels of sodium in the muscle, though the plasma concentrations were unchanged. Mammalian prolactin (a 'prolactin' has not been found in the Agnatha!) had a similar action. These effects may result from a decline in the intracellular sodium levels but they could also be due to a decrease in the extracellular space of the muscle. In another series of experiments CHESTER JONES et al. found that the extracellular space decreased following the injection of extracts of *Myxine* pituitaries into these fish. The site of such actions of the hormones is not clear.

The role of endocrines in the osmoregulation of the hagfishes seems to have been neglected in recent years. With the more refined techniques that have become available for measuring the permeability of fish it is high time that they received the attention to which they are entitled. The hagfishes could be exhibiting osmoregulation like that of the first vertebrates, but even if this is not so, it is a pattern that is unique among contemporary fishes.

References

ACEVES, J., D. ERLIJ, and C. EDWARDS: Na⁺ transport across the isolated skin of *Ambystoma mexicanus*. Biochim. biophys. Acta (Amst.) *150*, 744-746 (1968).

ACHER, R., R. BEAUPAIN, J. CHAUVET, M.T. CHAUVET, and D. CREPY: The neurohypophysial hormones of the amphibians: comparison of the hormones of *Rana esculenta* and *Xenopus laevis*. Gen. comp. Endocr. *4*, 596–601 (1964 a).

– J. CHAUVET, et M.T.CHAUVET: Les hormones neurohypophysaires des reptiles: Isolement de la mésotocine et de la vasotocine de la vipère *Vipera aspis*. Biochim. biophys. Acta (Amst.) *154*, 255-257 (1968).

– J. CHAUVET, and M.T. CHAUVET: The neurohypophysial hormones of reptiles: comparison of the viper, cobra, and elaph active principles. Gen. comp. Endocr. *13*, 357-360 (1969a).

– J. CHAUVET, and M.T. CHAUVET: Evolution of the neurohypophysial hormones with reference to amphibians. Nature (Lond.) *221*, 759–760 (1969 b).

– J. CHAUVET, M.T. CHAUVET, et D. CREPY: Isolement d'une nouvelle hormone neurohypophysaire, l'isotocine, présente chez les poissons osseux. Biochim. biophys. Acta (Amst.) *58*, 624–625 (1962).

– J. CHAUVET, M.T. CHAUVET, et D. CREPY: Phylogénie des peptides neurohypophysaires: isolement de la mésotocine (Ileu₈-ocytocine) de la grenouille, intermédiaire entre la Ser₄-Ileu₈-ocytocine des poissons osseux et l'ocytocine des mammifères. Biochim. biophys. Acta (Amst.) *90*, 613–614 (1964 b).

– J. CHAUVET, M.T. CHAUVET, et D.CREPY: Caractérisation des hormones neurohypophysaires d'un poisson osseux d'eau douce la carpe (*Cyprinus carpio*). Comparaison avec les hormones des poissons osseux marins. Comp. Biochem. Physiol. *14*, 245-254 (1965a).

– J. CHAUVET, M.T. CHAUVET, et D.CREPY: Phylogénie des peptides neurohypophysaires: isolement d'une nouvelle hormone la glumitocine (Ser₄-Gln₈-ocytocine) présente chez un poisson cartilagineux, la raie (*Raia clavata*). Biochim. Biophys. Acta (Amst.) *107*, 393-396 (1965b).

– J. CHAUVET, M.T. CHAUVET, et D.CREPY: Les hormones neurohypophysaires des amphibiens: isolement et caractérisation de la mésotocine et de la vasotocine chez le crapaud (*Bufo bufo*). Gen. comp. Endocr. *8*, 337-343 (1967).

– J. CHAUVET, M. T. LENCI, F. MOREL, et J. MAETZ: Présence d'une vasotocine dans la neurohypophyse de la grenouille (*Rana esculenta L.*). Biochim. biophys. Acta (Amst.) *42*, 379–380 (1960).

ADOLPH, E.F.: Exchanges of water in the frog. Biol. Rev. *8*, 224–240 (1933).

– Physiology of man in the desert. New York: Interscience 1947.

AKESTER, A.R., R.S. ANDERSON, K.J. HILL, and G.W. OSBALDISTON: A radiographic study of urine flow in the domestic fowl. Brit. Poultry Sci. *8*, 209–212 (1967).

ALVARADO, R. H. and A. MOODY: Sodium and chloride transport in tadpoles of the bullfrog *Rana catesbeiana*. Amer. J. Physiol. *218*, 1510–1516 (1970).

– S. R. JOHNSON: The effects of arginine vasotocin and oxytocin on sodium and water balance in *Ambystoma*. Comp. Biochem. Physiol. *16*, 531–546 (1965).

– S.R. JOHNSON: The effects of neurohypophysial hormones on water and sodium balance in larval and adult bullfrogs (*Rana catesbeiana*). Comp. Biochem. Physiol. *18*, 549–561 (1966).

– L.B. KIRSCHNER: Osmotic and ionic regulation in *Ambystoma tigrinum*. Comp. Biochem. Physiol. *10*, 55–67 (1963).

– L.B. KIRSCHNER: Effect of aldosterone on sodium fluxes in larval *Ambystoma tigrinum*. Nature (Lond.) *202*, 922–923 (1964).

AMATRUDA, T.T. and L.G. WELT: Secretion of electrolytes in thermal sweat. J. appl. Physiol. *5*, 759–771 (1953).

AMES, R.G. and H.B. VAN DYKE: Antidiuretic hormone in the urine and pituitary of the kangaroo rat. Proc. Soc. exp. Biol. (N.Y.) *75*, 417–420 (1950).

– H.B. Van Dyke: Antidiuretic activity in the serum or plasma of rats. Endocrinology *50*, 350–360 (1952).

Andre, R. and J. Crabbe: Stimulation by insulin of active sodium transport by toad skin: influence of aldosterone and vasopressin. Archives Internationales de Physiologie et de Biochimie *73*, 538–540 (1966).

Arai, R., H. Tajima, and B. Tamaoki: *In vitro* transformation of steroids by the head kidney, the body kidney and the corpuscles of Stannius of the rainbow trout (*Salmo gairdneri*). Gen. comp. Endocr. *12*, 99–109 (1969).

Arnold, J. and J. Shield: Oxygen consumption and body temperature of the Chuditch (*Dasyurus geoffroii*). J. Zool. (Lond.) *160*, 391–404 (1970).

Arvy, L., M. Fontaine, et M. Gabe: La voie neurosécrétrice hypothalamo-hypophysaire des téléostéens. J. Physiol. (Paris) *51*, 1031–1085 (1959).

Augee, M.L. and I.R. McDonald: The effect of cold stress on adrenalectomized echidnas (*Tachyglossus aculeatus*). Aust. J. exp. Biol. & med. Sci. *47*, PI (1969).

Ash, R.W.: Plasma osmolality and salt gland secretion in the duck. Q. Jl. exp. Physiol. *54*, 68–79 (1969).

– J.W. Pearce, and A. Silver: An investigation of the nerve supply to the salt gland of the duck. Q. Jl. exp. Physiol. *54*, 281–295 (1969).

Ausiello, D.A. and G.W.G. Sharp: Localization of physiological receptor sites for aldosterone in the bladder of the toad, *Bufo marinus*. Endocrinology *82*, 1163–1169 (1968).

Axelrad, A.A., C.P. Leblond, and H. Isler: Role of iodine deficiency in the production of goiter by the Remington diet. Endocrinology *56*, 387–403 (1955).

Baillien, M. and E. Schoffeniels: Origin of the potential difference in the intestinal epithelium of the turtle. Nature (Lond.) *190*, 1107–1108 (1961).

Baker, P.F., A.L. Hodgkin, and T.I. Shaw: Replacement of the protoplasm of a giant nerve fibre with artificial solutions. Nature (Lond.) *190*, 885–887 (1961).

– T.I. Shaw: A comparison of the phosphorus metabolism of intact squid nerve with that of isolated axoplasm and sheath. J. Physiol. (Lond.) *180*, 424–438 (1965).

Baldwin, E.: Dynamic Aspects of Biochemistry. 4th Ed. Cambridge: University Press 1963.

Balinsky, J.B., M.M. Cragg, and E. Baldwin: The adaptation of amphibian waste nitrogen excretion to dehydration. Comp. Biochem. Physiol. *3*, 236–244 (1961).

Ball, J.N.: Partial hypophysectomy in the teleost *Poecilia:* separate identities of teleostean growth hormone and prolactin-like hormone. Gen. comp. Endocr. *5*, 654–661 (1965).

– Effects of autotransplantation of different regions of the pituitary gland on fresh water survival in the teleost *Poecilia latipinna*. J. Endocr. *33*, v–vi (1965).

– D.M. Ensor: Specific action of prolactin on plasma sodium levels in hypophysectomized *Poecilia latipinna* (Teleostei). Gen. comp. Endocr. *8*, 432–440 (1967).

– M. Olivereau: Rôle de la prolactine dans la survie en eau douce de *Poecilia latipinna* hypophysectomisé et arguments en faveur de sa synthèse par les cellules érythrosinophiles de l'hypophyses des téléostéens. C.R. Acad. Sci. (Paris) *259*, 1443–1446 (1964).

– G.E. Pickford: Pituitary cytology and fresh water adaptation in the killifish, *Fundulus heteroclitus*. Anat. Rec. *148*, 358 (1964).

Bardack, D. and R. Zangerl: First fossil lamprey: a record from the Pennsylvania of Illinois. Science *162*, 1265–1267 (1968).

Bargmann, W.: Über die neurosekretorische Vernüpfung von Hypothalamus und Neurohypophyse. Z. Zellforsch. *34*, 610–634 (1949).

– Neurohypophysis. Structure and function. In: Handbuch d. exper. Pharmacol. Vol. 23, pp. 1–39. Ed. by B. Berde. Heidelberg and Berlin: Springer-Verlag 1968.

Barrell, J.: Dominantly fluviatile origin under seasonal rainfall of the old red Sandstone. Bull. geol. Soc. Amer. *27*, 345–386 (1916).

Barrington, E.J.W.: An Introduction to General and Comparative Endocrinology. Oxford: Clarendon 1963.

Bartholomew, G.A.: Temperature regulation in the macropod marsupial *Setonix brachyurus*. Physiol. Zool. *29*, 26–40 (1956).

– T.J. Cade: The water economy of land birds. The Auk *80*, 504–539 (1963).

– R.C. Lasiewski, and E.C. Crawford: Patterns of panting and gular flutter in cormorant, pelicans, owls and doves. The Condor *70*, 31–34 (1968).

256

BASTIDE, F. et S. JARD: Actions de la noradrénaline et de l'ocytocine sur le transport actif de sodium et la perméabilité a l'eau de la peau de grenouille. Rôle du 3', 5'-AMP cyclique. Biochim. biophys. Acta (Amst.) *150*, 113–123 (1968).

BAYLISS, L.E.: Principles of General Physiology. Vol. 2. pp. 724 London: Longmans, 1960.

BECK, R.R., K.A. BROWNELL, and P.K. BESCH: A further study of adrenal function in the Opossum (*Didelphis virginiana*). Gen. comp. Endocr. *13*, 165–172 (1969).

BELLAMY, D.: Movements of potassium, sodium and chloride in incubated gills from the silver eel. Comp. Biochem. Physiol. *3*, 125–135 (1961).

– I. CHESTER JONES: Studies on *Myxine glutinosa*. 1. The chemical composition of the tissues. Comp. Biochem. Physiol. *3*, 175–183 (1961).

– J.G.PHILLIPS: Effects of administration of sodium chloride solutions on the concentration of radioactivity in the nasal gland of ducks (*Anas platyrhynchos*) injected with [³H] corticosterone. J.Endocr. *36*, 97–98 (1966).

BENEDICT, F.G.: The physiology of large reptiles. Carnegie Inst. Wash. Publication No. 525 (1932).

BENNETT, M.V.L. and S.FOX: Electrophysiology of caudal neurosecretory cells in the skate and fluke. Gen. comp. Endocr. *2*, 77–95 (1962).

BENOIT, G.J. and E.R. NORRIS: Studies on trimethylamine oxide. II. The origin of trimethylamine oxide in young salmon. J. biol. Chem. *158*, 439–442 (1945).

BENTLEY, P.J.: Some aspects of the water metabolism of an Australian marsupial *Setonyx brachyurus*. J. Physiol. (Lond.) *127*, 1–10 (1955).

– The effects of neurohypophysial extracts on water transfer across the wall of the isolated urinary bladder of the toad *Bufo marinus*. J. Endocr. *17*, 201–209 (1958).

– Effects of elevated plasma sodium and potassium in the erythrocyte of the lizard *Trachysaurus rugosus*. Nature (Lond.) *184*, 1403 (1959a).

– Studies on the water and electrolyte metabolism of the lizard *Trachysaurus rugosus* (Gray). J. Physiol. (Lond.) *145*, 37–47 (1959 b).

– The effects of vasopressin on the short-circuit current across the wall of the isolated bladder of the toad *Bufo marinus*. J.Endocr. *21*, 161–170 (1960a).

– Evaporative water loss and temperature regulation in the marsupial *Setonyx brachyurus*. Aust. J. exp. Biol. *38*, 301–306 (1960b).

– Directional differences in the permeability to water of the isolated urinary bladder of the toad, *Bufo marinus*. J.Endocr. *22*, 95–100 (1961).

– Permeability of the skin of the cyclostome *Lampetra fluviatilis* to water and electrolytes. Comp. Biochem. Physiol. *6*, 95–97 (1962a).

– Studies on the permeability of the large intestine and urinary bladder of the tortoise (*Testudo graeca*) with special reference to the effects of neuro-hypophysial and adrenocortical hormones. Gen. comp. Endocr. *2*, 323–328 (1962 b).

– Composition of the urine of the fasting humpback whale (*Megaptera nodosa*). Comp. Biochem. Physiol. *10*, 257–259 (1963).

– Physiological properties of the isolated frog bladder in hyperosmotic solutions. Comp. Biochem. Physiol. *12*, 233–239 (1964).

– Rectification of water flow across the bladder of the toad; effect of vasopressin. Life Sciences *4*, 133–140 (1965).

– The physiology of the urinary bladder of Amphibia. Biol. Rev. *41*, 275–316 (1966a).

– Adaptations of Amphibia to arid environments. Science *152*, 619–623 (1966b).

– Mechanism of action of neurohypophysial hormones: actions of manganese and zinc on the permeability of the toad bladder. J. Endocr. *39*, 493–506 (1967).

– Neurohypophysial function in Amphibia: hormone activity in the plasma. J.Endocr. *43*, 359–369 (1969a).

– Neurohypophysial hormones in Amphibia: A comparison of their actions and storage Gen. comp. Endocr. *13*, 39–44 (1969 b).

– The permeability of the urinary bladder of a urodele amphibian (the mudpuppy, *Necturus maculosus*): studies with vasotocin and aldosterone. Gen. comp. Endocr. *16*, 356–362 (1971 a).

– Failure to detect levels of vasotocin similar to those in amphibians in the plasma of the eel, *Anguilla rostrata*. J. Endocr. *49*, 183–184 (1971 b).

– Inhibition by dopamine of the hydro-osmotic response of the toad bladder to vasopressin. (In preparation) (1971 c).
– W.F.C.Blumer: Uptake of water by the lizard *Moloch horridus*. Nature (Lond.) *194*, 699–700 (1962).
– W.L. Bretz, and K. Schmidt-Nielsen: Osmoregulation in the diamondback terrapin *Malaclemys terrapin centrata*. J. exp. Biol. *46*, 161–167 (1967 a).
– B.K. Follett: The action of neurohypophysial and adrenocortical hormones on sodium balance in the cyclostome, *Lampetra fluviatilis*. Gen. comp. Endocr. *2*, 329–335 (1962).
– B.K.Follett: Kidney function in a primitive vertebrate, the cyclostome *Lampetra fluviatilis*. J. Physiol. (Lond.) *169*, 902–918 (1963).
– B.K.Follett: The effects of hormones on the carbohydrate metabolism of the lamprey *Lampetra fluviatilis*. J. Endocr. *31*, 127–137 (1965).
– L. Greenwald: Neurohypophysial function in bullfrog (*Rana catesbeiana*) tadpoles. Gen. comp. Endocr. *14*, 412–415 (1970).
– H. Heller: The action of neurohypophysial hormones on the water and sodium metabolism of urodele amphibians. J. Physiol. (Lond.) *171*, 434–453 (1964).
– H.Heller: The water-retaining action of vasotocin on the fire salamander (*Salamandra maculosa*): The role of the urinary bladder. J. Physiol. (Lond.) *181*, 124–129 (1965).
– C.F.Herreid, and K.Schmidt-Nielsen: Respiration of a monotreme, the echidna, *Tachyglossus aculeatus*. Amer. J. Physiol. *212*, 957–961 (1967 b).
– A.K.Lee, and A.R.Main: Comparison of dehydration and hydration of two genera of frogs (*Heleioporus* and *Neobatrachus*) that live in areas of varying aridity. J. exp. Biol. *35*, 677–684 (1958).
– K. Schmidt-Nielsen: Permeability to water and sodium of the crocodilian *Caiman sclerops*. J. cell. comp. Physiol. *66*, 303–309 (1965).
– K. Schmidt-Nielsen: Cutaneous water loss in reptiles. Science *151*, 1547-1549 (1966).
– K. Schmidt-Nielsen: The role of the kidney in water balance of the echidna. Comp. Biochem. Physiol. *20*, 285-290 (1967).
– K. Schmidt-Nielsen: Comparison of water exchange in two aquatic turtles, *Trionyx spinifer* and *Pseudemys scripta*. Comp. Biochem. Physiol. *32*, 363–365 (1970).
– J.W. Shield: Metabolism and kidney function in the pouch young of the macropod marsupial *Setonix brachyurus*. J. Physiol. (Lond.) *164*, 127-137 (1962).
Bern, H.A.: Hormones and endocrine glands of fishes. Science, *158*, 455–462 (1967).
– F.G.W. Knowles: Neurosecretion. In: Neuroendocrinology, Vol. 1, pp. 139–186. Ed. by L. Martini and E.F. Ganong. New York: Academic Press 1966.
– J. Nandi: Endocrinology of poikilothermic vertebrates. In: The Hormones, Vol. 4, pp. 199–298. Ed. by G. Pincus, K.V. Thimann and E.B. Astwood. New York: Academic Press 1964.
– C.S. Nicoll: The comparative endocrinology of prolactin. Recent Prog. Horm. Res. *24*, 681–720 (1968).
– R.S. Nishioka, I. Chester Jones, D.K.O. Chan, J.C. Rankin, and S. Ponniah: The urophysis of teleost fish. J. Endocr. *37*, xl-xli (1967).
– C. deRoos, and E.G. Biglieri: Aldosterone and other corticosteroids from chondrichthyean interrenal glands. Gen. comp. Endocr. *2*, 490-494 (1962).
Berthet, J.: Pancreatic hormones: glucagon. In: Comparative Endocrinology, Vol. 1, pp. 410–427. Ed. by U.S. von Euler and H. Heller. New York: Academic Press 1963.
Blair-West, J.R., J.P. Coghlan, D.A. Denton, J.F. Nelson, E. Orchard, B.A. Scoggins, R.D. Wright, K. Meyers, and L.C.U. Junqueira: Physiological, morphological, and behavioural adaptation to a sodium deficient diet by wild native Australian and introduced species of animals. Nature (Lond.) *217*, 922–928 (1968).
– J.P. Coghlan, D.A. Denton, and R.D. Wright: Effects of endocrines on salivary glands. In: Handbook of Physiology Sec. 6. Alimentary canal, Vol. 2, pp. 633–664. Washington: Amer. Physiol. Soc. 1967.
Blanc-Livni, N. and M. Abraham: The influence of environmental salinity on the prolactin- and gonadotrophin-secreting regions in the pituitary of *Mugil* (Teleostei). Gen. comp. Endocr. *14*, 184–197 (1970).

BOGERT, C.M., and R.B. COWLES: Moisture loss in relation to habitat selection in some Floridian reptiles. Am. Mus. Novitates, *1358*, 1–34 (1947).

DU BOIS REYMOND, E.: Untersuchungen über tierische Elekricat. Berlin 1848.

BONDY, P.K., G.V. UPTON, and G.E. PICKFORD: Demonstration of cortisol in fish blood. Nature (Lond.) *179*, 1354-1355 (1957).

BONTING, S.L.: Studies on sodium-potassium-activated adenosinetriphosphatase-XV the rectal gland of the elasmobranchs. Comp. Biochem. Physiol. *17*, 953-966 (1966).

BOTT, E.J., R. BLAIR-WEST, J.P. COGHLAN, and D.A. DENTON: Action of aldosterone on the lacrimal gland. Nature (Lond.) *210*, 102 (1966).

BOURGUET, J., B. LAHLOUH, et J. MAETZ: Modifications experimentales de l'équilibre hydrominéral et osmorégulation chez *Carassius auratus*. Gen. comp. Endocr. *4*, 563–576 (1964).

– J. MAETZ: Arguments en faveur de l'indépendance des mécanismes d'action de divers peptides neurohypophysaires sur le flux osmotique d'eau et sur le transport actif de sodium au sein d'un même récepteur: études sur la vessie et sur la peau de *Rana esculenta* L. Biochim. biophys. Acta (Amst.) *52*, 552–565 (1961).

BOYD, E.M. and D.W. WHYTE: The effect of posterior hypophysial extract on retention of water and salt injected into frogs. Amer. J. Physiol. *125*, 415-422 (1939).

BOYLAN, J.W.: Gill permeability in *Squalus acanthias*. In: Sharks, Rays and Skates, pp. 197–206. Ed. by P.W. GILBERT, R.F. MATHEWSON, and D.P. RALL. Baltimore: The Johns Hopkins Press 1967.

BRADSHAW, S.D. and V.H. SHOEMAKER: Aspects of water and electrolyte changes in a field population of *Amphibolurus* lizards. Comp. Biochem. Physiol. *20*, 855-865 (1967).

BRODSKY, W.A. and T.P. SCHILB: Electrical and osmotic characteristics of the isolated turtle bladder. J. clin. Invest. *39*, 974 (1960).

– T.P. SCHILB: Osmotic properties of the isolated turtle bladder. Amer. J. Physiol. *208*, 46-57 (1965).

BROOKS, C.J.W., R.V. BROOKS, K. FOTHERBY, J.K. GRANT, A. KLOPPER, and W. KLYNE: The identification of steroids. J. Endocr. *47*, 265–272 (1970).

BROWN, G.W. and S.G. BROWN: Urea and its formation in coelacanth liver. Science. *155*, 570–572 (1967).

BROWN, G.W. Jr., J. JAMES, R.J. HENDERSON, W.N. THOMAS, R.O. ROBINSON, A.L. THOMPSON, E. BROWN, and S.G. BROWN: Uricotylic enzymes in liver of the dipnoan *Protopterus aethiopicus*. Science *153*, 1653–1654 (1966).

BRUNN, F.: Beitrag zur Kenntnis der Wirkung von Hypophysenextrakt auf den Wasserhaushalt des Frosches. Z. ges. exp. Med. *25*, 170–175 (1921).

BURDEN, C.E.: The failure of the hypophysectomized *Fundulus heteroclitus* to survive in fresh water. Biol. Bull. (Woods Hole) *110*, 8-28 (1956).

BURGER, J.W.: Further studies on the function of the rectal gland in the spiny dogfish. Physiol. Zool. *35*, 205–217 (1962).

– Roles of the rectal gland and the kidneys in salt and water excretion in the spiny dogfish. Physiol. Zool. *38*, 191-196 (1965).

– W.N. HESS: Function of the rectal gland in the spiny dogfish. Science *131*, 670-671 (1960).

– D.C. TOSTESON: Sodium influx and efflux in the spiny dogfish *Squalus acanthias*. Comp. Biochem. Physiol. *19*, 649-653 (1966).

BURGESS, W.W., A.M. HARVEY, and E.K. MARSHALL: The site of the antidiuretic action of pituitary extract. J. Pharmacol. exp. Ther. *49*, 237-248 (1933).

BURNS, T.W.: Endocrine factors in the water metabolism of the desert mammal, *G. gerbilus*. Endocrinology *58*, 243-254 (1956).

BUSH, I.E.: Species differences in adrenocortical secretion. J. Endocr. *9*, 95-100 (1953).

BUTLER, D.G.: Effect of hypophysectomy on osmoregulation in the European eel (*Anguilla anguilla* L.). Comp. Biochem. Physiol. *18*, 773-781 (1966).

– Corpuscles of Stannius and renal physiology in the eel (*Anguilla rostrata*). J. Fish. Res. Bd Can. *26*, 639–654 (1969).

– W.C. CLARKE, E.M. DONALDSON, and R.W. LANGFORD: Surgical adrenalectomy of a teleost fish (*Anguilla rostrata* Lesueur): effect on plasma cortisol and tissue electrolyte and carbohydrate concentrations. Gen. comp. Endocr. *12*, 503–514 (1969).

259

- E.M.Donaldson, and W.C.Clarke: Physiological evidence for a pituitary-adrenocortical feedback mechanism in the eel *(Anguilla rostrata).* Gen. comp. Endocr. *12,* 173–176 (1969).
- W.H. Knox: Adrenalectomy of the painted turtle (*Chrysemys picta belli*): effect on ion regulation and tissue glycogen. Gen. comp. Endocr. *14,* 551-566 (1970).

Buttle, J.M., R.L. Kirk, and H. Waring: The effects of complete adrenalectomy on the wallaby (*Setonyx brachyurus*). J. Endocr. *8,* 281–290 (1952).

Cade, T.J. and J.A. Dybas: Water economy of the budgerygah. The Auk *79,* 345-364 (1962).
- C.A. Tobin, and A. Gold: Water economy and metabolism of two estrilidine finches. Physiol. Zool. *39,* 9-33 (1965).

Calder, W.A.: Gaseous metabolism and water relations of the zebra finch, *Taeniopygia castanotis.* Physiol. Zool. *37,* 400–413 (1964).
- The diurnal activity of the roadrunner, *Geococcyx californianus.* The Condor *70,* 84-85 (1968).
- P.J.Bentley: Urine concentrations of carnivorous birds, the white pelican and the roadrunner. Comp. Biochem. Physiol. *22,* 607–609 (1967).

Caldwell, P.D., A.L. Hodgkin, R.D. Keynes, and T.I. Shaw: The effect of injecting energy rich phosphate compounds on the active transport of ions in the giant axons of *Loligo.* J. Physiol. (Lond.) *152,* 561–590 (1960).

Capelli, J.P., L.G. Wesson, and G.E. Aponte: A phylogenetic study of the renin-angiotensin system. Amer. J. Physiol. *218,* 1171-1178 (1970).

Carlson, I.H. and W.N. Holmes: Changes in the hormone content of the hypothalamo-hypophysial system of the rainbow trout (*Salmo gairdneri*). J. Endocr. *24,* 23–32 (1962).

Carstensen, H., A.C.J. Burgers, and C.H. Li: Isolation of aldosterone from incubates of adrenal glands of the American bullfrog and stimulation of its production by mammalian corticotrophin. J. Am. chem. Soc. *81,* 4109-4110 (1959).
- A.C.J. Burgers, and C.H. Li: Demonstration of aldosterone and corticosterone as the principal steroids formed in incubates of adrenals of the American bullfrog (*Rana catesbeiana*) and stimulation of their production by mammalian adrenocorticotrophin. Gen. comp. Endocr. *1,* 37-50 (1961).

Castel, M. and M. Abraham: Effects of a dry diet on the hypothalamic neurohypophyseal neurosecretory system in spiny mice as compared to the albino rat and mouse. Gen. comp. Endocr. *12,* 231-241 (1969).

Cereijido, M. and C.A. Rotunno: Transport and distribution of sodium across frog skin. J. Physiol. (Lond.) *190,* 481-497 (1967).

Chadwick, A.: Prolactin-like activity in the pituitary gland of fishes and amphibians. J. Endocr. *35,* 75-81 (1966).

Chadwick, C.S.: Identity of prolactin with water drive factor in *Triturus viridescens.* Proc. Soc. exp. Biol. (N.Y.) *45,* 335-337 (1940).
- Further observations on the water drive in *Triturus viridescens* II. Induction of the water drive with the lactogenic hormone. J. exp. Zool. *86,* 175–187 (1941).

Chamberlin, T.C.: On the habitat of the early vertebrates. J. Geol. *8,* 400-412 (1900).

Chan, D.K.O. and I. Chester Jones: Pressor effects of neurohypophysial peptides in the eel, *Anguilla anguilla* L., with some reference to their interaction with adrenergic and cholinergic receptors. J. Endocr. *45,* 161-174 (1969).
- I. Chester Jones, I.W. Henderson, and J.C. Rankin: Studies on the experimental alteration of water and electrolyte composition of the eel (*Anguilla anguilla* L.). J. Endocr. *37,* 297-317 (1967a).
- I. Chester Jones, and W. Mosley: Pituitary and adrenocortical factors in the control of the water and electrolyte composition of the freshwater European eel (*Anguilla anguilla* L.) J. Endocr. *42,* 91-98 (1968).
- I. Chester Jones, and S. Ponniah: Studies on the pressor substances of the caudal neurosecretory system of teleost fish: bioassay and fractionation. J. Endocr. *45,* 151-159 (1969).
- J.G. Phillips, and I. Chester Jones: Studies on electrolyte changes in the lip shark, *Hemiscyllium plagiosum* (Bennett), with special reference to the hormonal influence on the rectal gland. Comp. Biochem. Physiol. *23,* 185–198 (1967 b).

CHAN, S.T.H. and B.R. EDWARDS: Kinetic studies on the biosynthesis of corticosteroids *in vitro* from exogenous precursor by the interrenal glands of the normal, corticotrophintreated and adenohypophysectomized *Xenopus laevis* Daudin. J. Endocr. *47*, 183–195 (1970).

CHAUVET, J., M.T. CHAUVET, et R. ACHER: Les hormones neurohypophysaires des mammifères: Isolement et caracterisation de l'ocytocine et de la vasopressine de la baleine (*Balaenoptera physalus* L.). Bull. Soc. Chim. biol. *45*, 1369–1378 (1963).

– M.T. LENCI, et R. ACHER: Présence de deux vasopressines dans la neurohypophyses du poulet. Biochim. biophys. Acta (Amst.) *38*, 571–573 (1960).

CHESTER JONES, I.: The Adrenal Cortex. Cambridge: University Press 1957.

– D.K.O. CHAN, I.W. HENDERSON, W. MOSLEY, T. SANDOR, G.P. VINSON, and B. WHITEHOUSE: Failure of corpuscles of Stannius of the European eel (*Anguilla anguilla* L.) to produce corticosteroids *in vitro*. J. Endocr. *33*, 319–320 (1965).

– D.K.O. CHAN, and J.C. RANKIN: Renal function in the European eel (*Anguilla anguilla* L.): changes in blood pressure and renal function of the freshwater eel transferred to seawater. J. Endocr. *43*, 9-19 (1969a).

– D.K.O. CHAN, and J.C. RANKIN: Renal function in the European eel (*Anguilla anguilla* L.): effects of the caudal neurosecretory system, corpuscles of Stannius, neurohypophysial peptides and vasoactive substances. J. Endocr. *43*, 21–31 (1969 b).

– I.W. HENDERSON, and W. MOSLEY: Methods for the adrenalectomy of the European eel (*Anguilla anguilla*). J. Endocr. *30*, 155–156 (1964 b).

– I.W. HENDERSON, D.K.O. CHAN, J.C. RANKIN, W. MOSLEY, J.J. BROWN, A.F. LEVER, J.I.S. ROBERTSON, and M. TREE: Pressor activity in extracts of the corpuscles of Stannius from the European eel (*Anguilla anguilla* L.). J. Endocr. *34*, 393–408 (1966).

– J.G. PHILLIPS: Adrenocorticosteroids in fish. Symp zool. Soc. Lond. *1*, 17–32 (1960).

– J.G. PHILLIPS, and D. BELLAMY: The adrenal cortex throughout the vertebrates. Brit. med. Bull. *18*, 110-114 (1962a).

– J.G. PHILLIPS, and D. BELLAMY: Studies on water and electrolytes in cyclostomes and teleosts with special reference to *Myxine glutinosa* L. (the hagfish) and *Anguilla anguilla* L. (the European eel). Gen. comp. Endocr. Suppl. *1*, 36–47 (1962 b).

– J.G. PHILLIPS, and W.N. HOLMES: Comparative physiology of the adrenal cortex. In: Comparative Endocrinologyhon. pp. 582–612. Ed. by A. GORBMAN. New York: Wiley 1959.

– G.P. VINSON, I.G. JARRETT, and G.B. SHARMAN: Steroid components in the adrenal venous blood of *Trichosurus vulpecula* (Kerr). J. Endocr. *30*, 149–150 (1964 a).

CIVAN, M.M. and H.S. FRAZIER: The site of the stimulating action of vasopressin on sodium transport in toad bladder. J. gen. Physiol. *51*, 589-605 (1968).

CLAUSSEN, D.L.: Studies of water loss in two species of lizards. Comp. Biochem. Physiol. *20*, 115-130 (1967).

– Studies on water loss and rehydration in anurans. Physiol. Zool. *42*, 1-14 (1969).

COFRE, G. and J. CRABBE: Stimulation by aldosterone of active sodium transport by the isolated colon of the toad, *Bufo marinus*. Nature (Lond.) *207*, 1299-1300 (1965).

– J. CRABBE: Active sodium transport by the colon of *Bufo marinus:* stimulation by aldosterone and antidiuretic hormone. J. Physiol. (Lond.). *188*, 177–190 (1967).

COGHLAN, J.P. and B.A. SCOGGINS: The measurement of aldosterone, cortisol and corticosterone in the blood of the wombat (*Vombatus hirsutus* Perry) and the kangaroo (*Macropus giganteus*). J. Endocr. *39*, 445–448 (1967).

COLE, P.M., I. CHESTER JONES, and D. BELLAMY: Observations on the excretion of water and electrolytes in the desert rat (*Dipodomys spectabilis spectabilis* M.) and in the laboratory rat. J. Endocr. *25*, 515-532 (1963).

CONN, J.W., L.H. LOUIS, M.W. JOHNSTON, and B.J. JOHNSON: The electrolyte content of thermal sweat as an index of adrenal function. J. clin. Invest. *27*, 529-530 (1947).

CONTE, F.P. and D.H.Y. LIN: Kinetics of cellular morphogenesis in gill epithelium during sea water adaptation of *Oncorhynchus* (Walbaum). Comp. Biochem. Physiol. *23*, 945–957 (1967).

– H.H. WAGNER: Development of osmotic and ionic regulation in juvenile steelhead trout *Salmo gairdneri*. Comp. Biochem. Physiol. *14*, 603-620 (1965).

- H.H. WAGNER, J. FESSLER, and C. GNOSE: Development of osmotic and ionic regulation in juvenile coho salmon *Oncorhynchus kisutch*. Comp. Biochem. Physiol. *18*, 1-15 (1966).

COOPERSTEIN, I.L. and C.A.M. HOGBEN: Ionic transfer across the isolated frog large intestine. J. gen. Physiol. *42*, 461–473 (1959).

COPP, D.H.: Calcitonin-ultimobranchial hormone. Proc. int. Union of Physiol. Sci. *6*, 15-16 (1968).

- E.C. CAMERON, B.A. CHENEY, A.G.F. DAVIDSON, and K.G. HENZE: Evidence for calcitonin – a new hormone from the parathyroid which lowers blood calcium. Endocrinology *70*, 638–649 (1962).

CRABBE, J.: Stimulation of active sodium transport by the isolated toad bladder with aldosterone *in vitro*. J. clin. Invest. *40*, 2103–2110 (1961 a).

- Stimulation of active sodium transport across the isolated toad bladder after injection of aldosterone to the animal. Endocrinology *69*, 673-682 (1961b).

- The role of aldosterone in the renal concentration mechanism in man. Clin. Sci. *23*, 39–46 (1962).

- The Sodium Retaining Action of Aldosterone. pp. 65. Bruxelles: Academique Européennes 1963.

- Stimulation by aldosterone of active sodium transport across the isolated ventral skin of Amphibia. Endocrinology *75*, 809-811 (1964).

- La régulation de la sécrétion d'aldostérone chez *Bufo marinus*. Ann. Endocr. (Paris) *27*, 501-505 (1966).

- B. FRANCOIS: Stimulation par l'insuline du transport actif de sodium à travers les membranes épithéliales du crapaud, *Bufo marinus*. Ann. Endocr. (Paris) *28*, 713–715 (1967).

- P. DE WEER: Action of aldosterone on the bladder and skin of the toad. Nature (Lond.) *202*, 278-279 (1964).

CRANE, R.K.: Na^+-dependent transport in the intestine and other animal tissues. Fed. Proc. *24*, 1000-1006 (1965).

CRAWFORD, E.C. and K. SCHMIDT-NIELSEN: Temperature regulation and evaporative cooling in the ostrich. Amer. J. Physiol. *212*, 347-353 (1967).

CURRAN, P.F.: Ion transport in intestine and its coupling to other transport processes. Fed. Proc. *24*, 993-999 (1965).

CUTHBERT, A.W. and E. PAINTER: Mechanism of action of aldosterone. Nature (Lond.) *222*, 280-281 (1969).

CSAKY, T.Z.: Transport through biological membranes. Ann. Rev. Physiol. *27*, 415-450 (1965).

DANTZLER, W.H.: Renal response of chickens to infusion of hyperosmotic sodium chloride solution. Amer. J. Physiol. *210*, 640–646 (1966).

- Glomerular and tubular effects of arginine-vasotocin in water snakes (*Natrix sipedon*). Amer. J. Physiol. *212*, 83-91 (1967a).

- Stop-flow study of renal function in conscious water snakes (*Natrix sipedon*). Comp. Biochem. Physiol. *22*, 131-140 (1967b).

- B. SCHMIDT-NIELSEN: Excretion in fresh-water turtle (*Pseudemys scripta*) and desert tortoise (*Gopherus agassizii*). Amer. J. Physiol. *210*, 198–210 (1966).

DARLINGTON, P.J.: Zoogeography: the Geographical Distribution of Animals. New York: Wiley 1957.

DARWIN, C.: Journal of Researches into the Natural History and Geography of the Countries Visited during the Voyage of the H.M.S. Beagle around the World. London 1839.

- The Origin of Species by means of Natural Selection. London 1859.

DAVIS, J.O., C.I. JOHNSTON, P.M. HARTROFT, S.S. HOWARDS, and F.S. WRIGHT: The phylogenetic and physiologic importance of the renin angiotensin aldosterone system. Proc. 3rd Int. Congr. Nephrol. (Washington 1966) Vol. 1, pp. 215–225. Basel/New York: Karger 1967.

DAVSON, H. and J.F. DANIELLI: The Permeability of Natural Membranes. 2nd. ed., Cambridge: University Press 1952.

DAWSON, A.B.: Functional and degenerate or rudimentary glomeruli in the kidney of two species of Australian frog, *Cyclorana (Chiroleptes) platycephalus* and *Alboguttatus* (Gunther). Anat. Rec. *109*, 417–424 (1951).

DAWSON, T.J. and A.J. HULBERT: Standard energy metabolism of marsupials. Nature (Lond.) *221*, 383 (1969).

DAWSON, W.R. and K. SCHMIDT-NIELSEN: Terrestrial animals in dry heat: desert birds. In: Handbook of Physiology. Sec. 4. Adaptation of the Environment, pp. 481–492. Washington: Amer. Physiol. Soc. 1964.

DENTON, D.A.: Evolutionary aspects of the emergence of aldosterone secretion and salt appetite. Physiol. Rev. *45*, 245-295 (1965).

DEROOS, R.: The corticoids of the avian adrenal gland. Gen. comp. Endocr. *1*, 494–512 (1961).

– C.C. DEROOS: Angiotensin II: its effects on corticoid production by chicken adrenal *in vitro*. Science *141*, 1284 (1963).

– C. C. DEROOS: Presence of corticotropin activity in the pituitary gland of chondrichthyean fish. Amer. Zool. *4*, 393 (1964).

DHARMAMBA, M., R.I. HANDIN, J. NANDI, and H.A. BERN: Effect of prolactin on fresh water survival and on plasma osmotic pressure of hypophysectomized *Tilapia mossambica*. Gen. comp. Endocr. *9*, 295-302 (1967).

– R.S. NISHIOKA: Response of 'prolactin-secreting' cells of *Tilapia mossambica* to environmental salinity. Gen. comp. Endocr. *10*, 409-420 (1968).

DIBONA, D. R., M. M. CIVAN, and A. LEAF: Differential epithelial cell responses in toad urinary bladder to addition of vasopressin. Fed. Proc. *28*, 402 (1969).

DICKER, S.E.: Mechanisms of urine concentration and dilution in mammals. London: Arnold (1970).

– M.G. EGGLETON: Renal function in the primitive mammal *Aplodontia rufa*, with some observations on squirrels. J. Physiol. (Lond.) *170*, 186–194 (1964).

– A.B. ELLIOTT: Water uptake by *Bufo melanostictus*, as affected by osmotic gradients, vasopressin and temperature. J. Physiol. (Lond.) *190*, 359–370 (1967).

– A.B. ELLIOTT: Effects of neurohypophysial hormones on the skin permeability of *Rana cancrivora in vivo*. J. Physiol. (Lond.) *203*, 74–75p (1969).

– A.B. ELLIOTT: Water uptake by the crab-eating frog *Rana cancrivora*, as affected by osmotic gradients and by neurohypophysial hormones. J. Physiol. (Lond.) *207*, 119-132 (1970 a).

– A.B. ELLIOTT: Ultrastructure of urinary bladder of *Rana cancrivora*. J. Physiol. (Lond.) *209*, 23–24 P (1970 b).

– J. HASLAM: Water diuresis in the domestic fowl. J. Physiol. (Lond.) *183*, 225-235 (1966).

– J. NUNN: The role of the antidiuretic hormone during water deprivation in rats. J. Physiol. (Lond.) *136*, 235-248 (1957).

– C. TYLER: Vasopressor and oxytocic activities of the pituitary glands of rats, guinea pigs and cats and of human foetuses. J. Physiol. (Lond.) *121*, 205–214 (1953).

DONALDSON, E.M. and W.N. HOLMES: Corticosteroidogenesis in the fresh-water- and saline-maintained duck (*Anas platyrhynchos*). J. Endocr. *32*, 329-336 (1965).

– W.N. HOLMES, and J. STACHENKO: *In vitro* corticosteroidogenesis by the duck (*Anas platyrhynchos*) adrenal. Gen. comp. Endocr. *5*, 542-551 (1965).

– J.R. MCBRIDE: The effects of hypophysectomy in the rainbow trout *Salmo gairdnerii* (Rich) with special reference to the pituitary adrenal axis. Gen. comp. Endocr. *9*, 93–101 (1967).

DOUGLAS, D.S. and S.M. NEELY: The effect of dehydration on salt gland performance. Amer. Zool. *9*, 1095 (1969).

DOYLE, W.L.: Tubule cells of the rectal gland of *Urolophus*. Amer. J. Anat. *111*, 223-238 (1962).

DUNSON, W.A.: Sodium fluxes in fresh-water turtles. J. exp. Zool. *165*, 171-182 (1967).

– Salt gland secretion in the pelagic sea snake *Pelamis*. Amer. J. Physiol. *215*, 1512-1517 (1968).

– Electrolyte excretion by the salt gland of the Galapagos marine iguana. Amer. J. Physiol. *216*, 995-1002 (1969).

– E.G. BUSS: Abnormal water balance in a mutant strain of chickens. Science *161*, 167-169 (1968).

– A.M. TAUB: Extrarenal salt excretion in sea snakes (*Laticauda*). Amer. J. Physiol. *213*, 975-982 (1967).

– R.D. Weymouth: Active uptake of sodium by softshell turtles (*Trionyx spinifer*). Science *149*, 67-69 (1965).

Ealey, E.H.M., P.J. Bentley, and A.R. Main: Studies on water metabolism of the hill kangaroo, *Macropus robustus* (Gould) in Northwest Australia. Ecology *46*, 473–479 (1965).

Edelman, I.S., R. Bogoroch, and G.A. Porter: On the mechanism of action of aldosterone on sodium transport. The role of protein synthesis. Proc. nat. Acad. Sci. (Wash.). *50*, 1169-1177 (1963).

– H.L. Young, and J.B. Harris: Effects of corticosteroids on electrolyte metabolism during osmoregulation in teleosts. Amer. J. Physiol. *199*, 666-670 (1960).

Elizondo, R.S. and S.J. LeBrie: Adrenal-renal function in water snakes, *Natrix cyclopion*. Amer. J. Physiol. *217*, 419–425 (1969).

Elkington, J.R. and T.S. Danowski: The body fluids. Basic Physiology and Practical Therapeutics. pp. 77. Baltimore: Williams and Wilkins 1955.

Elliott, A.B.: Effect of adrenaline on water uptake in *Bufo*. J. Physiol. (Lond.) *197*, 87-88P (1968).

Emmart, E.W., G.E. Pickford, and A.E. Wilhelmi: Localization of prolactin within the pituitary of a cyprinodont fish, *Fundulus heteroclitus* (Linnaeus), by specific fluorescent antiovine prolactin globulin. Gen. comp. Endocr. *7*, 571–583 (1966).

– A.E. Wilhelmi: Immunochemical studies with prolactin-like fractions of fish pituitaries. Gen. comp. Endocr. *11*, 515-527 (1968).

Enami, M.: Caudal neurosecretory system in the eel (*Anguilla japonica*). Gunma J. med. Sci. *4*, 23-26 (1955).

Enemar, A. and B. Hanstrom: The relative sizes of the neural and glandular lobes in some rodents. Lunds Univ. Arsskr, N.F. Avd 2; 52, nr. 15.1–5 (1956).

Ensor, D.M. and J.N. Ball: A bioassay for fish prolactin (Paralactin). Gen. comp. Endocr. *11*, 104-110 (1968a).

– J.N. Ball: Prolactin and fresh water sodium fluxes in *Poecilia latipinna* (Teleostei). J. Endocr. *41*, xvi (1968b).

– D.H. Thomas, and J.G. Phillips: The possible role of the thyroid in extrarenal secretion following a hypertonic saline load in the duck, (*Anas platyrhynchos*). J. Endocr. *46*, x (1970).

Epstein, F.H., A.I. Katz, and G.E. Pickford: Sodium and potassium-activated adenosine triphosphatase of gills: Role in adaptation of teleosts to salt water. Science *156*, 1245-1247 (1967).

Evans, D.H.: Sodium, chloride and water balance of the intertidal teleost, *Xiphister atropurpureus* II. The role of the kidney and gut. J. exp. Biol. *47*, 519–524 (1967 a).

– Sodium, chloride and water balance of the intertidal teleost, *Xiphister atropurpureus*. II. The roles of simple diffusion, exchange diffusion and active transport. J. exp. Biol. *47*, 525–534 (1967 b).

– L.L. Sparks, and J.S. Dixon: The physiology and chemistry of adrenocorticotrophin. In: The Pituitary Gland, Vol. 1, pp. 317–373. Ed. by G.W. Harris and B.T. Donovan, Berkeley and Los Angeles: University of California Press 1966.

Ewer, R.F.: The effect of pituitrin on fluid distribution in *Bufo regularis* Reuss. J. exp. Biol. *29*, 173–177 (1952 a).

– The effects of posterior pituitary extracts on water balance in *Bufo carens* and *Xenopus laevis*. J. exp. Biol. *29*, 429-439 (1952b).

Fagerlund, U.H.M.: Plasma cortisol concentration in relation to stress in adult sockeye salmon during the freshwater stage of their life cycle. Gen. comp. Endocr. *8*, 197–207 (1967).

– J.R. McBride: Supression by dexamethasone of interrenal activity in adult sockeye salmon (*Oncorhynchus nerka*). Gen. comp. Endocr. *12*, 651-657 (1969).

Fanelli, G.M. and L. Goldstein: Ammonia excretion on the neotenous newt *Necturus maculosus* (Rafinesque). Comp. Biochem. Physiol. *13*, 193–204 (1964).

Fanestil, D.D. and I.S. Edelman: Characteristics of the renal nuclear receptors for aldosterone. Proc. nat. Acad. Sci. (Wash.) *56*, 872–879 (1966).

– G.A. Porter, and I.S. Edelman: Aldosterone stimulation of sodium transport. Biochim. biophys. Acta (Amst.) *135*, 74-88 (1967).

264

FÄNGE, R. and K. FUGELLI: Osmoregulation in Chimaeroid fish. Nature (Lond.) *196*, 689 (1962).

– J. KROG, and O. REITE: Blood flow in the avian salt gland studied with polarographic oxygen electrodes Acta. physiol. scand. *58*, 40–47 (1963).

– K. SCHMIDT-NIELSEN, and M. ROBINSON: Control of secretion from the avian salt gland. Amer. J. Physiol. *195*, 321–326 (1958).

FARNER, D.S.: The annual stimulus for migration: experimental and physiological aspects. In: Recent Studies in Avian Biology, pp. 198–237. Ed. by A. WOLFSON, Urbana: University of Illinois 1955.

– B.K. FOLLETT: Light and other environmental factors affecting avian reproduction. J. animal Sci. *25*, 90-118 (1966).

– A. OKSCHE: Neurosecretion in birds. Gen. comp. Endocr. *2*, 113-147 (1962).

FAVRE, L.C.: Recherches sur l'adaptation de Cyprins dorés à des milieux de salinité diverse et sur son déterminisme endocrinien. Diplôme d'Etudes Supérieures, Paris 1960.

FEAKES, M.J., E.P.HODGKIN, R. STRAHAN, and H. WARING: The effects of posterior lobe pituitary extracts on the blood pressure of *Ornithorhynchus* (duck-billed platypus). J. exp. Biol. *27*, 50-58 (1950).

FERGUSON, D.R.: The genetic distribution of vasopressins in the peccary *(Tayassu angulatus)* and warthog *(Phacochoerus aethiopicus)*. Gen. comp. Endocr. *12*, 609–613 (1969).

– H.HELLER: Distribution of neurohypophysial hormones in mammals. J. Physiol. (Lond.) *180*, 846-863 (1965).

FERREIRA, H.G. and M.W. SMITH: Effect of saline environment on sodium transport by the toad colon. J. Physiol. (Lond.) *198*, 329–343 (1968).

FERRERI, E., V.MAZZI, M.SOCINO, and F.SCALENGHE: Sodium and potassium metabolism in the newt *(Triturus cristatus carnifex* Laur.) VI. Effects of cortisol, and cortisol and aldosterone injected simultaneously. Gen. comp. Endocr. *9*, 10-16 (1967).

FIMOGNARI, G.M., G.A.PORTER, and I.S.EDELMAN: The role of the tricarboxylic acid cycle in the action of aldosterone on sodium transport. Biochim. biophys. Acta (Amst.) *135*, 89-99 (1967).

FINDLAY, J.D. and D.ROBERTSHAW: The role of the sympatho-adrenal system in the control of sweating in the ox *(Bos taurus)*. J. Physiol. (Lond.) *179*, 285–297 (1965).

FISHER, D.A.: Norepinephrine inhibition of vasopressin antidiuresis. J. clin. Invest. *47*, 540–547 (1968).

FITZPATRICK, R. J. and P. J. BENTLEY: The assay of neurohypophysial hormones in blood and other body fluids. In: Handbuch d. exper. Pharmacol. Vol. 23, pp. 190–285. Ed. by B. BERDE. Heidelberg and Berlin: Springer-Verlag 1968.

FLEMING, W.R. and F.I.KAMEMOTO: The site of sodium outflux from the gill of *Fundulus kansae*. Comp. Biochem. Physiol. *8*, 263-269 (1963).

– J.G. STANLEY: Effects of rapid changes in salinity on the renal function of a euryhaline teleost. Amer. J. Physiol. *209*, 1025–1030 (1965).

FLETCHER, G.L., I.M. STAINER, and W.N. HOLMES: Sequential changes in the adenosinetriphosphatase activity and the electrolyte excretory capacity of the nasal glands of the duck *(Anas platyrhynchos)* during the period of adaptation to hypertonic saline. J. exp. Biol. *57*, 375-392 (1967).

FOLLETT, B.K.: Mole ratios of the neurohypophysial hormones in the vertebrate neural lobe. Nature (Lond.) *198*, 693-694 (1963).

– Neurohypophysial hormones of marine turtles and of the grass snake. J. Endocr. *39*, 293-294 (1967).

– D.S. FARNER: The effects of daily photoperiod on gonadal growth, neurohypophysial hormone content, and neurosecretion in the hypothalamo-hypophysial system of the Japanese quail *(Coturnix coturnix japonica)*. Gen. comp. Endocr. *7*, 111-124 (1966).

– H. HELLER: The neurohypophysial hormones of lungfishes and amphibians. J. Physiol. (Lond.) *172*, 92-106 (1964a).

– H. HELLER: The neurohypophysial hormones of bony fishes and cyclostomes. J. Physiol. (Lond.) *172*, 72-91 (1964b).

FONTAINE, M.: Du déterminisme physiologique des migrations. Biol. Rev. *29*, 390-418 (1954).

– The hormonal control of water and salt-electrolyte metabolism in fish. Mem. Soc. Endocr. *5*, 69-82 (1956).
– Corpuscules de Stannius et régulation ionique (Ca, K, Na) du milieu intérieur de l'Anguille (*Anguilla anguilla* L.). C.R. Acad. Sci. (Paris) *259*, 875–878 (1964).
– Sur la nature de la fonction endocrine des corpuscules de Stannius. Gen. comp. Endocr. *5*, 678 (1965).
– O. Callamand, et M. Olivereau: Hypophyse et euryhalinité chez l'Anguille. C.R. Acad. Sci. (Paris) *228*, 513–514 (1949).
– J. Hatey: Sur la teneur en 17-OH-corticostéroides du plasma de saumon (*Salmo salar* L.) C.R. Acad. Sci. (Paris) *239*, 319–321 (1954).
– H. Koch: Les variations d'euryhalinité et d'osmorégulation chez les poissons. J. Physiol. (Paris) *42*, 287–318 (1950).
– M. Mazeaud, et F. Mazeaud: L'adrénalinémie du *Salmo salar* L. à quelques étapes de son cycle vital et de ses migrations. C.R. Acad. Sci. (Paris) *256*, 4562–4565 (1963).
Fontaine, Y.A.: Characteristics of the zoological specificity of some protein hormones from the anterior pituitary. Nature (Lond.) *202*, 1296–1298 (1964).
Forster, R.P. and L. Goldstein: Urea synthesis in the lungfish: relative importance of purine and ornithine cycle pathways. Science *153*, 1650-1652 (1966).
Fourman, P.: Parathyroids and tubular function. Mem. Soc. Endocr. *13*, 133-138 (1963).
Fowler, M.A. and I. Chester Jones: The adrenal cortex in the frog *Rana temporaria* and its relation to water and salt electrolyte metabolism. J. Endocr. *13*, vi-vii (1955).
Fregly, M.J. and R.E. Taylor: Effect of hypothyroidism on water and sodium exchange in rats. In: 'Thirst', Proceedings of the 1st Int. Symposium in the Regulation of Body Water, pp. 139–175. London: Pergamon 1964.
Fridberg, G. and H.A. Bern: The urophysis and the caudal neurosecretory system of fishes. Biol. Rev. *43*, 175-199 (1968).
– H.A. Bern, and R.S. Nishioka: The caudal neurosecretory system of the isospondylous teleost, *Albula vulpes*, from different habitats. Gen. comp. Endocr. *6*, 195-212 (1966).
Friedman, M., A. Kaplan, and E. Williams: Studies concerning the site of renin formation in the kidney. II. Absence of renin from glomerular kidneys in marine fish. Proc. Soc. exp. Biol. (N. Y.) *50*, 199-202 (1942).
Fromm, P.O.: Studies on renal and extra-renal excretion in a fresh-water teleost, *Salmo gairdneri*. Comp. Biochem. Physiol. *10*, 121-128 (1963).
Fuhrman, F.A. and H.H. Ussing: A characteristic response of the frog skin potential to neurohypophysial principles and its relations to the transport of sodium and water. J. cell. comp. Physiol. *38*, 109-130 (1951).
Ganong, W.F., E.G. Biglieri, and P.J. Mulrow: Mechanisms regulating adrenocortical secretion of aldosterone and glucocorticoids. Rec. Prog. horm. Res. *32*, 381–420 (1966).
Garcia Romeu, F. and J. Maetz: The mechanism of sodium and chloride uptake by the gills of a fresh-water fish, *Carassius auratus*. J. gen. Physiol. *47*, 1195-1207 (1964).
– R. Motais: Mise en evidence d'échanges Na^+/NH_4^+ chez l'anguille d'eau douce. Comp. Biochem. Physiol. *17*, 1201–1204 (1966).
– A. Salibian, and S. Pezzani-Hernandez: The nature of the *in vivo* sodium and chloride uptake mechanisms through the epithelium of the Chilean frog *Calyptocephalella gayi* (Dum. et Bibr., 1841). Exchanges of hydrogen against sodium and bicarbonate against chloride. J. gen. Physiol. *53*, 816–835 (1969).
Gerard, P.: Appareil excreteur. In: Traité de Zool. Vol. 12, pp. 974–1043. Ed. by P.P. Grasse. Paris: Masson 1954.
Geschwind, I.I.: Molecular variation and possible lines of evolution of peptide and protein hormones. Amer. Zool. *7*, 89–108 (1967).
– K. Lederis, H.A. Bern, and R.S. Nishioka: Purification of bladder-contracting principle from the urophysis of the teleost *Gillichthys mirabilis*. Amer. Zool. *8*, 758 (1968).
Ginsburg, M.: Production, release, transportation and elimination of the neurohypophysial hormones. In: Handbuch d. exper. Pharmacol., Vol. 23, pp. 286–371, Ed. by B. Berde. Heidelberg and Berlin: Springer-Verlag 1968.
– M. Ireland: The role of neurophysin in the transport and release of neurohypophysial hormones. J. Endocr. *35*, 289–298 (1966).

266

GISLEN, T. and H. KAURI: Zoogeography of the Swedish amphibians and reptiles with notes on their growth and ecology. Acta Vertebrat. *1*, 197–397 (1959).

GIST, D. H. and R. DEROOS: Corticoids of the alligator adrenal gland and the effects of ACTH and progesterone on their production *in vitro*. Gen. comp. Endocr. *7*, 304–313 (1966).

GODET, R.: Le problème hydrique et son contrôle hypophysaire chez le protoptère. Annales de la Faculté des Sciences de l'Université de Dakar. *6*, 183–201 (1961).

GOLDSTEIN, L.: Comparative physiology of ammonia excretion. Proc. Int. Union. Physiol. Sci. *6*, 25–26, Washington, 1968.

– R. P. FORSTER: Source of ammonia excreted by the gills of the marine teleost, *Myoxocephalus scorpius*. Amer. J. Physiol. *200*, 1116–1118 (1961).

– S. C. HARTMAN, and R. P. FORSTER: On the origin of trimethylamine oxide in the spiny dogfish, *Squalus acanthias*. Comp. Biochem. Physiol. *21*, 719–722 (1967).

– P. A. JANSSENS, and R. P. FORSTER: Lungfish *Neoceratodus forsteri:* activities of ornithine-urea cycle enzymes. Science *157*, 316–317 (1967).

GORBMAN, A.: Thyroid Hormones. In: Comparative Endocrinology Vol. 1, pp. 291–324. Ed. by U.S. VON EULER and H. HELLER. New York: Academic Press 1963.

– H. A. BERN: A Textbook of Comparative Endocrinology. New York: Wiley 1962.

GORDON, M. S.: Osmotic regulation in the green toad (*Bufo viridis*). J. exp. Biol. *39*, 261–270 (1962).

– Intracellular osmoregulation in skeletal muscle during salinity adaptation in two species of toads. Biol. Bull. (Woods Hole) *128*, 218–229 (1965).

– K. SCHMIDT-NIELSEN, and H. M. KELLY: Osmotic regulation on the crab-eating frog (*Rana cancrivora*). J. exp. Biol. *38*, 659–678 (1961).

– V. A. TUCKER: Osmotic regulation in the tadpoles of the crab-eating frog *(Rana cancrivora)*. J. exp. Biol. *42*, 437–445 (1965).

GOTTSCHALK, C. W. and M. MYLLE: Micropuncture study of the mammalian urinary concentrating mechanism: evidence for the countercurrent hypothesis. Amer. J. Physiol. *196*, 927–936 (1959).

GRANT, J. K.: Studies on the biogenesis of the adrenal steroids. Brit. med. Bull. *18*, 99–105 (1962).

GRANT, W. C. and J. A. GRANT: Water drive studies in hypophysectomized efts of *Diemyctilus viridescens:* Part. I. The Role of the lactogenic hormone. Biol. Bull. (Woods Hole) *114*, 1–9 (1958).

G. E. PICKFORD: Presence of the red eft water-drive factor prolactin in the pituitaries of teleosts. Biol. Bull. (Woods Hole) *116*, 429–435 (1959).

GREEN, E. E. and J. F. PERDUE: Membranes as expression of repeating units. Proc. nat. Acad. Sci. (Wash.) *55*, 1295–1302 (1966).

GREEN. J. D.: The comparative anatomy of the hypophysis with reference to its blood supply and innervation. Amer. J. Anat. 88, 225–312 (1951).

GREEN, K. and A. J. MATTY: Action of thyroxine on active transport in isolated membranes of *Bufo bufo*. Gen. comp. Endocr. *3*, 244–252 (1963)

GREEN, L. M. A.: Sweat glands in the skin of the quokka of Western Australia. Aust. J. exp. Biol. *39*, 481–486 (1961).

GREENWALD, L., W. B. STONE, and T. J. CADE: Physiological adjustments of the budgerygah (*Melopsittacus undulatus*) to dehydrating conditions. Comp. Biochem. Physiol. *22*, 91–100 (1967).

GREEP, R. O.: Parathyroid hormone. In: Comparative Endocrinology, Vol. 1, pp. 325–370 Ed. by U.S. VON EULER and H. HELLER. New York: Academic Press 1963.

GRENOT, C.: Observations physio-écologiques sur la régulation thermique chez le lézard Agamide *Uromastyx acanthinurus*, BELL. Bull. Soc. Zool. France *92*, 51–66 (1967).

– Sur l'excrétion nasale de sels chez le lézard saharien: *Uromastyx acanthinurus.* C.R. Acad. Sci. (Paris) *266*, 1871–1874 (1968).

GRIFFITHS, M.: Rate of growth and intake of milk in a suckling echidna. Comp. Biochem. Physiol. *16*, 383–392 (1965).

GRIMM, A. S., M. J. O'HALLORAN, and D. R. IDLER: Stimulation of sodium transport across the isolated toad bladder by 1α-hydroxycorticosterone from an Elasmobranch. J. Fish. Res. Bd Can. *26*, 1823–1835 (1969).

HAINES, H.: Salt tolerance and water requirements in the salt water harvest mouse. Physiol. Zool. *37*, 266–272 (1964).

HANDLER, J. S., R. BENSINGER, and J. ORLOFF: Effect of adrenergic agents on toad bladder response to ADH, 3', 5'-AMP, and theophylline. Amer. J. Physiol. *215*, 1024–1031 (1968).

– A. S. PRESTON, and J. ORLOFF: Effect of adrenal steroid hormones on the response of the toad's urinary bladder to vasopressin. J. clin. Invest. *48*, 823–833 (1969).

HANE, S. and O. H. ROBERTSON: Changes in the plasma 17-hydroxycorticosteroids accompanying sexual maturation and spawning of the Pacific salmon *(Oncorhynchus tschawytscha)* and rainbow trout *(salmo gairdnerii)*. Proc. nat. Acad. Sci. *45*, 886–893 (1959).

– O. H. ROBERTSON, B. C. WEXLER, and M. A. KRUPP: Adrenocortical response to stress and ACTH in pacific salmon *(Oncorhynchus tshawytscha)* and steelhead trout *(Salmo gairdnerii)* at successive stages in the sexual cycle. Endocrinology *78*, 791–800 (1966).

HANKE, W., K. BERGERHOFF, and D. K. O. CHAN: Histological observations on pituitary ACTH-cells, adrenal cortex, and the Corpuscles of Stannius of the European eel *(Anguilla anguilla* L.). Gen. comp. Endocr. *9*, 64–75 (1967).

HARDISTY, M. W.: Some aspects of osmotic regulation in lampreys. J. exp. Biol. *33*, 431–447 (1956).

HARGITAY, B. and W. KUHN: Das Multiplikationsprinzip als Grundlage der Harnkonzentrierung in der Niere. Zeitschr. f. Elektrochem. u. angewandte Physik. Che. *55*, 539–558 (1951).

HARRIS, E. J.: Transport and accumulation in biological systems. London: Butterworths 1960.

HARRIS, I.: The chemistry of the intermediate lobe hormones. In: The Pituitary Gland, Vol. 3, pp. 33–40. Ed. by G. W. HARRIS and B. T. DONOVAN. Berkeley and Los Angeles: University of California Press 1966.

HART, W. M. and H. E. ESSEX: Water metabolism of the chicken with special reference to the role of the cloaca. Amer. J. Physiol. *136*, 657–658 (1942).

HARTMAN, F. A. and K. A. BROWNELL: The Adrenal Gland. London: Kimpton 1949.

– L. A. LEWIS, K. A. BROWNELL, C. A. ANGERER, and F. F. SHELDEN: Effect of interrenalectomy on some blood constituents in the skate. Physiol Zool. *17*, 228–238 (1944).

HASAN, S. H. and H. HELLER: The clearance of neurohypophysial hormones from the circulation of non-mammalian vertebrates. Brit. J. Pharmacol. *33*, 523–530 (1968).

HAYS, R. M. and N. FRANKI: The role of water diffusion in the action of vasopressin. J. Membrane Biol. *2*, 263–276 (1970).

HELLER, H.: Über die Einwirkung von Hypophysenhinterlappenextrakten auf dem Wasserhaushalt des Frosches. Arch. expt. Path. Pharmak. *157*, 298–322 (1930).

– Differentiation of an (amphibian) water balance principle from the antidiuretic principle of the posterior pituitary gland. J. Physiol. (Lond.) *100*, 125–141 (1941).

– The posterior pituitary principles of a species of reptile *(Tripodonotus natrix)* with some remarks on the comparative physiology of the posterior pituitary gland generally. J. Physiol. (Lond.) *101*, 317–326 (1942).

– Neurohypophyseal hormones. In: Comparative Endocrinology, Vol. 1, pp. 26–80. Ed. by U. S. VON EULER and H. HELLER. New York: Academic Press 1963.

– P. J. BENTLEY: Comparative aspects of the actions of neurohypophysial hormones on water and salt metabolism. Mem. Soc. Endocr. *13*, 59–65 (1963).

– P. J. BENTLEY: Phylogenetic distribution of the effects of neurohypophysial hormones on water and sodium metabolism. Gen. comp. Endocr. *5*, 96–108 (1965).

– B. T. PICKERING: Neurohypophysial hormones of non-mammalian vertebrates. J. Physiol. (Lond.) *155*, 98–114 (1961).

– B. T. PICKERING, J. MAETZ, and F. MOREL: Pharmacological characterization of the oxytocic peptides in the pituitary of a marine teleost fish *(Pollachius virens)*. Nature (Lond.) *191*, 670–671 (1961).

HELLER, J.: The physiology of the antidiuretic hormone. VIII. The antidiuretic activity of the plasma of the mouse, guinea pig, cat, rabbit and dog. Physiol. bohemoslov. *10*, 167–172 (1961).

268

HENDERSON, I.W. and I. CHESTER JONES: Endocrine influences on the net extrarenal sodium fluxes of sodium and potassium in the European eel (*Anguilla anguilla* L.). J. Endocr. *37*, 319–325 (1967).

HERRERA, F.C.: Effect of insulin on short-circuit current and sodium transport across toad urinary bladder. Amer. J. Physiol. *209*, 819–824 (1965).

HICKMAN, C.P.: The osmoregulatory role of the thyroid gland in the starry flounder, *Platichthys stellatus*. Canad. J. Zool. *37*, 997–1060 (1959).

– Studies on renal function in freshwater teleost fish. Trans. roy. Soc. Can. *3*, 213–236 (1965).

HILL, C.W. and P.O.FROMM: Response of the interrenal gland of rainbow trout (*Salmo gairdnerii*) to stress. Gen. comp. Endocr. *11*, 69–77 (1968).

HIRANO, T.: Further studies on the neurohypophysial hormones in the avian median eminence. Endocr. japon. *11*, 87–95 (1964).

– Neurohypophysial hormones in the median eminence of the bullfrog, turtle and duck. Endocr. japon. *13*, 59–74 (1966).

– Effect of hypophysectomy on water transport in isolated intestine of the eel, *Anguilla japonica*. Proc. Jap. Acad. *43*, 793–796 (1967).

– Effects of hypophysectomy and salinity change on plasma cortisol concentration in the Japanese eel, *Anguilla japonica*. Endocr. japon. *16*, 557–560 (1969).

– S. UTIDA: Effects of ACTH and cortisol on water movement in isolated intestine of the eel, *Anguilla japonica*. Gen. comp. Endocr. *11*, 373–380 (1968).

HOAR, W.S.: Hormones in fish. Publ. Ont. Fisheries Research Lab. *71*, 1–51 (1951).

– Endocrine factors in the ecological adaptation of fishes. In: Comparative Endocrinology, pp. 1–23 Ed. by A. GORBMAN. New York: Wiley 1959.

– The endocrine regulation of migratory behaviour in anadromous teleosts. Proc. XVI int. Cong. Zool. *3*, 14–20 (1963).

– Hormonal activities of the pars distalis in cyclostomes, fish and Amphibia. In: The Pituitary Gland, Vol. 1, pp. 242–294. Ed. by G.W. HARRIS and B.T. DONOVAN. Berkeley and Los Angeles: University of California Press 1966.

HOHN, E.O.: Endocrine glands, thymus and pineal body. In: Biology and Comparative Physiology of Birds, Vol. 2, pp. 87–114. Ed. by A.J. MARSHALL. New York: Academic Press 1961.

HOLLANDS, B.C.S. and M.W. SMITH: Phosphatases of the goldfish intestine. J. Physiol. (Lond.) *175*, 31–37 (1964).

HOLMES, W.N.: Studies on the hormonal control of sodium metabolism in the rainbow trout (*Salmo gairdneri*). Acta endocr. (Kbh.) *31*, 587–602 (1959).

– B.W. ADAMS: Effects of adrenocortical and neurohypophysial hormones on the renal excretory pattern in the water loaded duck (*Anas platyrhynchos*). Endocrinology *73*, 5–10 (1963).

– D.G. BUTLER, and J.G. PHILLIPS: Observations on the effect of maintaining glaucous-winged gulls (*Larus glaucescens*) on fresh water and sea water for long periods. J. Endocr. *23*, 53–61 (1961).

– G.L.FLETCHER, and D.J.STEWART: The patterns of renal electrolyte excretion in the duck (*Anas platyrhynchos*) maintained on freshwater and on hypertonic saline. J. exp. Biol. *48*, 487–508 (1968),

– R.L.McBEAN: Studies on the glomerular filtration rate of rainbow trout (*Salmo gairdneri*). J. exp. Biol. *40*, 335–341 (1963).

– R.L.McBEAN: Some aspects of the electrolyte excretion in the green turtle, *Chelonia mydas mydas*. J. exp. Biol. *41*, 81–90 (1964).

– J.G.PHILLIPS: Adrenocortical hormones and electrolyte metabolism of birds. Proc. Sec. Int. Cong. Endocrin. Experta Medica Series *83*, 158–161 (1964).

– J.G.PHILLIPS, and D.G.BUTLER: The effect of adrenocortical steroids on the renal and extra-renal responses of the domestic duck (*Anas platyrhynchos*) after hypertonic saline loading. Endocrinology *69*, 483–495 (1961).

– J.G. PHILLIPS, and I. CHESTER JONES: Adrenocortical factors associated with adaptation of vertebrates to marine environments. Rec. Prog. Hormone Res. *19*, 619–672 (1963).

HORNBY, R. and S. THOMAS: Effect of prolonged saline-exposure on sodium transport across frog skin. J. Physiol. (Lond.) *200*, 321–344 (1969).

HOUSE, C.R.: Osmotic regulation in the brackish water teleost, *Blennius pholis*. J. exp. Biol. *40*, 87–104 (1963).

HOUSE, C.R. and K. GREEN: Sodium and water transport across isolated intestine of a marine teleost. Nature (Lond.) *199*, 1293–1294 (1963).

HOUSTON, A.H.: Osmoregulatory adaptation of steelhead trout (*Salmo gairdneri*, Richardson) to sea water. Can. J. Zool. *37*, 729–748 (1959).

HOWE, A. and P.A. JEWELL: Effects of water deprivation upon the neurosecretory material of the desert rat *(Meriones meriones)* compared with the laboratory rat. J. Endocr. *19*, 118–124 (1959).

HOWES, N.H.: The response of the water regulating mechanism of developmental stages of the common toad *Bufo bufo bufo* (L.) to treatment with extracts of the posterior lobe of the pituitary body. J. exp. Biol. *17*, 128–138 (1940).

HYMER, W.C. and W.H. McSHAN: Isolation of rat pituitary granules and the study of their biochemical properties and hormonal activities. J. Cell Biol. *17*, 67–86 (1963).

IDLER, D.R., M.J. O'HALLORAN, and D.A. HORNE: Interrenalectomy and hypophysectomy in relation to liver glycogen levels in the skate, (*Raja erinacea*). Gen. comp. Endocr. *13*, 303–306 (1969).

– B. TRUSCOTT: 1 α-hydroxycorticosterone from cartilaginous fish: a new adrenal steroid in blood. J. Fish. Res. Bd Can. *23*, 615–619 (1966).

– B. TRUSCOTT: 1α-hydroxycorticosterone and testosterone in body fluids of a cartilaginous fish *(Raja radiata)*. J. Endocr. *42*, 165–166 (1968).

ILETT, K.F.: Corticosteroids in the adrenal venous and heart blood of the quokka, *Setonix brachyurus* (Marsupialia: Macropodidae). Gen. comp. Endocr. *13*, 218–221 (1969).

IRELAND, M.P.: Effect of urophysectomy in *Gasterosteus aculeatus* on survival in fresh water and sea water. J. Endocr. *43*, 113–134 (1969).

ISHII, S., T. HIRANO, and H. KOBAYASHI: Neurohypophysial hormones in the avian median eminence and pars nervosa. Gen. comp. Endocr. *2*, 433–440 (1962).

ISLER, H., C.P. LEBLOND, and A.A. AXELRAD: Mechanism of the thyroid stimulation produced by sodium chloride in the mouse. Endocrinology *62*, 159–172 (1958).

JACKSON, D.C. and K. SCHMIDT-NIELSEN: Countercurrent heat exchanges in the respiratory passages. Proc. nat. Acad. Sci. (Wash.) *51*, 1192–1197 (1964).

JAMPOL, L.M. and F.H. EPSTEIN: Sodium-potassium-activated adenosinetriphosphatase and osmotic regulation by fishes. Amer. J. Physiol. *218*, 607–611 (1970).

JANSSENS, P.A.: The metabolism of the aestivating African lungfish. Comp. Biochem. Physiol. *11*, 105–117 (1964).

– G.P. VINSON, I. CHESTER JONES, and W. MOSLEY: Amphibian characteristics of the adrenal cortex of the African lungfish (*Protopterus* sp.). J. Endocr. *32*, 373–382 (1965).

JARD, S.: Etude des effets de la vasotocine sur l'excrétion de l'eau et des électrolytes par la rein de la grenouille *Rana esculenta* L.: analyse à l'aide d'analogues artificiels de l'hormone naturelle des caractères structuraux requis pour son activité biologique. J. Physiol. (Paris) Supp. 15. *58*, 1–124 (1966).

– J. MAETZ, et F. MOREL: Action de quelques analogues de l'ocytocine sur différents récepteurs intervenant dans l'osmorégulation de *Rana esculenta*. C.R. Acad. Sci. (Paris) *251*, 788–790 (1960).

– F. MOREL: Actions of vasotocin and some of its analogues on salt and water excretion by the frog. Amer. J. Physiol. *204*, 222–226 (1963).

– F. BASTIDE, et F. MOREL: Analyse de la relation 'Dose-effet Biologique' pour l'action de l'ocytocine et de la noradrénaline sur la peau et la vessie de la grenouille. Biochem. biophys. Acta (Amst.) *150*, 124–130 (1968).

JEWELL, P.A. and E.B. VERNEY: An experimental attempt to determine the site of neurohypophysial osmoreceptors in the dog. Phil. Trans. roy. Soc. B. *240*, 197–324 (1957).

JOHNSTON, C.I., J.O. DAVIS, F.S. WRIGHT, and S.S. HOWARDS: Effects of renin and ACTH on adrenal steroid production in the American bullfrog. Amer. J. Physiol. *213*, 393–399 (1967).

JONES, C.W. and B.T. PICKERING: Comparison of the effects of water deprivation and sodium chloride imbibition on the hormone content of the neurohypophysis of the rat. J. Physiol. (Lond.) *203*, 449–458 (1969).

JORGENSEN, C.B.: Influence of adenohypophysectomy on the transfer of salt across the frog skin. Nature (Lond.) *160*, 872 (1947).

– L.O. LARSEN: Effect of corticotrophin and growth hormone on survival in hypophysectomized toads. Proc. Soc. exp. Biol. (N.Y.) *113*, 94–96 (1963).

– L.O. LARSEN: Neuroendocrine mechanisms in lower vertebrates. In: Neuroendocrinology, Vol. 2, pp. 485–528. Ed. by L. MARTINI W.F. GANONG. New York: Academic Press 1967.

– H. LEVI, and H.H.USSING: On the influence of neurohypophysial principles on sodium metabolism in the axolotl *Ambystoma mexicanum*. Acta physiol. scand. *12*, 350–371 (1946).

– H. LEVI, and K. ZERAHN: On active uptake of sodium and chloride ions in anurans. Acta physiol. scand. *30*, 178–190 (1954).

– P. ROSENKILDE: Relative effectiveness of dehydration and neurohypophysial extracts in enhancing water absorption in toads and frogs. Biol. Bull. (Woods Hole) *110*, 306–309 (1956 a).

– P. ROSENKILDE: On regulation of concentration and content of chloride in goldfish. Biol. Bull. (Woods Hole) *110*, 300–305 (1956 b).

– P. ROSENKILDE: Chloride balance in hypophysectomized frogs. Endocrinology *60*, 219–224 (1957).

– P. ROSENKILDE, and K.G. WINGSTRAND: Role of the preoptic-neurohypophysial system in the water economy of the toad, *Bufo bufo* L.. Gen. comp. Endocr. *12*, 91–98 (1969).

– K.G. WINGSTRAND, and P. ROSENKILDE: Neurohypophysis and water metabolism in the toad *Bufo bufo* (L.). Endocrinology *59*, 601–610 (1956).

JUNQUEIRA, L.C.U., G. MALNIC, and C. MONGE: Reabsorptive function of the ophidian cloaca and large intestine. Physiol. Zool. *29*, 151–159 (1966).

KAMIYA, M: Changes in ion and water transport in isolated gills of the cultured eel during the course of salt adaptation. Ann. Zool. Japon. *40*, 123–129 (1967).

– S. UTIDA: Changes in activity of sodium-potassium-activated adenosinetriphosphatase in gills during adaptation of the Japanese eel to sea water. Comp. Biochem. Physiol. *26*, 675–685 (1968).

– S. UTIDA: Sodium-potassium-activated adenosine triphosphatase activity in gills of freshwater, marine and euryhaline teleosts. Comp. Biochem. Physiol. *31*, 671–674 (1969).

KAWADA, J., R.E. TAYLOR, and S.B. BARKER: Measurement of Na-K ATPase in the separated epidermis of *Rana catesbeiana* frogs and tadpoles. Comp. Biochem. Physiol. *30*, 965–975 (1969).

KEYNES, R.D. and R.C. SWAN: The effect of external sodium concentration on the sodium fluxes in frog muscle. J. Physiol. (Lond.) *147*, 591–625 (1959).

KEYS, A.: Chloride and water secretion and absorption by the gills of the eel. Z. vergl. Physiol. *15*, 364–388 (1931).

– The mechanism of adaptation to varying salinity in the common eel and the general problem of osmotic regulation in fishes. Proc. roy. Soc. B. *112*, 184–199 (1933).

– J.B. BATEMAN: Branchial responses to adrenaline and pitressin in the eel. Biol. Bull. (Woods Hole) *63*, 327–336 (1932).

– E.N. WILLMER: 'Chloride secreting cells' in the gills of fishes, with special reference to the common eel. J. Physiol. (Lond.) *76*, 368–378 (1932).

KHALIL, F. and G. HAGGAG: Ureotelism and uricotelism in reptiles. J. exp. Zool. *130*, 423–432 (1955).

– J. TAUFIC: Seasonal changes in the hypothalamo-hypophysial neurosecretory system of desert rodents. J. Endocr. *29*, 251–254 (1964).

KING, J.R. and D.S. FARNER: Terrestrial animals in humid heat: Birds. In: Handbook of Physiology. Sec. 4, Adaptation to the environment. pp. 603–624. Washington: Amer. Physiol. Soc. 1964.

KINNEAR, J.E., K.G. PUROHIT, and A.R. MAIN: The ability of the Tammar wallaby *Macropus eugenii*, Marsupialia) to drink sea water. Comp. Biochem. Physiol. *25*, 761–782 (1968).

KINSON, G. A. and B. SINGER: Sensitivity to angiotensin and adrenocorticotrophic hormone on the sodium deficient rat. Endocrinology *83*, 1108–1116 (1968).

KIRSTEN, E., R. KIRSTEN, A. LEAF, and G.W.G. SHARP: Increased activity of enzymes of the Tricarboxylic acid cycle in response to aldosterone in toad bladder. Pflugers Archiv. Ges. Physiol. *300*, 213–225 (1968).

KOBAYASHI, H., H.A.BERN, R.S.NISHIOKA, and Y.HYODO: The hypothalamo-hypophysial neurosecretory system of the parakeet, *Melopsittacus undulatus*. Gen. comp. Endocr. *1*, 545–564 (1961).

KORR, I.M.: The osmotic function of the chicken kidney. J. cell. comp. Physiol. *13*, 175–194 (1939).

KRAG, B. and E. SKADHAUGE: Renal salt and water excretion in the budgerygah (*Melopsittacus undulatus*). Abstract of 5th conference European Endocrinologists Utrecht 1969. E. SKADHAUGE: Personal communication in press 1970.

KRAULIS, I. and M.K.BIRMINGHAM: The conversion of 11 β-hydroxyprogesterone to corticosterone and other steroid fractions by rat and frog adrenal glands. Acta endocr. (Kbh.) *47*, 76–84 (1964).

KROGH, A.: Osmotic regulation in the frog *(R. esculenta)* by active absorption of chloride ions. Skand. Arch. Physiol. *76*, 60–74 (1937).

– Osmotic Regulation in Aquatic Animals. Cambridge: University Press 1939.

KRUHOFFER, P.: Handling of alkali metal ions by the kidney. In: Handbuch d. exper. Pharmakol. Vol. 13, pp. 233–423. Heidelberg and Berlin: Springer-Verlag 1960.

– J.H. THAYSEN: Intestinal absorption of alkali metal ions. In: Handbuch d. exper. Pharmakol. Vol. 13, pp. 508–515. Heidelberg and Berlin: Springer-Verlag 1960.

LACANILAO, F.: Teleostean urophysis: stimulation of water movement across the bladder of the toad *Bufo marinus*. Science *163*, 1326–1327 (1969).

LAHLOU, B.: Mise en évidence d'un 'recrutement glomerulaire' dans le rein des Téléostéens d'après la mesure du Tm glucose. C.R. Acad. Sci. (Paris) *262*, 1356–1358 (1966).

– Excrétion rénale chez un poisson euryhalin, le flet (*Platichthys flesus* L.): caractéristiques de l'urine normale en eau douce et en eau de mer et effets des changements de milieu. Comp. Biochem. Physiol. *20*, 925–938 (1967).

– A.GIORDAN: Le contrôle hormonal des échanges et de la balance de l'eau chez le Téléostéen d'eau douce *Carassius auratus*, intact et hypophysectomisé. Gen. comp. Endocr. *14*, 491–509 (1970).

– I.W.HENDERSON, and W.H.SAWYER: Sodium exchanges in goldfish (*Carassius auratus* L.) adapted to a hypertonic saline solution. Comp. Biochem. Physiol. *28*, 1427–1433 (1969a).

– I.W.HENDERSON, and W.H.SAWYER: Renal adaptations by *Opsanus tau*, a euryhaline aglomerular teleost, to dilute media. Amer. J. Physiol. *216*, 1266–1272 (1969b).

– W.H.SAWYER: Sodium exchanges in the toadfish, *Opsanus tau*, a euryhaline aglomerular teleost. Amer. J. Physiol. *216*, 1273–1278 (1969).

LAM, T. J. and J.F. LEATHERLAND: Effect of prolactin on freshwater survival of the marine form (*Trachurus*) of the Threespine Stickleback, *Gasterosteus aculeatus*, in the early winter. Gen. comp. Endocr. *12*, 385–394 (1969).

– W.S.HOAR: Seasonal effects of prolactin on freshwater osmoregulation of the marine form *(Trachurus)* of the stickleback *Gasterosteus aculeatus*. Can. J. Zool. *45*, 509-516 (1967).

LANDAU, R.L., D.M. BERGENSTAL, K. LUGIBIHL, and M.E. KASCHT: The metabolic effects of progesterone in man. J. clin. Endocr. *15*, 1194–1215 (1955).

LANDIS, E.M. and J.R. PAPPENHEIMER: Exchange of substances through the capillary walls. In: Handbook of Physiology Sect. 2. Circulation, Vol. 2, pp. 961–1034. Washington: Amer. Physiol. Soc. 1963.

LANGE, R. and K. FUGELLI: The osmotic adjustment in the euryhaline teleosts, the flounder *Pleuronectes flesus* L. and the three-spined stickleback *Gasterosteus aculeatus* L.. Comp. Biochem. Physiol. *15*, 283–292 (1965).

LAPOINTE, J. A.: Effects of ovarian steroids and neurohypophysial hormones on the oviduct of the viviparous lizard, *Klauberina riversiana*. J. Endocr. *43*, 197–205 (1969).

LARSEN, L. O.: Effects of hypophysectomy before and during sexual maturation in the cyclostome, *Lampetra fluviatilis* (L.) Gray. Gen. comp. Endocr. *12*, 200–208 (1969).

Leaf, A.: The mechanism of the asymmetrical distribution of endogenous lactate about the isolated toad bladder. J. cell. comp. Physiol. *54*, 103–108 (1959).

– J. Anderson, and L.B. Page: Active sodium transport by the isolated toad bladder. J. gen. Physiol. *41*, 657–668 (1958).

LeBrie, S.J. and R.S. Elizondo: Saline loading and aldosterone in water snakes *Natrix cyclopion*. Amer. J. Physiol. *217*, 426–430 (1969).

– I.D.W. Sutherland: Renal function in water snakes. Amer. J. Physiol. *203*, 995–1000 (1962).

Lederis, K.: Fine structure and hormone content of the hypothalamoneurohypophysial system of the rainbow trout (*Salmo irideus*) exposed to sea water. Gen. comp. Endocr. *4*, 638–661 (1964).

– Teleostean urophysis: Stimulation of contractions of bladder of the trout *Salmo gairdnerii*. Science *163*, 1327–1328 (1969).

Lee, A.K. and E.H. Mercer: Cocoon surrounding desert-dwelling frogs. Science *157*, 87–88 (1967).

Legait, H.: Contribution à l'étude morphologique et expérimentale du système hypothalamo-neurohypophysaire de la poule Rhode Island. Thèse, Louvain-Nancy 1959.

Leloup, J.: Absorption branchiale des iodures et son contrôle endocrinien chez l'Anguille. J. Physiol. (Paris) *58*, 560 (1966).

Leloup-Hatey, J.: Influence de l'hypophyse sur le taux des corticostéroides du plasma de l'Anguille (*Anguilla anguilla* L.). J. Physiol. (Paris) *53*, 405–406 (1961).

– Corpuscules de Stannius et équilibre minéral chez l'Anguille (*Anguilla anguilla* L.) J. Physiol. (Paris) *56*, 595–596 (1964).

Levinsky, N.G. and W.H. Sawyer: Influence of the adenohypophysis on the frog water-balance response. Endocrinology *51*, 110–116 (1952).

– W.H. Sawyer: Significance of neurohypophysis in regulation of fluid balance in the frog. Proc. Soc. exp. Biol. (N.Y.) *82*, 272–274 (1953).

Li, C.H., W.K. Liu, and J.S. Dixon: Human pituitary growth hormone. XII. The amino acid sequence of the hormone. J. Am. chem. Soc. *88*, 2050–2051 (1966).

Licht, P. and S.D. Bradshaw: A demonstration of corticotropic activity and its distribution in the Pars Distalis of the reptile. Gen. comp. Endocr. *13*, 226–235 (1969).

Lovatt Evans, C., D.F.G. Smith, and H. Weil-Malherbe: The relationship between sweating the catecholamine content of the blood in the horse. J. Physiol. (Lond.) *132*, 542–552 (1956).

Lynn, W.G., J.J. McCormick, and J.C. Gregorek: Environmental temperature and thyroid function in the lizard, *Anolis carolinensis*. Gen. comp. Endocr. *5*, 587–595 (1965).

Macchi, I.A. and J.G. Phillips: *In vitro* effect of adrenocorticotropin on corticoid secretion in the turtle, snake, and bullfrog. Gen. comp. Endocr. *6*, 170–182 (1966).

– J.G. Phillips, P. Brown, and M. Yasuna: Relationship between nasal gland function and plasma corticoid levels in birds. Fed. Proc. *24*, 575 (1965).

MacFarlane, W.V.: Terrestrial animals in dry heat: Ungulates. In: Handbook of Physiology. Sec. 5, Adaptation to the environment. pp. 509–539. Washington: Amer. Physiol. Soc. 1964.

MacIntyre, I.: Calcitonin: a review of its discovery and an account of purification and action. Proc. roy. Soc. B *170*, 49–60 (1968).

MacMillen, R.E. and A.K. Lee: Water metabolism of Australian hopping mice. Comp. Biochem. Physiol. *28*, 493–514 (1969).

Maetz, J.: Le contrôle endocrinien du transport actif de sodium à travers la peau de grenouille. 'la Méthode des Indicateurs Nucléaires dans l'Etude des transports Actifs d'Ions.' 185–196 Ed. by J. Coursaget. London: Pergamon Press 1959.

– Physiological aspects of neurohypophysial function in fishes with some reference to the Amphibia. Symp. Zool. Soc. Lond. No. 9, 107–140 (1963).

– Salt and water metabolism. In: Perspectives in Endocrinology, pp. 47–162, Ed. by E.J.W. Barrington and C.B. Jorgensen. London: Academic Press 1968.

– Observations of the role of the pituitary-interrenal axis in the ion regulation of the eel and other teleosts. (Symposium on Comp. Endocrinology New Delhi, 1967). Gen. comp. Endocr. Suppl. 2 pp. 299–316 (1969 a).

- Sea water teleosts: evidence for a sodium-potassium exchange in the branchial sodium excreting pump. Science *166*, 613–615 (1969b).
- Mechanisms of salt and water transfer across membranes in teleosts in relation to the aquatic environment. Mem. Soc. Endocr. *18*, 3–28 (1970).
- J. BOURGUET, et B. LAHLOUH: Action de l'urohypophyse sur les échanges de sodium (étudiés à l'aide ^{24}Na) et sur l'excrétion urinaire du téléostéen *Carassius auratus*. J. Physiol. (Paris) *55*, 159–160 (1963).
- J. BOURGUET, B. LAHLOUH, et J. HOURDRY: Peptides neurohypophysaires et osmorégulation chez *Carassius auratus*. Gen. comp. Endocr. *4*, 508–522 (1964).
- F. GARCIA ROMEU: The mechanism of sodium and water uptake by the gills of a fresh water fish, *Carassius auratus*. II. Evidence for NH^+_4/Na^+ and HCO_3^-/Cl^- exchanges. J. gen. Physiol. *47*, 1209–1227 (1964).
- B. LAHLOU: Les échanges de sodium et de chlore chez un Elasmobranche, *Scyliorhinus*, mesurés à l'aide des isotopes ^{24}Na et ^{36}Cl. J. Physiol. (Paris) *58*, 249 (1966).
- N. MAYER, et M. M. CHARTIER-BARADUC: La balance minérale du sodium chez *Anguilla anguilla* en eau de mer, en eau douce et au cours de transfert d'un milieu à l'autre: effets de l'hypophysectomie et de la prolactine. Gen. comp. Endocr. *8*, 177–188 (1967a).
- R. MOTAIS, and N. MAYER: Isotopic kinetic studies on the endocrine control of teleostean ion regulation. (Int. Cong. Endocrinol. Mexico City 1968). Exerpta Medica *184*, 187–193 (1969a).
- J. NIBELLE, M. BORNANCIN, et R. MOTAIS: Action sur l'osmorégulation de l'anguille de divers antibiotiques inhibiteurs de la synthèse de proteines ou du renouvellement cellulaire. Comp. Biochem. Physiol. *30*, 1125–1151 (1969b).
- J. C. RANKIN: Quelques aspects du rôle biologiques des hormones neurohypophysaires chez les poissons. Colloque du C. N. R. S. Paris No. 177 45–55 (1969).
- W. H. SAWYER, G. E. PICKFORD, et N. MAYER: Evolution de la balance minérale du sodium chez *Fundulus heteroclitus* au cours du transfert d'eau de mer en eau douce: effets de l'hypophysectomie et de la prolactine. Gen. comp. Endocr. *8*, 163–176 (1967b).
- E. SKADHAUGE: Drinking rates and gill ionic turnover in relation to external salinities in the eel. Nature (Lond.) *217*, 371–373 (1968).
- MAIN, A. R. and P. J. BENTLEY: Comparison of dehydration and hydration of burrowing desert frogs and tree frogs of the genus *Hyla*. Ecology *45*, 379–382 (1964).
- M. J. LITTLEJOHN, and A. K. LEE: Ecology of Australian frogs. Monographiae Biol. *8*, 396–411 (1959).
- MALNIC, G., R. M. KLOSE, and G. GIEBISCH: Micropuncture study of distal tubular potassium and sodium transport in rat nephrons. Amer. J. Physiol. *211*, 529–547 (1966a).
- R. M. KLOSE, and G. GIEBISCH: Microperfusion study of distal tubular potassium and sodium transfer in rat kidney. Amer. J. Physiol. *211*, 548–559 (1966b).
- MANWELL, C.: The structure of hemoglobin and hemorythrin as a source of taxonomic information. XVI Intern. cong. of Zool. Proc. Vol. 4, pp. 139–142 (1963).
- MARENZI, A. D. and O. FUSTINONI: El potasio de la sangre y de los tejidos de los sapos suprarenoprivos. Rev. Soc. Arg. Biol. *14*, 118–122 (1938).
- MARSHALL, A. J.: Breeding seasons and migration. In: The Biology and Comparative Physiology of Birds. Vol. 2, pp. 307–339. Ed. by A. J. MARSHALL. New York: Academic Press 1961.
- H. J. de S. DISNEY: Photostimulation of an equatorial bird (*Quelea quelea* Linneaus). Nature (Lond.) *177*, 143–144 (1957).
- MARSHALL, E. K.: The comparative physiology of the kidney in relation to theories of renal secretion. Physiol. Rev. *14*, 133–159 (1934).
- H. W. SMITH: The glomerular development of the vertebrate kidney in relation to habitat. Biol. Bull. (Woods Hole) *59*, 135–153 (1930).
- MARTIN, C. J.: Temperature and metabolism in monotremes and marsupials. Phil. Trans. B. 195, 1–37 (1902).
- MATTY, A. J.: Thyroidectomy and its effect upon oxygen consumption of a teleost fish, *Pseudoscarus gaucamaia*. J. Endocr. *15*, 1–8 (1957).
- MAURO, A.: Osmotic flow in a rigid porous membrane. Science *149*, 867–869 (1965).
- MAYER, N.: Nouvelles recherches sur l'adaption des grenouilles vertes *Rana esculenta* à des

milieux de salinité variée. Etude spéciale de l'excrétion rénale de l'eau et des électrolytes. Diplôme d'Etudes Supérieures. Paris. 1963.

– Adaptation de *Rana exculenta* à des milieux variés. Etude spéciale de l'excrétion rénale de l'eau et des électrolytes au cours des changements de milieux. Comp. Biochem. Physiol. *29*, 27–50 (1969).

– J. MAETZ: Axe hypophyso-interrénalien et osmorégulation chez l'anguille en eau de mer. Etude de l'animal intact avec note sur les perturbations de l'équilibre minéral produites par l'effet de choc. C.R. Acad. Sci. (Paris) *264*, 1632–1635 (1967).

– J. MAETZ, D. K. O. CHAN, M. E. FORSTER, and I. CHESTER JONES:Cortisol, a sodium ex creting factor in the eel *(Anguilla anguilla* L.) adapted to sea water. Nature (Lond.) *214*, 1118–1120 (1967).

MAYHEW, W. W.: Adaptations of the amphibian, *Scaphiopus couchi*, to desert conditions. Amer. Midland Naturalist *74*, 95–109 (1965).

MCBEAN, R. L. and L. GOLDSTEIN: Ornithine-urea cycle activity in *Xenopus laevis:* adaptation in saline. Science *157*, 931–932 (1967).

MCCANN, S.M.: Present status of hypothalmic hypophyseal stimulating and inhibiting hormones. Proc. Int. Union of Physiol. Sci. *6*, 202–203 (1968) XXIV International Congress, Washington, D.C. 1968).

MCCLANAHAN, L.: Adaptations of the spadefoot toad *Scaphiopus couchi*, to desert environments. Comp. Biochem. Physiol. *20*, 73–99 (1967).

– R.BALDWIN: Rate of water uptake through the integument of the desert toad, *Bufo cognatus*. Comp. Biochem. Physiol. *28*, 381–389 (1969).

MCDONALD, I. R. and M. L. AUGEE: Effects of bilateral adrenalectomy in the monotreme *Tachyglossus aculeatus*. Comp. Biochem. Physiol. *27*, 669–678 (1968).

. MCFARLAND, W.M. and F.W. MUNZ: Regulation of body weight and serum composition by hagfish in various media. Comp. Biochem. Physiol. *14*, 383–398 (1965).

MEIER, A. H. and W.R.FLEMING: The effects of pitocin and pitressin on water and sodium movements in the euryhaline killifish, *Fundulus kansae*. Comp. Biochem. Physiol. *6*, 215–231 (1962).

MIDDLER, S. A., C. R. KLEEMAN, and E. EDWARDS: Neurohypophysial function in the toad *Bufo marinus*. Gen. comp. Endocr. *9*, 38–48 (1967).

– C.R.KLEEMAN, E.EDWARDS, and D.BRODY: Effect of adenohypophysectomy on salt and water metabolism of the toad *Bufo marinus* studies in hormonal replacement. Gen. comp. Endocr. *12*, 290–304 (1969).

MILLER, R. A. and O.RIDDLE: The cytology of the adrenal cortex of normal pigeons and in experimentally induced atrophy and hypertrophy. Am. J. Anat. *71*, 311–341 (1942).

MIZOGAMI, S., M.OGURI, H.SOKABE, and H.NISHIMURA: Presence of renin in the glomerular and aglomerular kidney of marine teleosts. Amer. J. Physiol. *215*, 991–994 (1968).

MONOD, J.: On the mechanism of molecular interactions in the control of cellular metabolism. Endocrinology *78*, 412–425 (1966).

MOORHOUSE, D. E.: Experimental alteration of the peri-renal tissue of *Protopterus*. Q. Jl microsc. Sci. *97*, 519–534 (1956).

MOREL, F. and S.JARD: Inhibition of frog *(Rana esculenta)* antidiuretic action of vasotocin by some analogues. Amer. J. Physiol. *204*, 227–232 (1963).

– S. JARD: Actions and functions of the neurohypophysial hormones and related peptides in lower vertebrates. In: Handbuch d. exper. Pharmacol. Vol. 23, pp. 655–716 Ed. by B. BERDE. Heidelberg and Berlin: Springer-Verlag 1968.

MORITA, H., T. ISHIBASHI, and S. YAMASHITA: Synaptic transmission in neurosecretory cells. Nature (Lond.) *191*, 183 (1961).

MORRIS, R.: Some aspects of the structure and cytology of the gill of *Lampetra fluviatilis* L. Q. Jl microsc. Sci. *98*, 473–485 (1957).

– General problems of osmoregulation with special reference to cyclostomes. Symp. zool. Soc. Lond. *1*, 1–16 (1960).

– Studies on salt and water and water balance in *Myxine glutinosa* L. J. exp. Biol. *42*, 359–371 (1965).

MOTAIS, R.: Les échanges de sodium chez un Téléostéen euryhalin, *Platichthys flesus flesus*

Linné: Cinetique de ces échanges lors des passages d'eau de mer en eau douce et d'eau douce en eau de mer. C.R. Acad. Sci. (Paris) *253*, 724–726 (1961 a).

– Cinetique des échanges de sodium chez un Téléostéen euryhalin (*Platichthys flesus flesus* Linné.) au cours de passages successifs eau de mer – eau douce – eau de mer, en fonction du temps de séjour en eau douce. C.R. Acad. Sci. (Paris) *253*, 2609–2611 (1961 b).

– Effect of actinomycin D on the branchial Na-K dependent ATPase activity in relation to sodium balance of the eel. Comp. Biochem. Physiol. *34*, 497–501 (1970).

– F. GARCIA ROMEU, et J. MAETZ: Mécanisme de l'euryhalinité. Etude comparée du Flet (euryhalin) et du Serran (sténohalin) au cours du transfert en eau douce. C.R. Acad. Sci. (Paris) *261*, 801–804 (1965).

– F. GARCIA ROMEU, and J. MAETZ: Exchange diffusion effect and euryhalinity in teleosts. J. gen. Physiol. *50*, 391–422 (1966).

– J. ISAIA, J.C. RANKIN, and J. MAETZ: Adaptive changes of the water permeability of the teleostean gill epithelium in relation to external salinity. J. exp. Biol. *51*, 529–546 (1969).

– J. MAETZ: Action des hormones neurohypophysaires sur les échanges de sodium (mesurés a l'aide du radio-sodium Na^{24}) chez un téléostéen euryhalin: *Platichthys flesus* L. Gen. comp. Endocr. *4*, 210–224 (1964).

– J. MAETZ: Comparaison des échanges de sodium chez un Téléostéen euryhalin (le Flet) et un Téléostéen sténohalin (le Serran) en eau de mer. Importance relative du tube digestif et de la branchie dans ces échanges. C.R. Acad. Sci. (Paris) *261*, 532–535 (1965).

– J. MAETZ: Arginine vasotocine et évolution de la perméabilité branchiale au sodium au cours du passage d'eau douce en eau de mer chez le Flet. J. Physiol. (Paris) *59*, 271 (1967).

MULLINS, L. J.: Osmotic regulation in fish as studied with radioisotopes. Acta physiol. scand. *21*, 303–314 (1950).

MUNSICK, R.A.: Neurohypophysial hormones of chickens and turkeys. Endocrinology *75*, 104–112 (1964).

– Chromatographic and pharmacologic characterization of the neurohypophysial hormones of an amphibian and a reptile. Endocrinology *78*, 591–599 (1966).

– W.H. SAWYER, and H.B. VAN DYKE: The antidiuretic potency of arginine and lysine vasopressins in the pig with observations on porcine renal function. Endocrinology *63*, 688–693 (1958).

– W.H. SAWYER, and H.B. VAN DYKE: Avian neurohypophysial hormones; pharmacological properties and tentative identification. Endocrinology *66*, 860–871 (1960).

MUNZ, F.W. and W.N. McFARLAND: Regulatory function of as primitive vertebrate kidney. Comp. Biochem. Physiol. *13*, 381–400 (1964).

MURRAY, R.W. and W.T.W. POTTS: The composition of the endolymph, perilymph and other body fluids of elasmobranches. Comp. Biochem. Physiol. *2*, 65-75 (1961).

MURRISH, D.E. and K. SCHMIDT-NIELSEN: Water transport in the cloaca of lizards: Active or passive? Science *170*, 324–326 (1970).

MYERS, R., W.R. FLEMING, and B.T. SCHEER: Pituitary-adrenal control of sodium flux across frog skin. Endocrinology *58*, 674-676 (1956).

NAGRA, C.L., G.J. BAUM, and R.K. MEYER: Corticosterone levels in adrenal effluent blood of some gallinaceous birds. Proc. Soc. exp. Biol. (N. Y.) *105*, 68-70 (1960).

– J.G. BIRNIE, G.J. BAUM, and R.K. MEYER: The role of the pituitary in regulating steroid secretion by the avian adrenal. Gen. comp. Endocr. *3*, 274-280 (1963).

NANDI, J.: Comparative endocrinology of steroid hormones in vertebrates. Amer. Zool. *7*, 115–133 (1967).

– H.A. BERN: Chromatography of corticosteroids from teleost fishes. Gen. comp. Endocr. *5*, 1-15 (1965).

NECHAY, B., S. BOYARSKY, and P. CATACUTAN-LABAY: Rapid migration of urine into intestine of chickens. Comp. Biochem. Physiol. *26*, 369-370 (1968).

– B.D.C. LUTHERER: Handling of urine by cloaca and ureter in chickens. Comp. Biochem. Physiol. *26*, 1099-1105 (1968).

NEWCOMER, W.S.: Effects of hypophysectomy on some functional aspects of the adrenal gland of the chicken. Endocrinology *65*, 133–135 (1959).

NEWSOME, A.E.: The abundance of red kangaroos, *Megeleia rufa* (Demarest) in central Australia. Aust. J. Zool. *13*, 269–287 (1965 a).

- The distribution of red kangaroos *Megeleia rufa* (Demarest) about sources of persistent food and water in central Australia. Aust. J. Zool. *13*, 289–299 (1965 b).
NICOLL, C.S. and H.A. BERN: Further analysis of the occurrence of pigeon crop sac-stimulating activity (prolactin) in the vertebrate adenohypophysis. Gen. comp. Endocr. *11*, 5–20 (1968).
- H.A. BERN, and D. BROWN: Occurrence of mammotrophic activity (prolactin) in the vertebrate adenohypophysis. J. Endocr. *34*, 343-354 (1966).
NISHIMURA, H., M. OGURI, M. OGAWA, H. SOKABE, and M. IMAI: Absence of renin in kidneys of elasmobranchs and cyclostomes. Amer. J. Physiol. *218*, 911–915 (1970).
NOVELLI, A.: Lobulo posterior de hipofisis e imbibicion de los batrachios. II. Mecanismo de su accion. Rev. Soc. argent. Biol. *12*, 163–164 (1936).
O'CONNOR, W.J.: Renal Function. London: Arnold 1962.
- D.J. POTTS: The external water exchanges of normal laboratory dogs. Q. Jl exp. Physiol. *54*, 244-265 (1969).
- E.B. VERNEY: The effect of increased activity of the sympathetic system in the inhibition of water diuresis by emotional stress. Q. Jl exp. Physiol. *33*, 77–90 (1945).
OGURI, M.: Rectal glands of marine and fresh-water sharks: comparative histology. Science. *144*, 1151-1152 (1964).
OIDE, M.: Effects of inhibitors on transport of water and ions in isolated intestine and Na-K ATPase in intestinal mucosa of the eel. Ann. Zool. Japon. *40*, 130–135 (1967).
- S. UTIDA: Changes in water and ion transport in isolated intestines of the eel during salt adaptation and migration. Marine Biol. *1*, 102–106 (1967).
- S. UTIDA: Changes in intestinal absorbtion and renal excretion of water during adaptation to sea-water in the Japanese eel. Marine Biol. *1*, 172–177 (1968).
OKSCHE, A., D.S. FARNER, D.L. SERVENTY, F. WOLFF, and C.A. NICHOLLS: The hypothalamo-hypophysial neurosecretory system of the zebra finch, *Taeniopygia castanotis*. Zeitschrift für Zellforschung *58*, 846-914 (1963).
OLIVEREAU, M.: Influence d'une diminution de salinité sur l'activité de la glande thyroide de deux téléostéens marins: *Muraena helena* L. et *Labrus bergylta* Asc. C.R. Soc. Biol. (Paris) *142*, 176–177 (1948).
- Modifications d'l'interrénal du smolt (*Salmo salar* L.) au cours du passage d'eau douce en eau de mer. Gen. comp. Endocr. *2*, 565–573 (1962).
- Influence d'un séjour en eau déminéralisée sur le système hypophyso-surrénalien de l'anguille. Ann. Endocr. (Paris) *27*, 665–678 (1966).
- Réactions observées chez l'anguille maintenue dans un milieu privé d'électrolytes, en particulier au niveau du système hypothalamo-hypophysaire. Zeitschrift für Zellforschung *80*, 264–285 (1967).
- J.N. BALL: Contribution à l'histophysiologie de l'hypophyse de téléostéens, en particulier de celle de *Poecilia* species. Gen. comp. Endocr. *4*, 523–532 (1964).
- M. CHARTIER-BARADUC: Action de la prolactine chez l'anguille intacte et hypophysectomisée. Effets sur les électrolytes plasmatiques (sodium, potassium et calcium). Gen. comp. Endocr. *7*, 27–36 (1966).
- J. OLIVEREAU: Effets de l'interrénalectomie sur la structure histologique de l'hypophyse et de quelques tissus de l'anguille. Zeitschrift für Zellforschung *84*, 44-58 (1968).
VAN OORDT, G.J.: Male gonadal hormones. In: Comparative Endocrinology, Vol. 1, pp. 154–207. Ed. by U.S. VON EULER and H. HELLER. New York: Academic Press 1963.
ORLOFF, J. and J.S. HANDLER: The similarity of effects of vasopressin, adenosine-3'5' – phosphate (cyclic AMP) and theophylline on the toad bladder. J. clin. Invest. *41*, 702-709 (1962).
- J.S. HANDLER: The role of adenosine 3', 5'-phosphate in the action of antidiuretic hormone. Amer. J. Med. *42*, 757–768 (1967).
OVERACK, D.E. and R. DEROOS: Comparative effects of neurohypophysial hormones on body weight in the newt, *Notophthalmus viridescens*. Amer. Zool. *7*, 720 (1967).

PAPPENHEIMER, J.R.: Passage of molecules through capillary walls., Physiol. Rev. *33*, 387–423 (1953).
PARISI, M., P. RIPOCHE et J. BOURGUET: Mechanism by which hypertonic solutions increase

water net flux across the frog urinary bladder. IVth Int. Cong. Nephrology (abstracts) pp. 261 1969.

PARKES, A.S. and H.M. BRUCE: Olfactory stimuli in mammalian reproduction. Science *134*, 1049-1054 (1961).

PARKINS, W.M.: An experimental study of bilateral adrenalectomy in the fowl. Anat. Rec. 51, Suppl. 39 (1931).

PARRY, G.: Osmotic and ionic changes in blood and muscle of migrating salmonids. J. exp. Biol. *38*, 411-427 (1961).

− Osmotic adaptation in fishes. Biol. Rev. *41*, 392-444 (1966).

− F.G.T. HOLLIDAY, and J.H.S. BLAXTER: Chloride-secretory cells in the gills of teleosts. Nature (Lond.) *183*, 1248-1249 (1959).

PEAKER, M., J.G. PHILLIPS, and A. WRIGHT: The effect of prolactin on the secretory activity of the nasal salt-gland of the domestic duck (*Anas platyrhynchos*). J. Endocr. *47*, 123–127, (1970).

PETERSEN, M.J. and I.S. EDELMAN: Calcium inhibition of the action of vasopressin on the urinary bladder of the toad. J. clin. Invest. *43*, 583-594 (1964).

PETTUS, D.: Water relationships in *Natrix sipedon*. Copeia No. 3, 207-211 (1958).

PEYROT, A., E. FERRERI, V. MAZZI, and M. SOCINO: Il metabolismo del sodio e del potassio nel *Tretone crestato*. II. Effetti dell'ipofisectomia. Boll. Soc. Ital. Biol. Sper. *40*, 220–222 (1963).

PHILLIPS, J.G.: Adrenocorticosteroids in fish. J. Endocr. *18*, xxxvii-xxxix (1959).

− D. BELLAMY: Aspects of the hormonal control of nasal gland secretion in birds. J. Endocr. *24*, vi-vii (1962).

− I. CHESTER JONES: The identity of adrenocortical secretions in lower vertebrates. J. Endocr. *16*, iii (1957).

− I. CHESTER JONES, and D. BELLAMY: Biosynthesis of adrenocortical hormones by adrenal glands of lizards and snakes. J. Endocr. *25*, 233-237 (1962a).

− I. CHESTER JONES, D. BELLAMY, R.O. GREEP, L.R. DAY, and W.N. HOLMES: Corticosteroids in the blood of *Myxine glutinosa* L. (Atlantic hagfish). Endocrinology *71*, 329–331 (1962b).

− W.N. HOLMES, and D.G. BUTLER: The effect of total and sub-total adrenalectomy on the renal and extra-renal response of the domestic duck (*Anas platyrhynchos*) to saline loading. Endocrinology *69*, 958-969 (1961).

PICKERING, B.T.: The neurohypophysial hormones of a reptile species, the cobra (*Naja naja*). J. Endocr. *39*, 285–294 (1967).

− H. HELLER: Chromatographic and biological characteristics of fish and frog neurohypophysial extracts. Nature (Lond.,) *184*, 1463–1464 (1959).

− H. HELLER: Oxytocin as a neurohypophysial hormone in the holocephalian elasmobranch fish, *Hydrolagus collei*. J. Endocr. *45*, 597–606 (1969).

− S. McWATTERS: Neurohypophysial hormones of the South American lungfish *Lepidosiren paradoxa*. J. Endocr. *36*, 217-218 (1966).

PICKFORD, G.E. and F.B. GRANT: Serum osmolality in the coelacanth *Latimeria chalumnae*: urea retention and ion regulation. Science *155*, 568-570 (1967).

− P.K.T. PANG, and W.H. SAWYER: Prolactin and serum osmolality of hypophysectomized killifish, *Fundulus heteroclitus*, in fresh-water. Nature (Lond.) *209*, 1040–1041 (1966).

− P.K.T. PANG, J.G. STANLEY, and W.R. FLEMING: Calcium and fresh-water survival in the euryhaline cyprinodonts, *Fundulus kansae* and *Fundulus heteroclitus*. Comp. Biochem. Physiol. *18*, 503-509 (1966).

− P.K.T. PANG, E. WEINSTEIN, J. TORRETTI, E. HENDLER, and F.H. EPSTEIN: The response of the hypophysectomized cyprinodont, *Fundulus heteroclitus*, to replacement therapy with cortisol: Effects on blood serum and sodium-potassium activated adenosine triphosphatase in the gills, kidney and intestinal mucosa. Gen. comp. Endocr. *14*, 524-534 (1970).

− J.G. PHILLIPS: Prolactin, a factor in promoting survival of hypophysectomized killifish in fresh water. Science *130*, 454–455 (1959).

PORTER, G.A., R. BOGOROCH, and I.S. EDELMAN: On the mechanism of action of aldosterone on sodium transport. Proc. nat. Acad. Sci. (Wash.) *52*, 1326–1333 (1964).

POTTS, W.T.W. and D.H. EVANS: The effects of hypophysectomy and bovine prolactin on salt fluxes in fresh-water adapted *Fundulus heteroclitus*. Biol. Bull. (Woods Hole) *131*, 362-368 (1966).

– D.H. EVANS: Sodium and chloride balance in the killifish *Fundulus heteroclitus*. Biol. Bull. (Woods Hole) 133, 411-425 (1967).

– M.A. FORSTER, P.P. RUDY, and G.P. HOWELLS: Sodium and water balance in the cichlid teleost, *Tilapia mossambica*. J. exp. Biol. *47*, 461–470 (1967).

POULSEN, T.L.: Countercurrent multipliers in avian kidneys. Science *148*, 389–391 (1965).

– G.A. BARTHOLOMEW: Salt balance in the savannah sparrow. Physiol. Zool. *35*, 109–119 (1962).

PRANGE, H.D. and K. SCHMIDT-NIELSEN: Evaporative water loss in snakes. Comp. Biochem. Physiol. *28*, 973–975 (1969).

PRICE, K.S. Jr.: Fluctuations in two osmoregulatory components, urea and sodium chloride, of the clearnose skate, *Raja eglanteria*, Bosc 1802–II. Natural variation of the salinity of the external medium. Comp. Biochem. Physiol. *23*, 77–82 (1967).

– E.P. CREASER: Fluctuations in two osmoregulatory components, urea and sodium chloride, of the clearnose skate, *Raja eglanteria* Bosc 1802–I. upon laboratory modification of external salinities. Comp. Biochem. Physiol. *23*, 65–76 (1967).

RACE, G.J. and H.M. WU: Adrenal cortex functional zonation in the whale (*Physeter catadon*). Endocrinology *68*, 156–158 (1961).

RASQUIN, P.: Effects of carp pituitary and mammalian ACTH on the endocrine and lymphoid systems of the teleost *Astyanax mexicanus*. J. exp. Zool. *117*, 317–357 (1951).

– Cytological evidence for a role of the corpuscles of Stannius in the osmoregulation of teleosts. Biol. Bull. (Woods Hole) *111*, 399–409 (1956).

REID, I.A. and I.R. MCDONALD: Bilateral adrenalectomy and steroid replacement in the marsupial *Trichosurus vulpecula*. Comp. Biochem. Physiol. *26*, 613-625 (1968).

– I.R. MCDONALD: The renin-angiotensin system in a marsupial (*Trichosurus vulpecula*). J. Endocr. *44*, 231-240 (1969).

REINKE, E.E. and C.S. CHADWICK: Inducing land stage of *Triturus viridescens* to assume water habitat by pituitary implantations. Proc. Soc. exp. Biol. (N.Y.) *40*, 691–693 (1939).

RENKIN, E.M. and B.D. ZAUN: Effects of adrenal hormones on capillary permeability in perfused rat tissues. Amer. J. Physiol. *180*, 498-502 (1955).

REY, P.: Recherches expérimentales sur l'économie de l'eau chez les Batraciens. Annls. Physiol. Physiochem. biol. *13*, 1081–1144 (1937).

RICHARDS, B.D. and P.O. FROMM: Patterns of blood flow through the filaments and lamellae of isolated perfused rainbow trout (*Salmo gairdneri*) gills. Comp. Biochem. Physiol. *29*, 1063-1070 (1969).

RIDDLE, O. and P.F. BRAUCHER: Studies on the physiology of the reproduction in birds. XXX. Control of the special secretion of the crop gland in pigeons by an anterior pituitary hormone. Amer. J. Physiol. *97*, 617–625 (1931).

RIDLEY, A.: Effects of osmotic stress and hypophysectomy on ion distribution in bullfrogs. Gen. comp. Endocr. *4*, 481-485 (1964).

RINFRET, A.P. and S. HANE: Presence of ACTH in pituitary gland of Pacific salmon (*O. keta*). Proc. Soc. exp. Biol. (N.Y.) *90*, 508–510 (1955).

ROBERTS, J.S. and B. SCHMIDT-NIELSEN: Renal ultrastructure and excretion of salt and water by three terrestrial lizards. Amer. J. Physiol. *211*, 476–486 (1966).

ROBERTSHAW, D. and C.R. TAYLOR: A comparison of sweat gland activity in eight species of East African bovids. J. Physiol. (Lond.) *203*, 135-143 (1969).

ROBERTSON, J.D.: The chemical composition of the blood of some aquatic chordates, including members of the Tunicata, Cyclostomata and Osteichthyes. J. exp. Biol. *31*, 424-442 (1954).

– The habitat of early vertebrates. Biol. Rev. *32*, 156-187 (1957).

ROBERTSON, J. DAVID: Structural alterations in nerve fibers produced by hypotonic and hypertonic solutions. J. biophys. biochem. Cytol. *4*, 349–364 (1958).

– Unit membranes: A review with recent new studies of experimental alterations and a new subunit structure in synaptic membranes. In: Cellular Membranes in Development, pp. 1–81 Ed. by M. LOCKE New York: Academic Press 1964.

279

ROBERTSON, O.H., M.A. KRUPP, S.F. THOMAS, C.B. FAVOUR, S. HANE, and B.C. WEXLER: Hyperadrenocorticism in spawning migratory and nonmigratory rainbow trout (*Salmo gairdnerii*); comparison with pacific salmon (Genus *Oncorhynchus*). Gen. comp. Endocr. *1*, 473–484 (1961).

- B.C. WEXLER: Hyperplasia of the adrenal cortical tissue in Pacific salmon (Genus *Oncorhynchus*) and rainbow trout (*Salmo gairdnerii*) accompanying sexual maturation and spawning. Endocrinology *65*, 225-238 (1959).

ROBINSON, K.W. and W.V. MACFARLANE: Plasma antidiuretic activity in marsupials during exposure to heat. Endocrinology *60*, 679–680 (1957).

ROCHE, J., G. SALVATORE, et I. COVELLI: Métabolisme de ^{131}I et fonction thyroidienne chez la larve (ammoncoete) d'un cyclostome *Petromyzon planeri*. Comp. Biochem. Physiol. *2*, 90–99 (1961).

ROMER, A.S.: Vertebrate Paleontology 2nd ed. Chicago: University Press 1955.

- Major steps in vertebrate evolution. Science *158*, 1629-1637 (1967).

ROY, B.P.: Distribution of neurohypophyseal hormones in some elasmobranch fishes. Gen. comp. Endocr. *12*, 326-338 (1969).

RUIBAL, R.: The adaptive value of bladder water in the toad, *Bufo cognatus*. Physiol. Zool. *35*, 218-223 (1962).

SALADINO, A.J., P.J. BENTLEY, and B.F. TRUMP: Ion movements in cell injury. Effect of amphotericin B on the ultrastructure and function of the epithelial cells of the toad bladder. Amer. J. Path. *54*, 421-466 (1969).

SANDOR, T., J. LAMOUREUX, and A. LANTHIER: Adrenal cortical function in reptiles. The *in vitro* biosynthesis of adrenal cortical steroids by adrenal slices of two common North American turtles, the slider turtle (*Pseudemys scripta elegans*) and the painted turtle (*Chrysemys picta picta*). Steroids *4*, 213–227 (1964).

SAWYER, W.H.: Effect of posterior pituitary extract on permeability of frog skin to water. Amer. J. Physiol. *164*, 44-48 (1951).

- Differences in antidiuretic responses of rats to the intravenous administration of lysine and arginine vasopressin. Endocrinology *63*, 694-698 (1958).

- Oxytocic activity in the neural complex of two ascidians, *Chelyosoma productum* and *Pyura haustor*. Endocrinology *65*, 520-523 (1959).

- Increased water permeability of the bullfrog (*Rana catesbeiana*) bladder *in vitro* in response to synthetic oxytocin and arginine vasotocin and to neurohypophysial extracts from non-mammalian vertebrates. Endocrinology *66*, 112-120 (1960).

- Active neurohypophysial principles from a cyclostome (*Petromyzon marinus*) and two cartilaginous fishes (*Squalus acanthias* and *Hydrolagus collei*). Gen. comp. Endocr. *5*, 427-439 (1965).

- Diuretic and natriuretic responses of lungfish (*Protopterus aethiopicus*) to arginine vasotocin. Amer. J. Physiol. *210*, 191-197 (1966a).

- Biological assays for neurohypophysial principles in tissues and in blood. In: The Pituitary Gland, Vol. 3, pp. 288–306, Ed. by G.W. HARRIS and B.T. DONOVAN. London: Butterworths 1966 b.

- Evolution of antidiuretic hormones and their functions. Amer. J. Med. *42*, 678-686 (1967a).

- Chromatographic and pharmacological characteristics of the active neurohypophysial principles of the spiny dogfish *Squalus acanthias*. Gen. comp. Endocr. *9*, 303-311 (1967b).

- Phylogenetic aspects of the neurohypophysial hormones. In: Handbuch d. exp. Pharmacol. Vol. 23, pp. 717–747, Ed. by B. BERDE. Heidelberg and Berlin: Springer-Verlag 1968.

- The active neurohypophysial principles of two primitive bony fishes, the bichir (*Polypterus senegalis*) and the African lungfish (*Protopterus aethiopicus*). J. Endocr. *44*, 421-435 (1969).

- Vasopressor, diuretic and natriuretic responses by lungfish to arginine vasotocin. Amer. J. Physiol. *218*, 1789–1794 (1970).

- H.A. BERN: Examination of the caudal neurosecretory system of *Tilapia mossambica* (Teleostei, Cichlidae) for the presence of neurohypophysial-like activity. Amer. Zool. *3*, 334 (1963).

- R.J. FREER, and T-C. TSENG: Characterization of a principle resembling oxytocin in the pituitary of the holocephalian ratfish (*Hydrolagus colliei*) by partition chromatography on sephadex columns. Gen. comp. Endocr. *9*, 31–37 (1967).

– M. MANNING, E. HEINICKE, and A.M. PERKS: Elasmobranch oxytocin-like principles: comparisons with synthetic glumitocin. Gen. comp. Endocr. *12*, 387-390 (1969).

– R.A. MUNSICK, and H.B. VAN DYKE: Pharmacological evidence for the presence of arginine vasotocin and oxytocin in neurohypophysial extracts from cold blooded vertebrates. Nature (Lond.) *184*, 1464-1465 (1959).

– R.A. MUNSICK, and H.B. VAN DYKE: Pharmacological characteristics of neurohypophysial hormones from a marsupial (*Didelphys virginiana*) and a monotreme (*Tachyglossus (Echidna) aculeatus*). Endocrinology *67*, 137-138 (1960).

– R.A. MUNSICK, and H.B. VAN DYKE: Evidence for the presence of arginine vasotocin (8-arginine oxytocin) and oxytocin in neurohypophyseal extracts from amphibians and reptiles. Gen. comp. Endocr. *1*, 30-36 (1961).

– G.E. PICKFORD: Neurohypophyseal principles of *Fundulus heteroclitus*: characteristics and seasonal changes. Gen. comp. Endocr. *3*, 439-445 (1963).

– M.K. SAWYER: Adaptive responses to neurohypophyseal fractions in vertebrates. Physiol. Zool. *25*, 84-98 (1952).

– R.M. SCHISGALL: Increased permeability of the frog bladder to water in response to dehydration and neurohypophysial extracts. Amer. J. Physiol. *187*, 312-314 (1956).

– H. VALTIN and H.W. SOKOL: Neurohypophysial principles in rats with familial hypothalamic diabetes insipidus (Brattleboro strain). Endocrinology *74*, 153-155 (1964).

SCHALLY, A.V. and R. GUILLEMIN: Isolation and chemical characterization of a ß-CRF from pig posterior pituitary glands. Proc. Soc. exp. Biol. (N.Y.) *112*, 1014–1017 (1963).

– M. SAFFRAN, and P. ZIMMERMAN: A corticotrophin-releasing factor: partial purification and amino acid composition. Biochem. J. *70*, 97–103 (1958).

SCHARRER, E. and B. SCHARRER: Neurosecretion. Physiol. Rev. *25*, 171–181 (1945).

SCHEER, B.T. and R.P. MARKEL: The effect of osmotic stress and hypophysectomy on blood and urine urea levels in frogs. Comp. Biochem. Physiol. *7*, 289–297 (1962).

SCHIMKE, R.T.: Studies on factors affecting the levels of urea cycle enzymes in rat liver. J. biol. Chem. *238*, 1012–1018 (1963).

SCHMID, W.D.: Some aspects of the water economies of nine species of amphibians. Ecology *46*, 261–269 (1965).

– R.E. BARDEN: Water permeability and lipid content of amphibian skin. Comp. Biochem. Physiol. *15*, 423–427 (1965).

SCHMIDT-NIELSEN, B.: Organ systems in adaptation: the excretory system. In: Handbook of Physiology Sec. 4 Adaptation to the environment, pp. 215–243. Washington: Amer. Physiol. Soc. 1964.

– L.É. DAVIS: Fluid transport and tubular intercellular spaces in reptilian kidneys. Science *159*, 1105–1108 (1968).

– K. SCHMIDT-NIELSEN, T.R. HOUPT, and S.A. JARNUM: Water balance in the camel. Amer. J. Physiol. *185*, 185–194 (1956).

– E. SKADHAUGE: Function of the excretory system in the crocodile *(Crocodylus acutus)*. Amer. J. Physiol. *212*, 973–980 (1967).

SCHMIDT-NIELSEN, K.: The salt secreting glands in marine birds. Circulation *21*, 955–967 (1960).

– Desert animals. Physiological Problems of Heat and Water. Oxford: Clarendon Press 1964 a.

– Terrestrial animals in dry heat: Desert rodents. In: Handbook of Physiology Sec. 4 Adaptation to the environment, pp. 493–507. Washington: Amer. Physiol. Soc. 1964 b.

– Physiology of salt glands. In: Funktionelle und Morphologische Organisation der Zelle. pp. 269–288. Heidelberg and Berlin: Springer-Verlag 1965.

– P.J. BENTLEY: Desert tortoise *Gopherus agassizii:* cutaneous water loss. Science *154*, 911 (1966).

– A. BORUT, P. LEE, and E.C. CRAWFORD: Nasal salt excretion and the possible function of the cloaca in water conservation. Science *142*, 1300–1301 (1963).

– E.C. CRAWFORD, and P.J. BENTLEY: Discontinuous respiration in the lizard *Sauromalus obesus*. Fed. Proc. *25*, 506 (1966).

– E.C. CRAWFORD, A.E. NEWSOME, K.S. RAWSON, and H.T. HAMEL: Metabolic rate of camels; effect of body temperature and dehydration. Amer. J. Physiol. *212*, 341–346 (1967).
– T.J. DAWSON, and E.C. CRAWFORD: Temperature regulation in the Echidna (*Tachyglossus aculeatus*). J. Cell. Physiol. *67*, 63–71 (1966).
– R. FÄNGE: Salt glands in marine reptiles. Nature (Lond.) *182*, 783–785 (1958).
– H.B. HAINES, and D.B. HACKEL: Diabetes mellitus in the sand rat induced by standard laboratory diets. Science *143*, 689–690 (1964).
– C.B. JORGENSEN, and H. OSAKI: Extrarenal salt excretion in birds. Amer. J. Physiol. *193*, 101–107 (1958).
– Y.T. KIM: The effect of salt intake on the size and function of the salt gland of ducks. The Auk *81*, 160–172 (1964).
– P. LEE: Kidney function in the crab-eating frog *(Rana cancrivora)*. J. exp. Biol. *39*, 167–177 (1962).
– B. SCHMIDT-NIELSEN, S.A. JARNUM, and T.R. HOUPT: Body temperature of the camel and its relation to water economy. Amer. J. Physiol. *188*, 103–112 (1957).
SCHREIBMAN, M.P. and K.D. KALLMAN: Endocrine control of freshwater tolerance in teleosts. Gen. comp. Endocr. *6*, 144–155 (1966).
– K.D. KALLMAN: The effect of hypophysectomy on freshwater survival in teleosts of the order Atherinformes. Gen. comp. Endocr. *13*, 27–38 (1969).
SEXTON, A.W.: Factors influencing the uptake of sodium against a diffusion gradient in the goldfish gill. Diss. Abstr. *15*, 2270–2271 (1955).
SHARP, G.W.G., C.H. COGGINS, N.S. LICHTENSTEIN, and A. LEAF: Evidence for a mucosal effect of aldosterone on sodium transport in the toad bladder. J. clin. Invest. *45*, 1640–1647 (1966).
– A. LEAF: On the stimulation of sodium transport by aldosterone. J. gen. Physiol. *51*, 271–279 (1968).
SHARRATT, B.M., I. CHESTER JONES, and D. BELLAMY: Water and electrolyte composition of the body and renal function of the eel (*Anguilla anguilla* L.). Comp. Biochem. Physiol. *11*, 9–18 (1964 a).
– D. BELLAMY, and I. CHESTER JONES: Adaptation of the silver eel (*Anguilla anguilla* L.) to seawater and to artificial media together with observations on the role of the gut. Comp. Biochem. Physiol. *11*, 19–30 (1964 b).
SHIELD, J.: The development of certain external characters in the young of the macropod marsupial *Setonix brachyurus*. Anat. Rec. *140*, 289–294 (1961).
SHIRLEY, H.V. and A.V. NALBANDOV: Effects of neurohypophysectomy in domestic chickens. Endocrinology *58*, 477–483 (1956).
SHOEMAKER, V.H.: The effects of dehydration on electrolyte concentrations in a toad, *Bufo marinus*. Comp. Biochem. Physiol. *13*, 261–271 (1964).
– The stimulus for the water-balance response to dehydration in toads. Comp. Biochem. Physiol. *15*, 81–88 (1965).
– P. LICHT, and W.R. DAWSON: Effects of temperature on kidney function in the lizard *Tiliqua rugosa*. Physiol. Zool. *39*, 244–252 (1966).
– H. WARING: Effect of hypothalamic lesions on the water-balance response of a toad *(Bufo marinus)*. Comp. Biochem. Physiol. *24*, 47–54 (1968).
SIGLER, M.H., J.N. FORREST, and J.R. ELKINGTON: Renal concentrating ability in adrenalectomized rats. Clin. Sci. *28*, 29–37 (1965).
SILVETTE, H. and S.W. BRITTON: Renal function in opossum and the mechanism of corticoadrenal and post-pituitary action. Amer. J. Physiol. *123*, 630–639 (1938).
SKADHAUGE. E.; Effects of unilateral infusion of arginine-vasotocin into the portal circulation of the avian kidney. Acta endocr. (Kbh.) *47*, 321–330 (1964).
– *In vivo* perfusion studies of the cloacal water and electrolyte resorption in the fowl *(Gallus domesticus)*. Comp. Biochem. Physiol. *23*, 483–501 (1967).
– The cloacal storage of urine in the rooster. Comp. Biochem. Physiol. *24*, 7–18 (1968).
– Activités biologiques des hormones neuro-hypophysaires chez les oiseaux et les reptiles. Colloques C.N.R.S. Paris No. 177. 63–68 (1969 a).
– The mechanism of salt and water absorption in the intestine of the eel (*Anguilla anguilla*) adapted to waters of various salinities. J. Physiol. (Lond.) *204*, 135–158 (1969 b).

– J. MAETZ: Etude *in vivo* de l'absorption intestinale d'eau et d'électrolytes chez *Anguilla anguilla* adapté à des milieux de salinités diverses. C.R. Acad. Sci. (Paris) *265*, 347–350 (1967).

– B. SCHMIDT-NIELSEN: Renal function in domestic fowl. Amer. J. Physiol. *212*, 793–798 (1967).

SKOU, J.C.: The influence of some cations on an adenosine triphosphatase from peripheral nerves. Biochim. biophys. Acta. (Amst.) *23*, 394–401 (1957).

SMITH, D.C.: The role of the endocrine organs in the salinity tolerance of the trout. Mem. Soc. Endocr. *5*, 83–101 (1956).

SMITH, H.W.: Metabolism of the lungfish *Protopterus aethiopicus*. J. biol. Chem. *88*, 97–130 (1930a).

– The absorption and secretion of water and salts by marine teleosts. Amer. J. Physiol. *93*, 480–505 (1930b).

– The absorption and excretion of water and salts by the elasmobranch fishes. II. Marine elasmobranchs. Amer. J. Physiol. *98*, 296–310 (1931).

– The retention and physiological role of urea in the elasmobranchii. Biol. Rev. *11*, 49–82 (1936).

– The Kidney. New York: Oxford University Press 1951.

– Kamongo, or the Lungfish and the Padre. New York: Viking Press 1956.

– From Fish to Philosopher. New York: Doubleday 1961.

SMITH, L.F.: Species variation in the amino acid sequence of insulin. Amer. J. Med. *40*, 662–666 (1966).

SMITH, M.W.: The *in vitro* absorption of water and solutes from the intestine of goldfish, *Carassius auratus*. J. Physiol. (Lond.) *175*, 38–49 (1964).

– Sodium-glucose interactions in the goldfish intestine. J. Physiol. (Lond.) *182*, 559–573 (1966).

– Influences of temperature acclimatization on the temperature-dependence and ouabain-sensitivity of goldfish intestinal adenosine triphosphatase. Biochem J. *105*, 65–71 (1967).

– V.W. COLOMBO, and E.A. MUNN: Influence of temperature acclimatization on the ionic activation of goldfish intestinal adenosine triphosphatase. Biochem. J. *107*, 691–698 (1968).

SMYTH, M. and G.A. BARTHOLOMEW: Effects of water deprivation and sodium chloride on blood and urine of the mourning dove. The Auk *83*, 597–602 (1966).

SOKABE, H., S. MIZOGAMI, T. MURASE, and F. SAKAI: Renin and euryhalinity in the Japanese eel, *Anguilla japonica*. Nature (Lond.) *212*, 952–953 (1966).

SPEIDEL, C.C.: Gland-cells of intestinal secretion in the spinal cord of skates. Papers Dep. mar. Biol. Carnegie Inst. Wash. *13*, 1–31 (1919).

SPERBER, I.: Studies on the mammalian kidney. Zool. bidrag fran Upsala *22*, 249–431 (1944).

SPIGHT, T.M.: The water economy of salamanders: exchange of water with the soil. Biol. Bull. (Woods Hole) *132*, 126–132 (1967a).

– The water economy of salamanders; water uptake after dehydration. Comp. Biochem. Physiol. *20*, 767–771 (1967b).

– The water economy of salamanders: evaporative water loss. Physiol. Zool. *41*, 195–203 (1968).

STAALAND, H.: Anatomical and physiological adaptations of the nasal glands in charadriiformes birds. Comp. Biochem. Physiol. *23*, 933–944 (1967).

STANLEY, J.G. and W.R. FLEMING: The effect of urophysectomy on sodium metabolism of *Tilapia mossambica*. Amer. Zool. *4*, 406 (1964).

– W.R. FLEMING: Effect of hypophysectomy on the function of the kidney of the euryhaline teleost, *Fundulus kansae*. Biol. Bull. (Woods Hole) *130*, 430–441 (1966a).

– W.R. FLEMING: The effect of hypophysectomy on sodium metabolism of the gill and kidney of *Fundulus kansae*. Biol. Bull. (Woods Hole) *131*, 155–165 (1966b).

– W.R. FLEMING: The effect of hypophysectomy on the electrolyte content of *Fundulus kansae* held in freshwater and seawater. Comp. Biochem. Physiol. *20*, 489–497 (1967a).

– W.R. FLEMING: Effect of prolactin and ACTH on the serum and urine levels of *Fundulus kansae*. Comp. Biochem. Physiol. *20*, 199–208 (1967b).

STANNIUS, H.: Die Nebennieren bei Knochenfischen. Arch. Anat. Physiol. *97*, 97–101 (1839).

STEEN, J.G. and A. KRUYSSE: The respiratory function of teleostean gills. Comp. Biochem. Physiol. *12*, 127–142 (1964).

STEGGERDA, F.R.: Comparative study of water metabolism in amphibians injected with pituitrin. Proc. Soc. exp. Biol. (N.Y.) *36*, 103–106 (1937).

STEIN, W.D.: The movement of molecules across cell membranes. New York: Academic Press 1967.

STEWART, A.D.: Genetic variation in the neurohypophysial hormones of the mouse (*Mus musculus*). J. Endocr. *41*, xix-xx (1968).

STEWART, D.J., W.N. HOLMES, and G. FLETCHER: The renal excretion of nitrogenous compounds by the duck (*Anas platyrhynchos*) maintained on freshwater and on hypertonic saline. J. exp. Biol. *50*, 527–539 (1969).

STOICOVICI, F. and E.A. PORA: Comportarea la variatiuni de salinitate. Nota XXX Stud. Cercet Stiint., Acad. Rep. Pop. Romậne, Fil. Cluj, *2*, 159–219 (1951).

STRAHAN, R.: Pituitary hormones of *Myxine* and *Lampetra*. Trans. Asia and Oceania Reg. Cong. Endocrinol. Vol. 1 No. 24 Kyoto Japan 1959.

STURKIE, P.D. and Y. LIN: Release of vasotocin and oviposition in the hen. J. Endocr. *35*, 325–326 (1966).

SZARSKI, H.: The origin of the Amphibia. Quart. Rev. Biol. *37*, 189–241 (1962).

TAKASUGI, N. and H.A. BERN: Experimental studies on the caudal neurosecretory system of *Tilapia mossambica*. Comp. Biochem. Physiol. *6*, 289–303 (1962).

TAUBENHAUS, M., I.B. FRITZ, and J.V. MORTON: *In vitro* effects of steroids upon electrolyte transfer through frog skin. Endocrinology *59*, 458–462 (1956).

TAYLOR, A.A., J.O. DAVIS, R.P. BREITENBACH, and P.M. HARTCROFT: Adrenal steroid secretion and a renal-pressor system in the chicken (*Gallus domesticus*). Gen. comp. Endocr. *14*, 321–333 (1970).

TAYLOR, C.R.: The Eland and the Oryx. Scientific American *220*, 88–95 (1969).

– C.A. SPINAGE, and C.P. LYMAN: Water relations of the water buck and East African antelope. Amer. J. Physiol. *217*, 630–634 (1969).

TAYLOR, R.E. and S.B. BARKER: Transepidermal potential difference: development in anuran larvae. Science *148*, 1612–1613 (1965).

– S.B. BARKER: Absence of an *in vitro* thyroxine effect on oxygen consumption and sodium or water transport by anuran skin and bladder. Gen. comp. Endocr. *9*, 129–134 (1967).

– T. TU, S.B. BARKER, and E.C. JORGENSEN: Thyroxine-like actions of 3'-isopropyl-3, 5-dibromo-L-thyronine, a potent iodine free-analog. Endocrinology *80*, 1143–1147 (1967).

TECHNAU, G.: Die Nasendrüse der Vögel. J.f. Orn. *84*, 511–617 (1936).

TEMPLETON, J.R.: Nasal salt excretion in terrestrial lizards. Comp. Biochem. Physiol. *11*, 223–229 (1964).

– Responses of the lizard nasal salt gland to chronic hypersalemia. Comp. Biochem. Physiol.. *18*, 563–572 (1966).

– Nasal salt excretion and adjustment to sodium loading in the lizard, *Ctenosaura pectinata*. Copeia No. *1*, 136–140 (1967).

– D. MURRISH, E. RANDALL, and J. MUGAAS: The effect of aldosterone and adrenalectomy on nasal salt excretion of the desert iguana, *Dipsosaurus dorsalis*. Amer. Zool. *8*, 818–819 (1968).

TERCAFS, R.R.: Phénomènes de perméabilité au niveau de la peau des Reptiles. Arch. Int. Physiol. Biochim. *71*, 318–320 (1963).

THAYSEN, J.H.: Handling of alkali metals by exocrine glands other than the kidney. In: Handbuch d. exper. Pharmacol. Vol. 13, pp. 424–507. Heidelberg and Berlin: Springer-Verlag 1960.

THORN, N.: The total body contents of sodium and potassium. Total exchangeable sodium and potassium. In: Handbuch d. exper. Pharmakol. Vol. 13, pp. 226–232. Heidelberg and Berlin: Springer-Verlag 1960.

THORSON, T.B.: The relationship of water economy to terrestrialism in amphibians. Ecology *36*, 100-116 (1955).

– Partitioning of body water in sea lamprey. Science *130*, 99-100 (1959).

– The partitioning of body water in osteichthyes: Phylogenetic and ecological implications of aquatic vertebrates. Biol. Bull. (Woods Hole) *120*, 238-254 (1961).

- Partitioning of body fluids in the lake Nicaragua shark and three marine sharks. Science *138*, 688-690 (1962).
- A. SVIHLA: Correlation of the habitats of amphibians with their ability to survive the loss of body water. Ecology *24*, 374–381 (1943).

TOWNSON: On the absorption of the Amphibia. In: Tracts and observations in natural history 1799. p. 65 London. Quoted by C. G. CARUS in: An introduction to the comparative anatomy of animals. Vol. 2, p. 126 London 1827.

TOYOFUKU, H. and P.J. BENTLEY: The osmotic composition of the ocular fluids of an amphibian: the toad *Bufo marinus*. Comp. Biochem. Physiol. *35*, 263–272 (1970).

TRUSCOTT, B. and D.R. IDLER: The widespread occurrence of a cortiocosteroid 1 α-hydroxylase in the interrenals of elasmobranchii. J. Endocr. *40*, 515–526 (1968).

TUCHMANN-DUPLESSIS, H.: Dêvelopment des caractères sexuels du Triton traité par des hormones hypophysaires gonadotropes et lactogènes. C.R. Soc. Biol. *142*, 629–630 (1948).

TUCKER, V.A.: Respiratory exchange and evaporative water loss in the flying budgerygah. J. exp. Biol. *48*, 67-87 (1968).

UEMURA, H.: Effects of water deprivation on the hypothalamohypophysial neurosecretory system in the grass parakeet, *Melopsittacus undulatus*. Gen. comp. Endocr. *4*, 193-198 (1964).

ULICK, S. and E. FEINHOLTZ: Metabolism and rate of secretion of aldosterone in the bullfrog. J. clin. Invest. *47*, 2523-2529 (1968).

URANGA, J. and W.H. SAWYER: Renal responses of the bullfrog to oxytocin, arginine-vasotocin and frog neurohypophysial extract. Amer. J. Physiol. *198*, 1287–1290 (1960).

USSING, H.H.: Interpretation of the exchange of radiosodium in isolated muscle. Nature (Lond.) *160*, 262-263 (1947).
- The alkali metal ions in isolated systems and tissues. In: Handbuch d. exper. Pharmakol. Vol. 13 pp. 1–195 Heidelberg and Berlin: Springer-Verlag 1960.
- B. ANDERSEN: The relation between solvent drag and active sodium transport of ions. Proc. 3rd Int. Cong. Biochem. Bruxelles pp. 434–441 (1955).
- K. ZERAHN: Active transport of sodium as the source of electric current in the short-circuited isolated frog skin. Acta physiol. scand. *23*, 110-117 (1951).

UTIDA, S., T. HIRANO, M. OIDE, M. KAMIYA, S. SAISYU, and H. OIDE: Na$^+$-K$^+$ activated adenosinetriphosphatase in gills and Cl$^-$-activated alkaline phosphatase in intestinal mucosa with special reference to salt adaptation of eels. Proc. XI Pacific Sci. Congress, (Tokyo) *7*, 5 (1966).
- N. ISONO, and T. HIRANO: Water movement in the isolated intestine of the eel adapted to fresh water or seawater. Zool. Magazine (Debutsugaku Zasshi) *76*, 203–204 (1967).

VALTIN, H., H.A. SCHROEDER, K. BENIRSCHKE, and H.W. SOKOL: Familial hypothalamic diabetes insipidus in rats. Nature (Lond.) *196*, 1109-1110 (1962).

VERNEY, E.B.: The antidiuretic hormone and the factors which determine its release. Proc. roy. Soc. B. *135*, 25-106 (1947).

Du VIGNEAUD, V., H.C. LAWLER, and E.A. POPENOE: Enzymic cleavage of glycinamide from vasopressin and a proposed structure for this pressor antidiuretic hormone of the posterior pituitary. J. Amer. chem. Soc. *75*, 4880-4881 (1953a).
- C. RESSLER, J.M. SWAN, C.W. ROBERTS, P.G. KATSOYANNIS, and S. GORDON: The synthesis of an octapeptide amide with the hormonal activity of oxytocin. J. Amer. chem. Soc. *75*, 4879-4880 (1953b).

WALKER, E.P.: Mammals of the World. Baltimore: Johns Hopkins 1964.

WALKER, A.M., C.L. HUDSON, T.J. FINDLEY, and A.N. RICHARDS: The total molecular concentration and the chloride concentration of fluid from different segments of the renal tubule of amphibia. The site of chloride resorption. Amer. J. Physiol. *118*, 121-129 (1937).

WARBURG, M.R.: The influence of ambient temperature and humidity on the body temperature and water loss from two Australian lizards, *Tiliqua rugosa* (Gray) (Scincidae) and *Amphibolurus barbatus* (Cuvier) (Agamidae). Aust. J. Zool. *13*, 331–350 (1965 a).
- Studies on the water economy of some Australian frogs. Aust. J. Zool. *13*, 317-330 (1965b).
- On thermal and water balance of three central Australian frogs. Comp. Biochem. Physiol. *20*, 27-43 (1967).

WARING, H.: Color change mechanisms of cold blooded vertebrates. New York: Academic Press 1963.
– R.J. MOIR, and C.H. TYNDALE-BISCOE: Comparative physiology of marsupials. In: Advances in Comparative Physiology and Biochemistry, Vol. 2, pp. 237–376. New York: Academic Press 1966.
– L. MORRIS, and G. STEPHENS: The effect of pituitary posterior lobe extracts on the blood pressure of the pigeon. Aust. J. exp. Biol. med. Sci. 34, 235–238 (1956).
De WEER, P. and J. CRABBE: The role of nucleic acids in the sodium retaining action of aldosterone. Biochim. biophys. Acta (Amst.) 155, 280–289 (1968).
WEISBART, M. and D.R. IDLER: Reexamination of the presence of corticosteroids in two cyclostomes, the Atlantic hagfish (Myxine glutinosa L.) and the sea lamprey (Petromyzon marinus L.). J. Endocr. 46, 29-43 (1970).
WEISS, M. and I.R. McDONALD: Corticosteroid secretion in the monotreme Tachyglossus aculeatus. J. Endocr. 33, 203–210 (1965).
– I.R. McDONALD: Adrenocortical secretion in the wombat Vombatus hirsutus Perry. J. Endocr. 35, 207–208 (1966 a).
– I.R. McDONALD: Corticosteroid secretion in the Australian Phalanger (Trichosurus vulpecula). Gen. comp. Endocr. 7, 345–351 (1966 b).
– I.R. McDONALD: Corticosteroid secretion in kangaroos (Macropus cangurus Major and M. (Megaleia) rufus). J. Endocr. 39, 251–261 (1967).
WIKGREN, B.: Osmotic regulation in some animals with special reference to temperature. Acta zool. Fenn. 71, 1-102 (1953).
WINGSTRAND, K.G.: Comparative anatomy and evolution of the hypophysis. In: The Pituitary gland, Vol. 1, pp. 58–126. Ed. by G.W. HARRIS and B.T. DONOVAN. Berkeley and Los Angeles: University of California 1966.
WOODHEAD, A.D.: The presence of adrenocorticotrophic hormone in the pituitary of the arctic cod, Gadus morhua L. J. Endocr. 21, 295–301 (1960).
WOOLLEY, P.: The effect of posterior lobe pituitary extracts on blood pressure in several vertebrate classes. J. exp. Biol. 36, 453-458 (1959).
WRIGHT, A. and I. CHESTER JONES: The adrenal gland in lizards and snakes. J. Endocr. 15, 83-99 (1957).
– I. CHESTER JONES, and J.G. PHILLIPS: The histology of the adrenal gland in prototheria. J. Endocr. 15, 100–107 (1957).
– J.G. PHILLIPS, and D.P. HUANG: The effect of adenohypophysectomy on the extrarenal and renal excretion of the saline loaded duck (Anas platyrhynchos). J. Endocr. 36, 249-256 (1966).
YAGI, K. and H.A. BERN: Electrophysiologic indications of the osmoregulatory role of the teleost urophysis. Science 142, 491-493 (1963).
– H.A. BERN: Electrophysiologic analysis of the response of the caudal neurosecretory system of Tilapia mossambica to osmotic manipulation. Gen. comp. Endocr. 5, 509–526 (1965).
YAPP, W.B.: Two physiological considerations in bird migration. Wilson Bull. 68, 312-319 (1956).
YOUNG, T.: Depletion of antidiuretic hormone in the pituitary of rats with diabetes mellitus. Endocrinology 84, 960-962 (1969).
ZADUNAISKY, J.A. and O.A. CANDIA: Active transport of sodium and chloride by the isolated skin of the South American frog Leptodactylus ocellatus. Nature (Lond.) 195, 1004 (1962).
ZARROW, M.X. and J.T. BALDINI: Failure of adrenocorticotrophin and various stimuli to deplete the ascorbic acid content of the adrenal gland of the quail. Endocrinology 50, 555-561 (1952).

Subject Index

Italics (5) refer to references in Figures or tables
Boldface (5) indicate more substantial parts of the text

289

Systematic and Species Index